ASTRONOMY THROUGH THE
TELESCOPE

ASTRONOMY THROUGH THE TELESCOPE

Richard Learner

VNR VAN NOSTRAND REINHOLD COMPANY
NEW YORK CINCINNATI TORONTO LONDON MELBOURNE

First published in 1981 by

Van Nostrand Reinhold Company
A division of Litton Educational Publishing, Inc
135 West 50th Street, New York, NY 10020, USA

Van Nostrand Reinhold Limited
1410 Birchmount Road
Scarborough, Ontario M1P 2E7, Canada

Library of Congress Catalog Card Number 81-4059
ISBN 0-442-25839-9

Edited, designed and produced by
Harrow House Editions Limited
7a Langley Street, Covent Garden,
London WC2H 9JA

Filmset in Melior by
Oliver Burridge & Co Ltd,
Crawley, Sussex, England
Illustrations originated by
Reprocolor Llovet S.A., Barcelona, Spain
Printed and bound by
Artes Graficas, Toledo, Spain

16 15 14 13 12 11 10 9 8 7 6 5 4 3 2 1

Library of Congress cataloging
in Publication Data
Learner, Richard.
Astronomy through the Telescope

1. Telescope I. Title.
QB88.L4 1981 522'.s 81-4059
ISBN 0-442-25839-9 AACR2
D.L. TO. 706-1981

CONTENTS

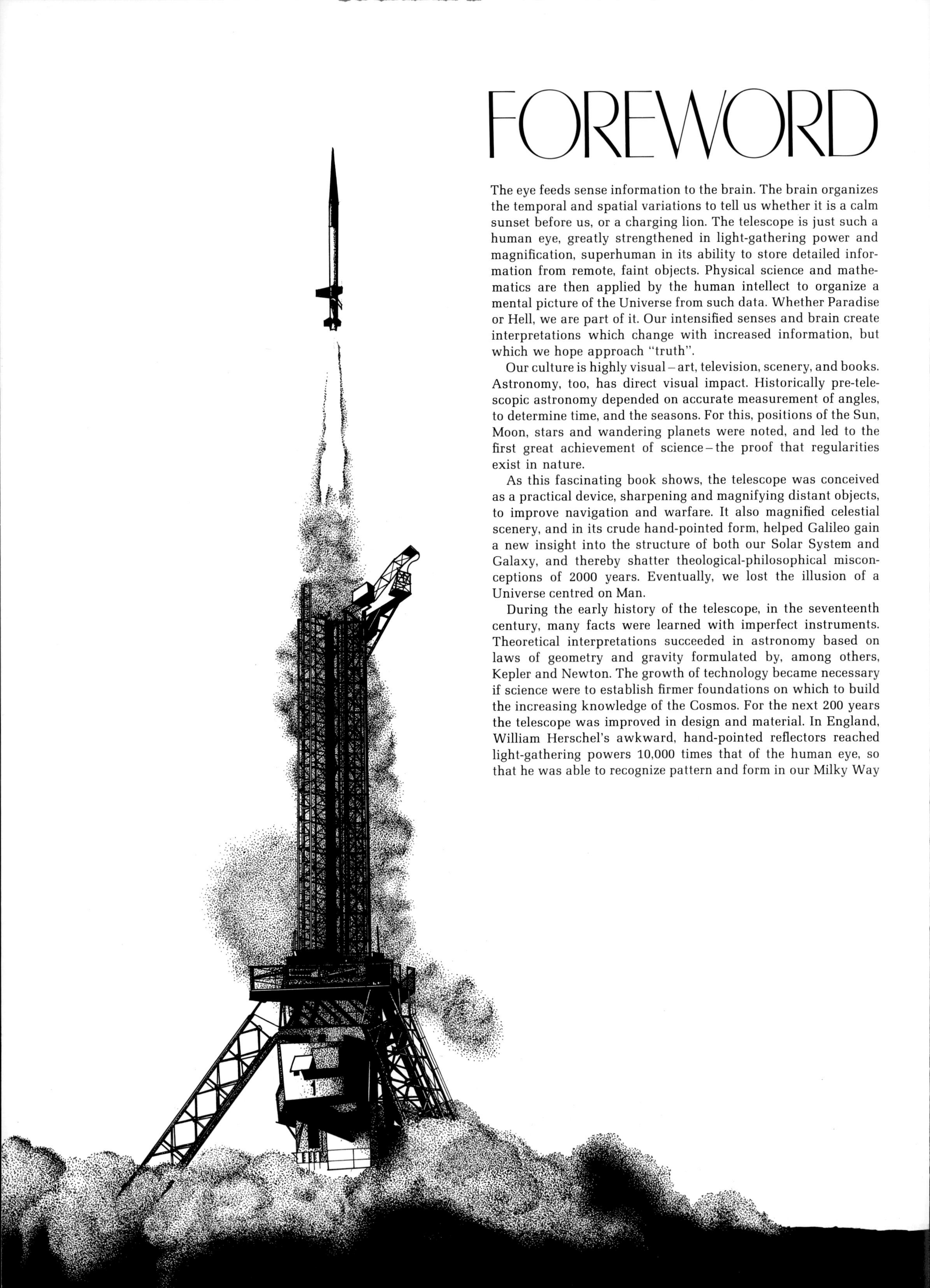

FOREWORD

The eye feeds sense information to the brain. The brain organizes the temporal and spatial variations to tell us whether it is a calm sunset before us, or a charging lion. The telescope is just such a human eye, greatly strengthened in light-gathering power and magnification, superhuman in its ability to store detailed information from remote, faint objects. Physical science and mathematics are then applied by the human intellect to organize a mental picture of the Universe from such data. Whether Paradise or Hell, we are part of it. Our intensified senses and brain create interpretations which change with increased information, but which we hope approach "truth".

Our culture is highly visual – art, television, scenery, and books. Astronomy, too, has direct visual impact. Historically pre-telescopic astronomy depended on accurate measurement of angles, to determine time, and the seasons. For this, positions of the Sun, Moon, stars and wandering planets were noted, and led to the first great achievement of science – the proof that regularities exist in nature.

As this fascinating book shows, the telescope was conceived as a practical device, sharpening and magnifying distant objects, to improve navigation and warfare. It also magnified celestial scenery, and in its crude hand-pointed form, helped Galileo gain a new insight into the structure of both our Solar System and Galaxy, and thereby shatter theological-philosophical misconceptions of 2000 years. Eventually, we lost the illusion of a Universe centred on Man.

During the early history of the telescope, in the seventeenth century, many facts were learned with imperfect instruments. Theoretical interpretations succeeded in astronomy based on laws of geometry and gravity formulated by, among others, Kepler and Newton. The growth of technology became necessary if science were to establish firmer foundations on which to build the increasing knowledge of the Cosmos. For the next 200 years the telescope was improved in design and material. In England, William Herschel's awkward, hand-pointed reflectors reached light-gathering powers 10,000 times that of the human eye, so that he was able to recognize pattern and form in our Milky Way and clouds of stars in other galaxies, millions of light years distant.

The deeper aspects of the physical world became understandable with the growth, in the late nineteenth century, of the physics of heat and the structure of atoms and atomic nuclei. Only then could we understand the Sun or a star, make accurate telescopic measurements, and ask how far, how big, how hot, and why they shine. Later came the development of auxiliary instruments such as the spectroscope, which permitted identification of chemical elements in the Sun and stars. Proof of large-scale uniformity in nature was now possible.

The advent of photography, which was developed first as a visual art, had a major impact on astronomy. It was soon applied to photography of stars and their spectra. The mechanical structure of telescopes and their driving mechanism were perfected for long exposures. The eye sees all it can in less than 0·1 seconds; the photographic plate can accumulate information for 100,000 times longer, and record it with geometric accuracy. But nature evolved a retina for our eyes more sensitive than chemists could perfect for the photographic emulsion. Light comes in tiny packages, called photons, or quanta. The brain recognizes a dim light when the eye receives as little as five quanta. The photographic plate needs several hundred quanta to blacken each spot. However, twentieth-century technology has seen the invention of surfaces which emit electrical pulses when irradiated with light, and these are more efficient than the eye and are sensitive to ultraviolet and infra-red light which neither the eye nor the photographic plate can see. The realization of television detectors, which can also integrate signals for a long time, increased a hundredfold the response to weak signals. This was equivalent to a tenfold increase in telescope diameter at minimal cost. Finally, the Next Generation Telescope will be large indeed, and use electronic light detectors of nearly 100 per cent efficiency. Such improvements require information-storage devices and computers at the telescope. Electronic control systems point at and track the sky nearly perfectly, while television-viewing devices make faint targets visible. The road on which Galileo started now approaches an end, with detailed study of targets 100 million times fainter than he could ever see.

In the last few decades new telescope construction has been rapid. The total area of large telescope mirrors in use increased ten-fold from 1910 to 1970 and will double again by the early 1980s. The analysing instruments fed by these mirrors, such as electronic spectrometers and data-handling computers, are complex. Astronomy has become a big and expensive science on Earth; in space, everything costs at least ten times as much. The large Space Telescope to be launched in the Shuttle by 1985, together with associated research (with which I am involved personally), will cost nearly $1000 million; a 4-meter ground-based telescope costs $10 million. The incredible pictures obtained with small television cameras on the Voyager spacecraft to Jupiter and Saturn show how much the telescope has grown and travelled since Galileo. But, in this epoch of large ground and space optical telescopes few recognizably great names and personalities emerge, but instead, great teams. Groups of scientists, instrument designers, engineers, managers and budgeteers must work together ten years or more to see a giant new eye or ear on the Universe start work.

New major sensory-enhancing instruments are radio telescopes. These differ conspicuously in appearance from optical devices. Some are 100-meter dishes (like those used for radar). Others consist of grids of wires spread over many kilometers. One, the Very Large Array, consists of twenty-seven dishes on tracks covering 50 kilometers. They reveal unfamiliar aspects of the Universe, to be explained in this book. (Although we might associate radio signals from space with sound, rather than light, a radio wave is merely very long wavelength light.) Radio astronomy is barely fifty years old, growing explosively after the Second World War. The typical collecting area of a radio telescope grew a hundredfold, and the image sharpness a millionfold, in those decades. From the beginning, the most advanced electronic and computer technology was necessary. One result is that radio equipment in particular requires large team efforts, large expenditures, and becomes rapidly obsolescent.

The special province of radio astronomy is the new insight into violent events requiring a new source of energy. Stellar explosions and distant quasi-stellar radio sources, or quasars, both have so large an energy output that nuclear energy seems insufficient. Transcending Newton's laws of gravity, Einstein's theory of general relativity allows large stellar masses to collapse under their own weight into a Black Hole, releasing 10 to 100 times as much energy as a nuclear explosion of the same mass. Magnetic forces have been emphasized, in space and in collapsed stars, so that a link has been established with cosmic-ray physics. Clouds of magnetic field and high-energy particles are found to have been blown repeatedly out of active galaxies, emitting strong radio waves, detectable halfway across our Universe. Radio astronomy has given us such exotic objects as quasars and pulsars. It has also provided a less well-known technique for chemical analysis of cold clouds of interstellar matter in our Galaxy: complex molecular compounds of carbon have been identified for the first time, which seem to be precursors to the birth of stars. But radio astronomy's greatest triumph has been in the detection and study of the most distant galaxies. Short wavelength radio telescopes have detected emissions coming from all directions that mark the birth of our Universe, 15,000 million years ago.

I must end this preamble with a personal note. This book introduces you to the many complex forms of one of Man's most important scientific tools. It will help you share in the first glimpses of progressively deeper mysteries. I have been lucky enough to observe with the greatest telescopes for nearly a thousand and one nights. I viewed a sky which was a storyteller. It was like Scheherazade, who entertained the Sultan with stories that never ended, needed to be continued another night, stories full of surprising reversals of fortune, encounters with new friends, each with a new story. This magical entertainment was matched, for me, by high aesthetic and intellectual adventure. You, too, I hope will find astronomy and the history of the telescope as presented herein to be just such a stirring exercise for your intellect and imagination.

Jesse L. Greenstein
Lee A. DuBridge Professor of Astrophysics,
California Institute of Technology,
and Staff Member, Palomar Observatory.

CHAPTER ONE

LOOKING

Descriptive astronomy with the newly invented telescope, 1609–70

"We are here so on fire with thes things that I must render my request and your promise to send mee of all sortes of thes cylinders. My man shal deliver you monie for anie charge requisite, and content your man for his paines and skill."

Letter from Sir William Lower to Thomas Harriot, June 1610.

The telescope was first described in October 1608, and it was first used for astronomy in July 1609. Within a year, arguments began as to who had invented the new instrument, and who had made the first astronomical observations with it. Historians of science still debate these questions, but quite regardless of the minutiae of the arguments, Galileo is the dominating figure in the telescope's early history. His position at the forefront of early seventeenth-century science is so secure that he is still referred to by only his Christian name, a familiarity denied even to the comparable Newton and Einstein.

Galileo was not the first to use the telescope for astronomy; he was not the most accurate of observers, nor the best at sorting out his observations – and he was not always right. He himself stated that he did not invent the instrument which posterity insisted on christening the Galilean telescope. Why then was he so preeminent?

For a start, Galileo was the first to publish his findings, although the rapid appearance of his pamphlet *The Starry Messenger* in March 1610 is more a symptom of his greatness than one of its causes. His achievements rest on his immense intellectual confidence, even arrogance. He was confident enough to accept that in eight months he had accumulated sufficient evidence to reject the picture of the universe that had been built up by 2000 years of endeavour by pre-telescopic astronomers: the Book of Genesis was wrong, the philosopher Aristotle was wrong, the great Greek astronomer Ptolemy was wrong, even St Thomas Aquinas was wrong – but Galileo Galilei was not. In his support Galileo had few allies. Similar controversial ideas had first been proposed by an ecclesiastical administrator in Poland, Nicolas Copernicus, in a book published seventy years earlier, but it contained a preface stating the ideas in it were not to be taken too seriously; and a mathematician in Prague, Johannes Kepler, had just published some calculations that supported Copernicus but only on the strength of minute discrepancies between observation and the established theory. It is this immediate acceptance, not of his new results but of their implications and importance, that puts Galileo in a separate class from the other astronomers who were making similar observations at the same time: Scheiner in Bavaria, Marius in central Germany, Fabricius in Holland and Harriot in England.

The speed with which Galileo reacted also unsettled his colleagues at the University of Padua. The professor of philosophy refused even to look through a telescope, and the professor of mathe-

Galileo Galilei revolutionized astronomy by using a telescope to study the skies. The implications of his observations opposed the general beliefs of the time and led to a change in much of seventeenth-century thought.

Galileo's classic work, Sidereus Nuncius, The Starry Messenger, *was published in 1610. In it Galileo described his astronomical observations – four satellites of Jupiter, mountains and craters on the Moon and a multitude of stars that had never before been seen.*

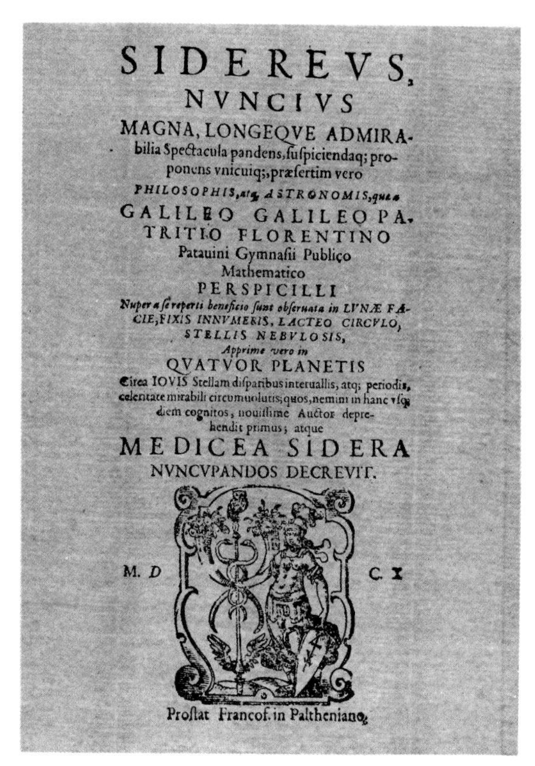
SIDEREVS,
NVNCIVS
MAGNA, LONGEQVE ADMIRA-
bilia Spectacula pandens, suspiciendaq; pro-
ponens vnicuiq;, præsertim vero
PHILOSOPHIS, atq; ASTRONOMIS, quæ
GALILEO GALILEO PA-
TRITIO FLORENTINO
Patauini Gymnasii Publico
Mathematico
PERSPICILLI
Nuper a se reperti beneficio sunt obseruata in LVNÆ FA-
CIE, FIXIS INNVMERIS, LACTEO CIRCVLO,
STELLIS NEBVLOSIS,
Apprime vero in
QVATVOR PLANETIS
Circa IOVIS Stellam disparibus interuallis, atq; periodis,
celeritate mirabili circumuolutis; quos, nemini in hanc vsq;
diem cognitos, nouissime Auctor depre-
hendit primus; atque
MEDICEA SIDERA
NVNCVPANDOS DECREVIT.
M. D C. X
Prostat Francof. in Paltheniano.

matics stated that he would put forth his best endeavours to see the new results discredited. Galileo and his telescope were probably none too popular with the faculty in any case: as soon as he had a working instrument in his hand, Galileo had hurried to Venice, the source of all local power and wealth. The Venetian senate were much impressed by the possible military application of the new device and promptly gave him life tenure of his professorship and a handsome pay increase.

These measures did not suffice to keep Galileo in Venice. He dedicated his new book of astronomical observations to the Florentine Grand Duke Cosimo II de Medici, and in July 1610 took up a new appointment as court mathematician at Pisa. He was already familiar with the town, having been born there in 1564 and brought up in nearby Florence. From 1581 he had been a medical student at Pisa University, although due to lack of funds he withdrew without completing the course. Galileo then turned to mathematics, and in 1589 he had been appointed to the mathematics staff at the university. His early work concerned errors inherent in the traditional Greek view of the basic concepts of force, energy and momentum. As was to happen in Padua, the rest of the faculty at Pisa were none too keen to be deprived of their cherished illusions. To make matters worse, Galileo had achieved this in a style as sarcastic as it was polished, and in 1591 the thoroughly unpopular genius had left for Padua.

Galileo's new job on his return to Pisa nearly twenty years later left him free to research rather than teach, but in the longer term it proved unwise for the proponent of such controversial ideas to move from a university within the liberal Venetian republic back to the more orthodox air of Florence.

The origin of the telescope used by Galileo and his contemporaries is curious. Unlike most technical devices, it was not invented. Its birth has little in common with, say, Samuel Crompton's invention of the spinning mule or Eli Whitney's cotton gin, and is more reminiscent of Monsieur Jourdain's discovery that all his life he had been speaking prose. In the autumn of 1608 one of the spectacle-lens makers working in the shadow of the glass factory at Middelburg in Holland realized that the lenses he had in his hands would make a useful telescope. Arguments over which of

Hans Lippershey, a Dutch spectacle maker, constructed the first effective telescope by mounting two lenses at opposite ends of a tube.

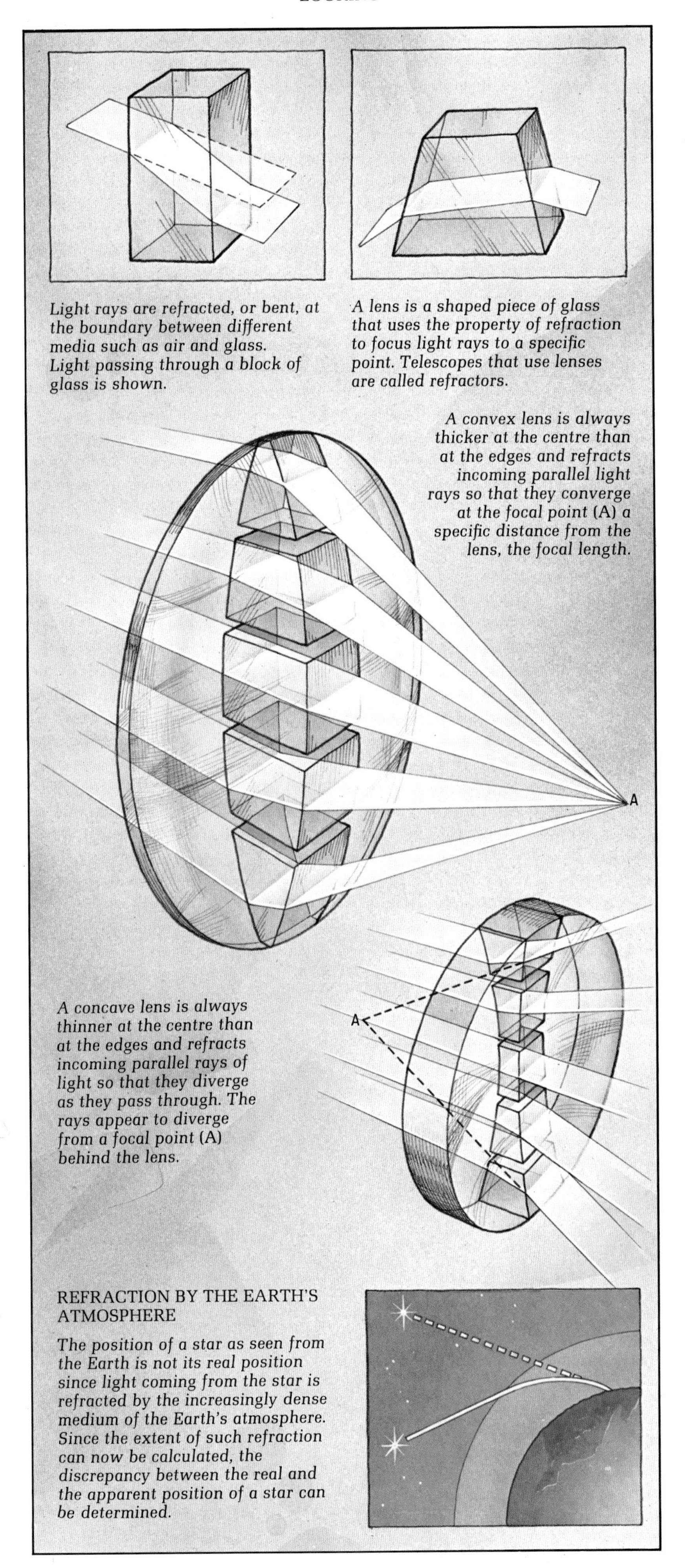

Light rays are refracted, or bent, at the boundary between different media such as air and glass. Light passing through a block of glass is shown.

A lens is a shaped piece of glass that uses the property of refraction to focus light rays to a specific point. Telescopes that use lenses are called refractors.

A convex lens is always thicker at the centre than at the edges and refracts incoming parallel light rays so that they converge at the focal point (A) a specific distance from the lens, the focal length.

A concave lens is always thinner at the centre than at the edges and refracts incoming parallel rays of light so that they diverge as they pass through. The rays appear to diverge from a focal point (A) behind the lens.

REFRACTION BY THE EARTH'S ATMOSPHERE

The position of a star as seen from the Earth is not its real position since light coming from the star is refracted by the increasingly dense medium of the Earth's atmosphere. Since the extent of such refraction can now be calculated, the discrepancy between the real and the apparent position of a star can be determined.

the Middelburg opticians it was lasted for 300 years: the palm has now been awarded to Hans Lippershey. Neither he nor his rival, Zacharias Janssen, was native to Middelburg. Lippershey came from Wesel in Westphalia, and Janssen from The Hague. Janssen was not only a lens maker but a thoroughgoing rogue, who had been forging Spanish money. This had been an acceptable, even patriotic, act for a Dutchman while Spain and Holland were at war, but Janssen forgot to stop when the war ended. As the punishment was, in the words of W. S. Gilbert's *Mikado*, "something lingering with boiling oil in it", Janssen left Middelburg rather hurriedly.

Lippershey's telescope consisted of two lenses. At the front end was a weak convex lens – a lens which is a little thicker in the middle than at the edges – and at the eye end a rather strong concave lens, a lens that in contrast is a good bit thinner in the middle than at the edges. The first combination of these lenses was probably just a result of holding one lens in each hand, with the convex at arm's length and the concave immediately in front of the eye. According to tradition, it was Lippershey's apprentice who first noticed the effect of this lens combination.

Convex lenses had been known by then for about 300 years, and concave lenses for 150 years. Once the construction of such a telescope was known to be readily possible – and to be worth while – any optician could manufacture the pair of lenses needed and fit them at opposite ends of a lead or paper tube; and so the device spread across Europe from Middelburg in only a few months. Galileo made his own; in England, Thomas Harriot's man, Christopher Tooke, made dozens; while Johannes Kepler in Prague, an archetypal theorist, regretted that he was all thumbs and had to borrow a telescope that Galileo had presented to the Duke of Bavaria.

Lippershey's patent application to the government for his new device was rejected on the grounds that he had done nothing new – everyone knew how to make telescopes. The government did not, however, answer the obvious question: why, if everyone knew how to combine the two simple component lenses of the new telescope, there had been no telescope the year before, or even the century before. The answer involves a mixture of philosophical, intellectual and commercial factors. The telescope had been foretold in the thirteenth century, by the polymath Roger Bacon. His version was, however, more magical than rational, allowing Bacon to sit in Oxford with a device "in which men myght see thynges that were doon in other places". This was so far

During the Dutch-Spanish wars of the early sixteenth century the telescope was exploited by the military. This painting by Velásquez portrays the surrender of Breda to the Spanish in 1625.

Galileo's representations of the surface of the Moon as viewed through his telescope portray his discovery that the Moon is not smooth like a mirror, as the early Greeks had thought, but rough with valleys and mountains.

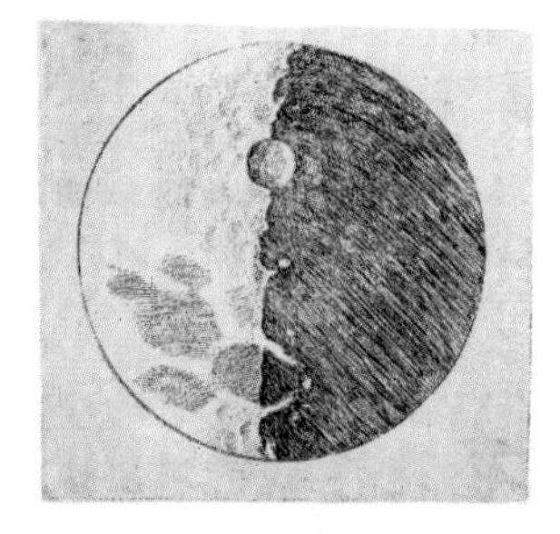

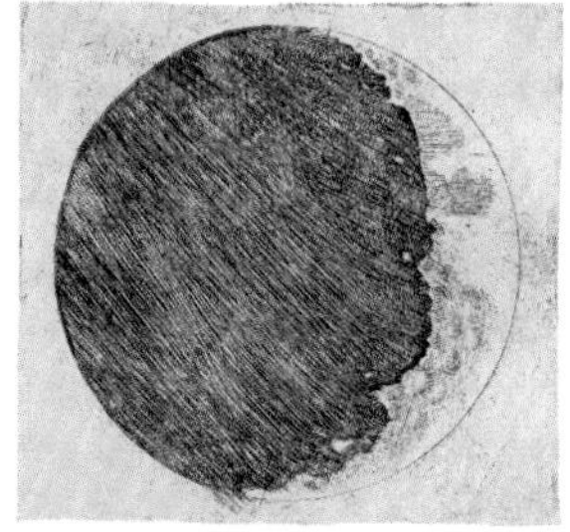

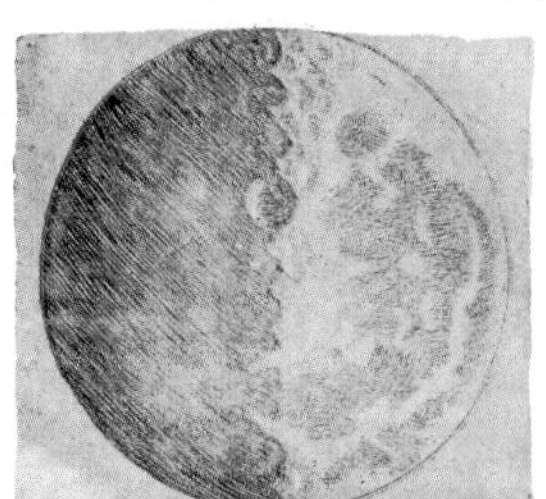

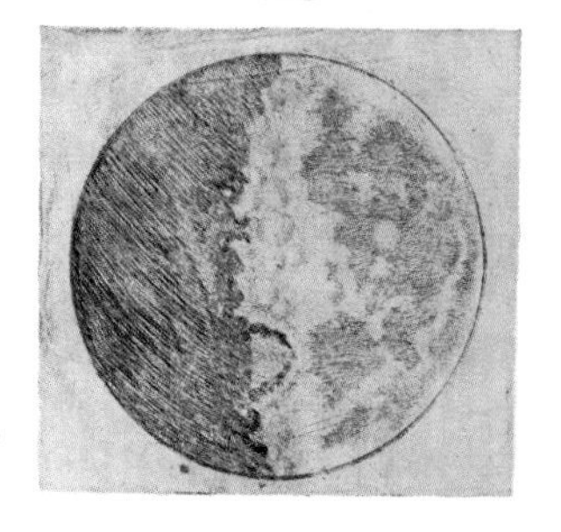

from the instruments that could be made even in the sixteenth century, when the poor optical quality of the lenses still more than offset any small gain in magnification, and none of the primitive trials were recognized even to be telescopes.

Intellectually, the lenses of the Galilean telescope are a most unlikely pair. To make a telescope that magnifies a distant object, the optician must use as the objective (front lens) the convex lens that, of all those in his drawer, is the weakest and worst magnifying glass. This is then combined with a strong concave lens; and a concave lens on its own powerfully diminishes the distant scene. In the sixteenth century, it was known that two weak magnifying glasses placed close together would give a higher magnification than either lens alone. No one would have expected that good magnification could be achieved by combining two components neither of which was itself a powerful magnifying glass. Indeed, if the two lenses needed for the Galilean telescope are placed in contact, rather than at the opposite ends of a tube, and then held up in front of the eye, they do not magnify but diminish the apparent size of a distant object.

Finally, the two lenses needed are those which are of least application as spectacle lenses, and therefore were the least likely to be found in the working optician's stockpile – it was the opposite types, strong convex and weak concave lenses, that were the commercially successful items for correcting poor vision.

Lippershey did receive some reward for his recognition that a suitable combination of well-made lenses would give an instrument which, if not magical, was certainly useful. The Zeeland provincial council recommended his device to the Dutch commander-in-chief, Prince Maurice, Count of Nassau, who was at the time negotiating the treaty that marked the end of the Dutch wars for independence from Spain. Prince Maurice recognized the military use of the new telescope, tried to keep it secret and ordered Lippershey to make a better instrument for 900 crowns. To put his act in modern terms, Maurice, at a peace conference, issued a fat defence development contract on the basis of classified information. The telescope worked, the security did not, and soon all the other delegates at the conference were writing home describing the new device.

With the news came argument. Jacob Metius wrote from northern Holland to say that he had thought of it first, and in Naples Giambaptista della Porta said that he had tried the experiment years earlier and it did not work. Both were probably right. Metius's patent application, filed a fortnight after Lippershey's, indicates that he had been working on the instrument for some time, while Porta would have used lenses of poorer quality and less suitable power than Lippershey in his earlier and unsuccessful experiment.

What then did Galileo and the others see through their new instruments, and why did it cause so much trouble? The

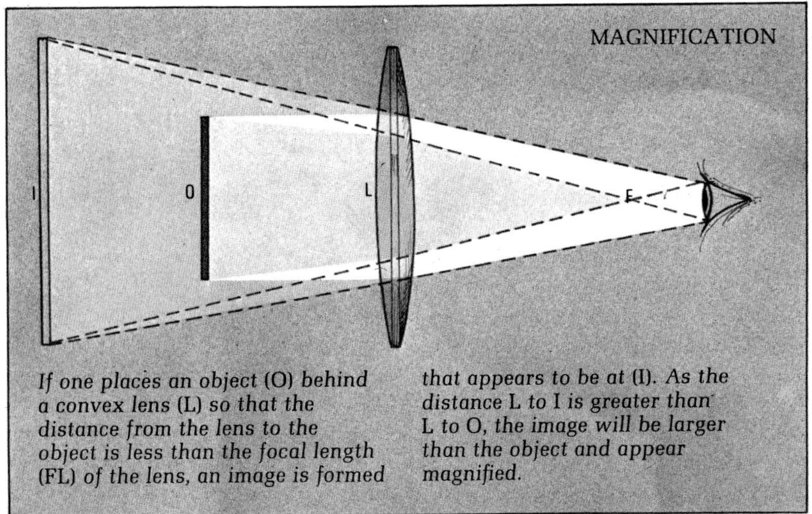

If one places an object (O) behind a convex lens (L) so that the distance from the lens to the object is less than the focal length (FL) of the lens, an image is formed that appears to be at (I). As the distance L to I is greater than L to O, the image will be larger than the object and appear magnified.

The Copernican model of the universe (right) as drawn by Galileo incorporates the latter's discovery of the moons of Jupiter. Copernicus proposed that the planets revolved round the central Sun and showed that this greatly simplified the mathematical treatment of planetary motion.

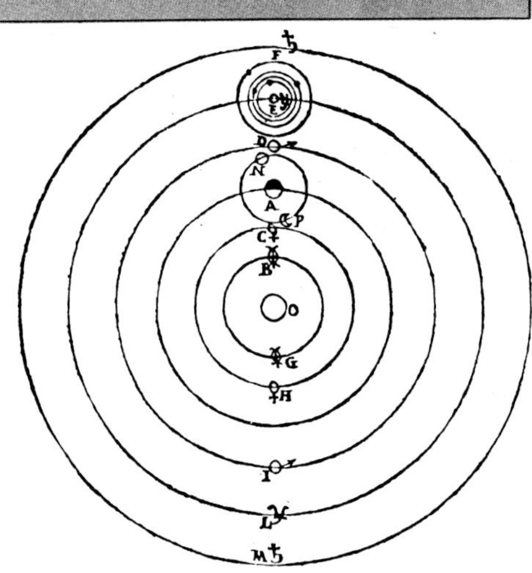

Christopher Scheiner's method for observing the Sun with a telescope without damage to his eyes was to project the image on to a white screen.

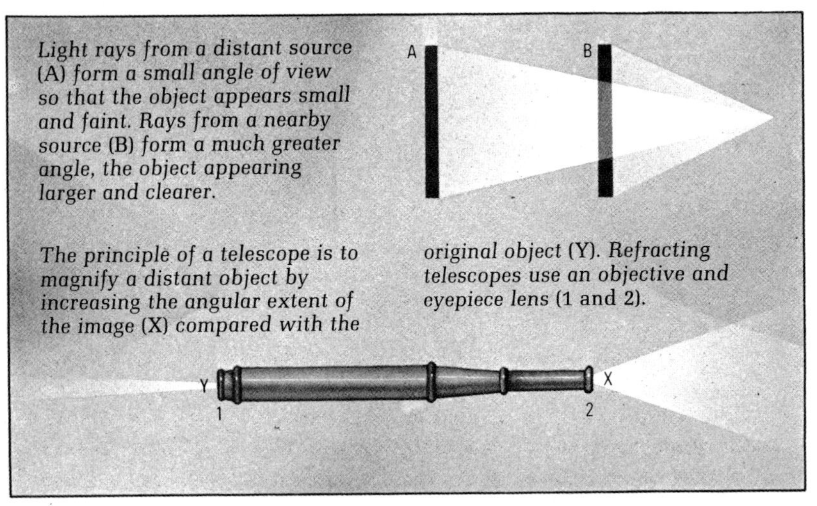

Light rays from a distant source (A) form a small angle of view so that the object appears small and faint. Rays from a nearby source (B) form a much greater angle, the object appearing larger and clearer.

The principle of a telescope is to magnify a distant object by increasing the angular extent of the image (X) compared with the original object (Y). Refracting telescopes use an objective and eyepiece lens (1 and 2).

observations presented in *The Starry Messenger* covered four topics. First, the Milky Way is not a continuous band of light, as it appears to the unaided eye, but is composed of myriad separate stars. Second, wherever else you look, more stars are visible through the telescope than without it; not just a few but hundreds and hundreds. Since previous explanations of the universe had been based entirely on eye observation Galileo's demonstration that things are not as they seem was the first step towards a new explanation. Next was the observation that the Moon is not a perfect sphere, but rough and mountainous. Its evident imperfections conflicted with the established belief at the time that the sphere is the perfect form and that the heavenly bodies must therefore be perfectly spherical.

The last and most important of the four topics concerned the planet Jupiter as the centre of a set of four satellites, called by Galileo the "Medicean stars". These four specks of light fatally wounded the accepted picture of the universe, which required *everything* to go round the Earth. The observation did not, however, conflict with the alternative view, proposed by Copernicus, that the Earth revolves about the Sun. Indeed, it eased an embarrassing inconsistency in Copernicus's theory. Although Copernicus placed the Sun at the centre of the universe, he had to admit that not everything, but only *almost* everything, went round the Sun. Copernicus's opponents made much of the exceptional status of the Moon, which even in the new theory still went round the Earth. Their argument was based on the implausibility, even impossibility, of a moon moving in a serene and stable orbit, not round a fixed centre but round a planet that was travelling at high speed round the Sun. Surely the Moon would get left behind. Jupiter's moons showed that it is indeed quite possible to have regular motion around a centre that is itself moving.

If Jupiter's moons were a fatal wound to the theory of an Earth-centred universe, Galileo dealt the death blow soon after moving to Pisa in the autumn of 1610. The object of his studies was the planet Venus, and his conclusion was that the planet showed phases like those of the Moon. The crucial point here is that in the Earth-centred model of the solar system, Venus is always between the Earth and Sun, and so the view from Earth must always show Venus as an illuminated crescent. In the Copernican model, however, the Earth and Venus can be on opposite sides of the Sun, and at these times Venus can be seen with more than half the planet illuminated, looking like a nearly full Moon in miniature.

After looking at Venus, Galileo turned his telescopes on Saturn and found a puzzle: the planet appeared to have two moons, but, unlike Jupiter, the moons did not move. As he had found a puzzle, he set a puzzle. He did not publish his findings as such, nor even a brief summary, but an anagram of the brief summary. He sent Kepler a list of thirty-seven letters with a characteristic message to the effect that, never mind what it was, he saw it first. The letters were: SMAISMRMILMEPOETALEUMIBUNENVGTTAVIRAS.

Kepler picked out MARTIA and tried to put the remainder into a message about the moons of Mars. Harriot's friends in London had a go. One of them made rather tortured sentences based on Jupiter (I = J in Latin), on Montibus . . . Martem, and on Ignis Lunarem. Another of Harriot's group took a more earthy view of this deliberate obfuscation: he picked out the letters GALILEI and then made rude words with the rest. The correct solution was "Altissimum planetam tergeminum observavi" – I have observed the furthest planet to be triple.

The last discovery to come out of the initial surge of observation in 1610 was the discovery of sunspots. Once again, several observers made the discovery virtually simultaneously. The first to publish was the youngest, the Dutchman Johann Fabricius, who had the rare advantage of being the son of another astronomer, David Fabricius. A Jesuit, Christoph Scheiner, made the longest and most careful study of the new spots, first in Ingoldstadt and then in Rome, but he had to publish his early results under a pseudonym, as his superior was sure that the spots did not exist. Harriot's observations probably were the first and typically he never published at all. Galileo was sure he was first, and rather suspiciously, when twenty years later he retold the story of "his" discovery, he adjusted the evidence in his favour.

Among the early observers, Scheiner had the best equipment and Harriot the worst. Scheiner built the first properly mounted telescope and arranged to project the Sun's image on to a sheet of paper, the safest way of observing the Sun. Thomas Harriot, who had as a young man been tutor to Sir Walter Raleigh and in 1585 had gone as geographer on the second expedition to Virginia, was now a member of the household of Henry Percy, Duke of Northumberland. In contrast to Scheiner, Harriot relied on misty mornings beside the river at Syon House to look through his telescope directly at the Sun. He sometimes, most unwisely, did this when it was not misty. In March 1612 his observation notes record, "At 12^{h} all the sky being cleare and the Sonne I saw the

THE PHASES OF VENUS

According to the Ptolemaic model of the solar system (above), Mercury (M), Venus (V), the Sun (S) and Moon (Mo) followed paths round the central and stationary Earth (E). But Galileo noticed that Venus exhibits phases similar to those of the Moon, disproving Ptolemy's theory because the different phases could only be seen if Venus and the Earth were on opposing sides of the Sun.

Recent photographs of Venus (right), both shown at the same scale, illustrate the extremes of the different phases. The crescent phase appears when Venus is close to the Earth and the full phase is visible when Venus is at its furthest point from Earth.

The present-day model of the solar system (below) is based on the Sun-centred Copernican model. Venus is shown in a complete cycle of its phases. (For comparison with the diagram above, the orbits of the planets are shown as being circular; they are, in fact, elliptical.)

Galileo's trial by the Inquisition, held in Rome in 1633, was instigated by Pope Urban VIII. Galileo was sentenced to house arrest in his declining years as a punishment for his unbalanced presentation of orthodox views on the solar system. Only as recently as 1979 did the Vatican officially admit that Galileo's ideas were indeed true.

great spot. . . . My sight was after dim for an houre."

Once again, regardless of priority, the theory of pure spherical form took a battering. More important, Fabricius and Galileo noticed that the spots change position from day to day, showing that the Sun rotates on its axis. Hitherto the only evidence supporting the theory that the Earth rotates was rather shaky. It certainly does not feel as if the Earth is going round, and indirect evidence from the Moon is unconvincing: one does not need a telescope to see that the "Man in the Moon" always faces the Earth, meaning that the Moon rotates only once a month, and this made one rotation a day for the much larger Earth rather hard to accept.

By Christmas 1612 the rush of discoveries was over and there was a pause while the new data were considered. Apart from an intemperate sermon against Galileo in 1614, for which the preacher's superior wrote and apologized, it was late 1615 before the establishment decided that the new ideas were so unsettling that something had to be done. This took the form of a firm but friendly warning to Galileo to keep away from doctrinal controversy. Galileo took the view that he had enough evidence to convince the entire College of Cardinals, and repaired to Rome to do just that. He was wrong, and in February 1616 he was formally forbidden to believe in Copernicus's theory. A week later Copernicus's book, published seventy-three years before, was put on the list of prohibited books, the *Index Librorum Prohibitorum*, where it stayed until 1835.

With the benefit of hindsight, this most famous misjudgement has become part of every modern schoolchild's knowledge, with Galileo cast as hero, and the Inquisition as villains. At the time, things were much less clear cut. Galileo, like all great experimental scientists, was working at the absolute limit of his apparatus. The crucial observation of Venus when it was nearly "full" was made with a telescope that gave a blurred image, and the diameter of the blur is almost exactly the same as the apparent diameter of the image of Venus. As Giorgio Abetti put it after he had tested the original instruments in 1923, Galileo saw the phases of Venus "with the sharpness of his intuition".

Other points were not as telling as Galileo expected. His observation of many more stars than could be seen with the naked eye had been predicted by Clavius, the papal astronomer, who had pointed out that the Bible describes "innumerable" stars (Hebrews xi, 12) and so there must be many more than the few thousand that can be seen on a dark night. Many telescopes were poorly made and even a good telescope was hard to use. Its field of view was tiny and the skill needed to use it was acquired only by long practice. If a telescope was not properly made, accurately

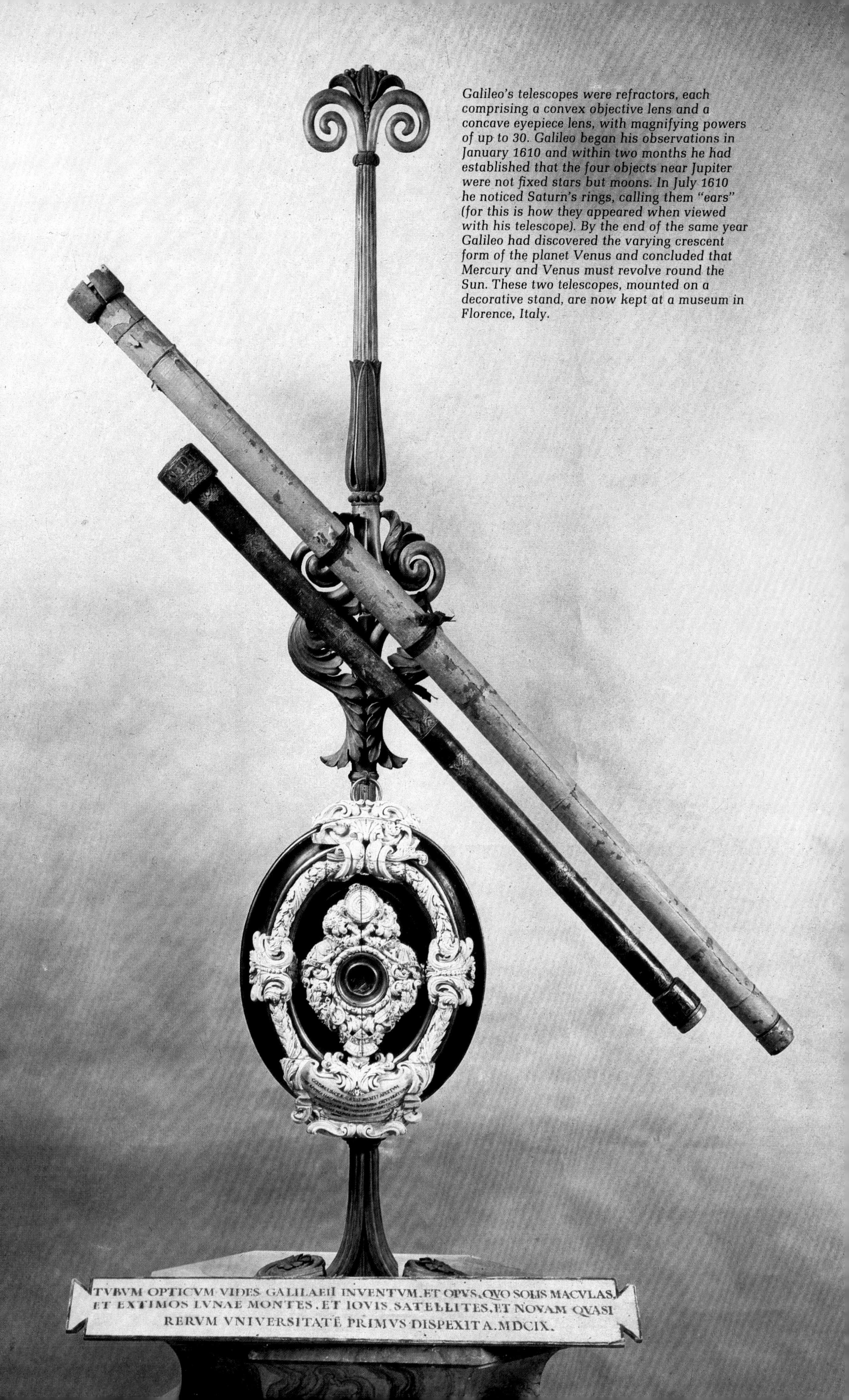

Galileo's telescopes were refractors, each comprising a convex objective lens and a concave eyepiece lens, with magnifying powers of up to 30. Galileo began his observations in January 1610 and within two months he had established that the four objects near Jupiter were not fixed stars but moons. In July 1610 he noticed Saturn's rings, calling them "ears" (for this is how they appeared when viewed with his telescope). By the end of the same year Galileo had discovered the varying crescent form of the planet Venus and concluded that Mercury and Venus must revolve round the Sun. These two telescopes, mounted on a decorative stand, are now kept at a museum in Florence, Italy.

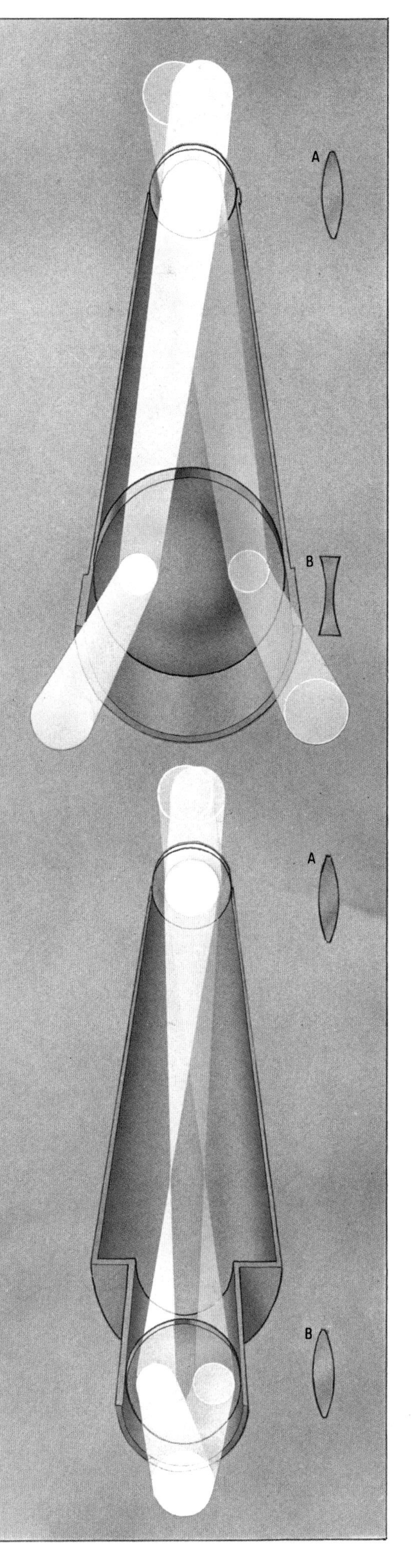

REFRACTING TELESCOPES

Galileo made a simple telescope by placing a combination of two lenses, one convex and one concave, at opposite ends of a tube. In 1610 he demonstrated the new invention to an assembly of nobles in Venice. The image is formed by the objective lens (A), which is convex, and is magnified by a concave lens forming the eyepiece (B). As the focal length of the objective lens is large compared with that of the eyepiece lens, an enlarged image is produced. This system of lenses is so successful that it is still used in modern opera glasses. Its disadvantage is that the field of view is very small if the telescope is required to give high magnification. In addition, Galileo's lens-making technique was far from perfect, so that the telescopes always gave blurred images.

Kepler, although he was inept at making telescopes, in his Dioptrice *published in 1611, he suggested using convex lenses for both the objective and eyepiece. The image in this telescope is formed by the objective convex lens (A) and is then magnified by the other convex lens (B) placed at the opposite end of the tube. The idea of using two convex lenses has the advantage of giving a wider field of view than the Galilean telescope, but the image is inverted, rendering Kepler's telescope only suitable for astronomical purposes. It still suffered from the combined defects caused by the spherical shaped lenses and the dispersive effect of glass lenses on light of different colours. None the less, the telescope gave much sharper images than the original "Galilean" instrument.*

pointed and held steady, then all the observer saw was a random collection of rapidly moving blurs. Many would-be observers saw nothing, and even those who could see were sometimes unsure. Galileo claimed two moons for Saturn in 1610; a few years later even he had to agree that they were not there. "Does Saturn devour his children?" he wrote. It was forty years before the telescope was good enough to show Saturn's rings as they really are, and thus explain Galileo's puzzle.

Most important, and the underlying factor to all our hindsight, is the date – 1616. It was to be another four years before Francis Bacon's *Novum Organum* spelled out the new philosophy of knowledge that science has ascribed to since: that the test of truth must be experiment and observation. In 1616 the attitude of the medieval schoolmen still held sway, the doctrine that the only way to get things right was to think about them and to read what others had thought. It was almost inconceivable that some arrogant and abrasive astronomer could be right as a result of mere observation.

For the next twenty years development was slow. The pace of discovery could not be maintained and the Galilean telescope was hard to improve. Two important astronomical discoveries were however made. Their discoverers did not know what they had found, but clearly they were neither stars nor planets. Simon Marius, in Ansbach, saw something he described as "like a candle shining through horn" – an object now known as the Andromeda Galaxy – while Cysat, a pupil of Scheiner, saw that one of the stars in Orion was not a bright point of light but a tiny luminous patch, known today as the Orion Nebula. These distant diffuse patches of light were to become two of the most thoroughly studied objects in the sky.

While the pace of astronomy slowed, improvements in optics continued. The Galilean telescope has the serious drawback that when it is used at high magnification the "field of view" – the amount of sky visible through it – is very small. The telescope's optical arrangement means that light from objects just off the telescope axis (away from the centre of the field of view) emerges from the eyepiece at such a steep angle that it misses the pupil of the observer's eye, and such objects cannot be seen without moving the telescope. Kepler found a way round this by inventing a new type of telescope. He described how to combine two convex lenses to give a telescope that would have both good magnification and a reasonable field of view. In Kepler's telescope the first lens forms an image of the star, and this

René Descartes, the French philosopher and mathematician, made a determined effort to overcome the optical difficulties created by lenses in focusing light. He outlined his theories on the subject in a book Dioptrique, *published in 1637.*

image is then examined using the second lens as a magnifying glass.

Another step forward in optical science occurred in 1621, when Willebrord Snell of Leiden discovered the law which describes the refraction, or bending, of light at a surface; and this was publicized by René Descartes in 1637. Once again the tacit genius of Harriot had been there before – using "Snell's law" in 1597. As Sir William Lower, writer of the send-telescopes-regardless-of-expense missive quoted at the beginning of the chapter, said in 1610: "Al these were your deues and manie others that I could mention; and yet to(o) great reservednesse hath robd you of those glories. . . . Onlie let this remember you, that it is possible by to(o) much procrastination to be prevented in the honor of some of your rarest inventions and speculations. Let your countrie and frends injoye the comforts they would purchase your selfe by publishing some of your choise works."

Another reason for the lull in astronomy in the 1620s is that it was a strife-torn field to enter. To be an astronomer at all needed an enquiring and independent mind, and it inevitably drew the observer into bitter theological confrontations. This was emphasized when Galileo's friend, Maffeo Barberini, became Pope Urban VIII, an appointment which encouraged Galileo to write up his work. Friendship was not enough. Four years later the seventy-two-year-old Galileo was under arrest and forced to recant. Scheiner, who had himself been refused permission to

Ingolstadt Observatory was the site of Scheiner's major observations of the Sun.

Galileo's drawing of a star cluster in Orion (above) *showed stars that were not visible to the naked eye.*

Scheiner's drawing of the Sun (left) *portrayed his observation of sunspots.*

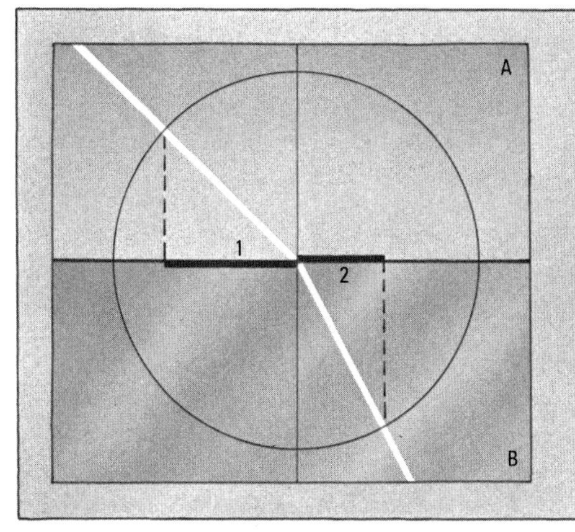

SNELL'S LAW

Light rays passing from one medium to another, such as from air (A) *to glass* (B), *are refracted at the boundary. The degree of refraction is a constant for any two given media and can be represented as a ratio of the two distances 1 and 2, which can be measured from the two triangles shown in the drawing.*

Huygens developed an aerial telescope in which the objective lens was mounted in a short iron tube placed atop a high pole.

In his Selenographia, *published in 1647, Hevelius drew a chart of the Moon* (left).

Hevelius's unwieldy 46-metre (150-foot) "celestial machine" was erected in Danzig. It was the extreme of seventeenth-century attempts to reduce the blurring of an image by using an objective lens of a very long focal length.

The standard and accuracy of astronomical measurements using telescopes was greatly improved with the invention of the pendulum clock (right) *in 1657. It was originally designed by Christian Huygens.*

publish his early sunspot work, wrote a book which attacked Galileo. Fabricius was murdered, though not for being an astronomer, and Kepler ran into religious persecution.

Before Harriot even started observing he was strongly accused of atheism in the row that followed the fall of his first employer, Sir Walter Raleigh. In November 1604 Harriot had unwittingly made the mistake of being at a dinner at Syon House with the Percys, the Duke of Northumberland's family. The meal was interrupted by an urgent message for Thomas Percy, brought by a certain Guy Fawkes. The next day the famous Gunpowder Plot to blow up Parliament was uncovered and Fawkes arrested in a cellar full of gunpowder – a cellar that had been rented by Thomas Percy. A short while later Harriot was signing a letter to the Privy Council as "Your honors' humble petitioner: a poore prisoner in the gatehouse [of the Tower of London]."

The only real step forward in observational astronomy during the 1620s was a systematic set of observations of the Moon by Langren, mathematician and cosmographer to Philip II of Spain. His intention was to make a map of the Moon, and this was shown in draft to Queen Isabel in 1628 but not published until 1645. The pace picked up in the late 1630s. A leading figure was Francisco Fontana, who worked in Naples, then one of the three greatest cities in Europe and controlling a revenue more than three times that of the whole of England. Fontana started to use Kepler's style of telescope for astronomical observation. This is easier to make and to use than the Galilean design,

The ever-changing appearance of Saturn puzzled early astronomers – Hevelius's drawings of the planet in 1656 are shown below. In that year Huygens discovered the presence of rings around the planet and by 1675 Cassini had noticed a large gap between them. A recent photograph of Saturn (left), taken by the visiting spacecraft Voyager 1 *in 1980, reveals that the rings are subdivided into hundreds and, to add further complexities, the outer ones are braided like a rope. The rings are composed of smallish chunks of ice but their very existence remains a mystery.*

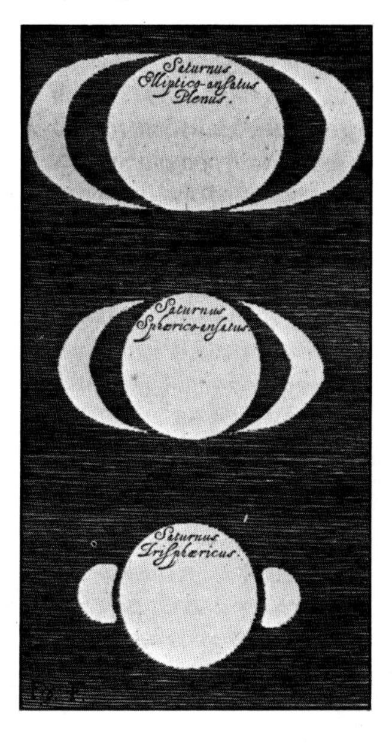

but it did not have the same commercial impact as the earlier instrument because the resulting view is upside down. The inverted view does not matter in astronomy, but it prevented the Keplerian telescope from becoming popular for everyday use. A version of this telescope using two additional convex lenses (making four altogether) to produce an image the right way up did, however, appear a few years later. This "terrestrial" telescope has never been a rival to the "astronomical" telescope for observation of the night sky because the extra lenses reduce the brightness of the final image, and a bright image is all-important in astronomy.

Fontana's improved telescope gave better images than the earlier makers had achieved. He saw the phases of Mars (due to the differing angle of illumination as it and the Earth orbit the Sun) and he drew a much improved map of the Moon. Both of these observations attracted the attention of other workers. Chief among these were Johannes Hevelius and Christian Huygens. These two, one in Danzig and the other first in Holland and then in Paris, led the efforts to improve the new "Keplerian" telescope. Hevelius's experience and the optical theories of Descartes and Kepler showed that lenses with strongly curved surfaces gave bad images. Weak objective lenses were better. The result was the first of a generation of very long telescopes, the length being necessary because a convex lens with only slightly curved surfaces brings light to a focus a long way from the lens itself. Hevelius's early telescopes used in 1647 for the compilation of his *Selenographia*, the first proper atlas of the Moon, were from 2·5 to 3·5 metres (8 to 12 feet) long. With such instruments he could repeat the early observations and confirm the discovery of the phases of Mercury made by Ionnes Zupo in 1639. Zupo was one of a dozen or more Jesuits, including Scheiner, Fontana, Riccioli and Grimaldi, who responded to the need for their Church to be better informed if further embarrassment was to be avoided. Mercury, even at its nearest to the Earth, looks smaller than Venus when at her most distant, and to make matters worse, is usually only visible in the twilight glow. These observations of Mercury clearly established the superiority of the long Keplerian telescope.

Christian Huygens in Holland was the first to pursue Hevelius's use of weak lenses. With his brother Constantine, Huygens developed techniques for polishing lenses with small curvatures and consequently a long focal length. In addition he invented an improved eyepiece, in which the magnification was shared between two thin lenses rather than one fat one. The traditional method for grinding and polishing spectacle lenses relied on the use of a crude metal tool and a rod, in addition to the disc of glass that would become the lens, and abrasives. For a convex lens the metal tool was cast or forged with a hollow of approximately the right radius for the lens surface. The glass blank was fixed to the end of the rod and then ground against the tool using successively finer abrasives until both tool and lens acquired an accurate spherical shape. The radius was controlled by the length of the rod, the other end of which was held in a loose socket; and all the grinding was done by hand. These methods are only

Isaac Newton, besides being the builder of the first working telescopes that used mirrors rather than lenses, was also a renowned authority on optics. Even after his epoch-making discoveries in mathematics, mechanics and gravitation, he was still described to the French academy as "the celebrated telescope-maker, Monsieur Newton".

Giovanni Cassini, 1625–1712, used refracting telescopes of ever-increasing length in his discovery of four of Saturn's moons, Iapetus, Rhea, Tethys and Dione.

practicable for small lenses of modest focal length; a lens with a focal length of 3 metres (10 feet) requires a rod at least as long. The Huygens brothers developed a grinding and polishing machine that not only needed no rod but also used gears to rotate both lens blank and polishing tool, and as a result could produce lenses with much longer focal lengths.

The first of these lenses to be built into a telescope had a focal length of a little over 3·5 metres (12 feet). Christian Huygens discovered Titan, the first of Saturn's genuine moons, with this instrument in 1655. Later that year, with a 7-metre (23-foot) telescope, he had partially unravelled the detail of the rings, the feature that had so puzzled Galileo. In 1656, with a 37-metre (123-foot) giant, he was sure, and in 1659 Huygens published his description of the planet as having "a ring, thin, plane, nowhere attached". During this interval he had rediscovered the Orion Nebula, and, far more significant, had invented that most important of astronomical instruments, the pendulum clock. Astronomers need an accurate clock to keep track of the positions of the stars as the Earth turns, and the two devices, a good telescope and a good clock, have been the fundamental tools of astronomy ever since.

The improved telescope assisted others as well as triggering off a series of ever-longer instruments. The very long focus lenses were not a great success. Hevelius built a 46-metre (150-foot) telescope in Danzig, but the difficulty of pointing it at a moving star or planet, combined with the way that a puff of wind would set the whole thing shaking, offset any gain that the vast length promised. Huygens tried an alternative approach, replacing the telescope tube with a length of string with which the observer could pull the telescope objective to face the right way. With this style of telescope Huygens discovered an unwelcome fact: the quality of the image seen was affected by the air currents in the Earth's atmosphere. This problem, called "seeing", still bedevils astronomy. It is the reason why modern observatories are to be found on mountains in Chile or Arizona or on oceanic islands like Hawaii and the Canaries, above the worst atmospheric turbulence (and far from the dazzle of city lights).

The most successful astronomer among those using the very long instruments was the professor at Bologna, Giovanni Cassini. With his colleague, Giuseppe Campani, who made the lenses, Cassini was able to see surface markings on the planetary discs. The ability to study the other planets as worlds in their own right set in motion the next phase of the astronomy of the solar system. The surface features appeared to move, showing that the other planets rotate, and it was seen that Venus, Mars and Jupiter all rotate about axes roughly parallel to that of the Earth. The observation of Venus's rotation marked the start of a long controversy, only really resolved a few years ago, for astronomers from Cassini onwards have only been able to see the top of cloud layers which hide the solid planet beneath. Jupiter was found to be spinning rapidly on its axis, going round once every ten hours. The new telescopes were able to show the resulting distortion of the planet, which has an equatorial diameter seven per cent greater than the distance from pole to pole.

Others were at work elsewhere. In London the restoration of the monarchy led, in 1660, to the formation of the Royal Society. Several of the members had an interest in astronomy, including Christopher Wren, better known as the architect of St Paul's cathedral, and the quarrelsome Robert Hooke. Both of these worked on problems of lens manufacture, and Hooke made observations of the cloud belts of Jupiter.

In spite of Hooke's experimental genius, the main influence on telescope design came from experiments in a darkened room in Cambridge. These were carried out by a recent graduate, Isaac Newton, and investigated the coloured spectrum produced when sunlight passed through a prism. First, Newton showed that the "rainbow" of colours was a property of the light, not of the prism, and that white light was a combination of light of all colours. Second, he found that Snell's law of refraction applied to each individual colour; this meant that the behaviour of a lens focusing a beam of white light could be calculated by working out how it focused each colour separately. Third and last, he found that the extent to which the colours were spread out into a spectrum did depend on the prism. If a weak prism was used, so that the beam of light was bent only by a modest amount from its original path, then the colours in the beam were dispersed into a spectrum of only modest length. If the prism produced a larger deviation of the beam, then the dispersion of colours was greater too. He concluded, rather too hastily, that deviation and dispersion would always be related to one another, and that the relation would be the same regardless of the material of the prism.

Newton's experiments led to two conclusions which were relevant to Huygens's and Hevelius's telescopes. First, there was a fundamental limit to their performance that was set by the properties of the glass used for the objective. This limit also depended on length. A 15-metre (50-foot) telescope was twice as good as one of only 7 metres (25 feet); a 30-metre (100-foot) telescope twice as good again. Second, no ingenuity in design, or skill in lens making, could avoid the need to increase the length yet further to gain better optical performance – neither replacing the objective lens with several less powerful lenses close together, nor polishing the lens surface to a different shape would help. Increase in length was the only solution, and every increase in length made the mechanical problems of supporting, pointing and steadying the telescope more severe. As Newton said: "The improvement of telescopes by refractions is desperate." Newton's conclusions were published in 1672 in a newly invented format, a paper in a scientific journal, a method that has now become standard for communicating scientific results. The full title of the new journal was *The Philosophical Transactions of the Royal Society of London for Improving Natural Knowledge*; it is still being published and has now reached volume 290.

Meanwhile Cassini had become Chief Astronomer at Paris, where Huygens was now president of the Academy. Cassini continued to make discoveries, using telescopes of considerable but reasonable length, suspended on either a mast or a government surplus water tower, which had been moved to the new French Royal Observatory. In 1675 he was able to see that Saturn has two rings. The single ring seen in poorer telescopes is split by a narrow dark gap, still known as Cassini's division. Cassini published a map of the Moon in 1680, a map that overshadowed all previous efforts. Other members of Cassini's family joined in. His nephew, Maraldi, studied Mars and discovered the white polar ice caps. Cassini started a notation still in use for Jupiter's moons, calling them J-I, J-II, J-III and J-IV, a notation that was later of use in identifying the members of the Cassini family. Hence, Giovanni Cassini (pére) is Cassini I, Jacques was the next director, Cassini II. Then came Cesar, and finally Jacques Dominique, Cassini IV, who was the Observatory director more than a century later, when he resigned in order to write the family history. The four great astronomers, Cassini I to Cassini IV, cover a remarkable span of more than two centuries, dating from 1625 to 1845.

The names now used for Jupiter's moons are a compromise. Galileo, their discoverer, named his "Medicean stars" after individual members of the Medici family, but his proposal was too parochial for the growing international scientific community. Simon Marius proposed the names of four victims of Jupiter's extramarital promiscuity – Io, Callisto, Europa and, for the brightest of the moons, the beautiful boy Ganymede. A fifth moon, discovered in 1892, was called Amalthea after Jupiter's nurse. Later discoveries came at a time when improving astronomical technique was associated with declining classical education, and the unromantic Jupiter VI to XIII were only belatedly named after mythological figures as recent as 1976.

Newton did not only point out the limitation of the telescope in use at the time. He proposed a solution to the problem, and built two examples to demonstrate the concept. His answer was to use a concave mirror, rather than a lens, to collect and focus light. While a simple lens focuses different colours at different points, a mirror focuses all the colours of white light at the same point. In other words, the newly discovered limitation on lenses (called chromatic aberration), which produces a coloured haze around the image seen in a telescope, does not affect images produced by a mirror.

There had been previous attempts to build a telescope using mirrors rather than lenses. The Italian physicist Nicolas Zucchi had tried a simple system, in 1616, and the French mathematicians Marin Mersenne and René Descartes had written on the theory of the various forms such an instrument could take. James Gregory, a Scot, had designed a telescope based on two concave mirrors, but after he came to London in 1665 he was unable to get mirrors of sufficiently precise workmanship. A Frenchman, N. Cassegrain, invented yet another design using one concave mirror and one convex mirror, though this did not influence Newton, who was completing his second telescope when Cassegrain's design was announced.

Newton proposed and built a design that avoided the cardinal weakness of Zucchi's device. Because a mirror reflects light back the way it has come, the observer has to look down into the mirror and in simple designs his head obstructs the incoming light from the sky. The answer is to use a second, smaller mirror to deflect the light again, and Newton's design was simpler than Gregory's as the second mirror was flat, rather than concave, and so was easier to make. After some experiments to find the best metal alloy for his mirrors (silvered glass was nearly 200 years in the future) he made two small telescopes; and these, the first reflecting telescopes ever made, caused a stir. The Royal Society exhibited the better of them to its members in January 1672, at the meeting at which Newton was elected a Fellow.

Newton's telescope was small, scarcely 30 centimetres (12 inches) long, but it foreshadowed the close of the first chapter of the telescope's history. In the previous sixty-three years astronomers had taken the refracting telescope to its apparent limit. In search of better instruments to study the sky, they had extended the small telescopes of Galileo's day to vast lengths and found further improvement was unpracticable for structural rather than optical reasons. But now, Newton's compact reflecting telescope would render obsolete Huygens's and Hevelius's and Cassini's optical dinosaurs, whose tiny heads had needed such enormous bodies.

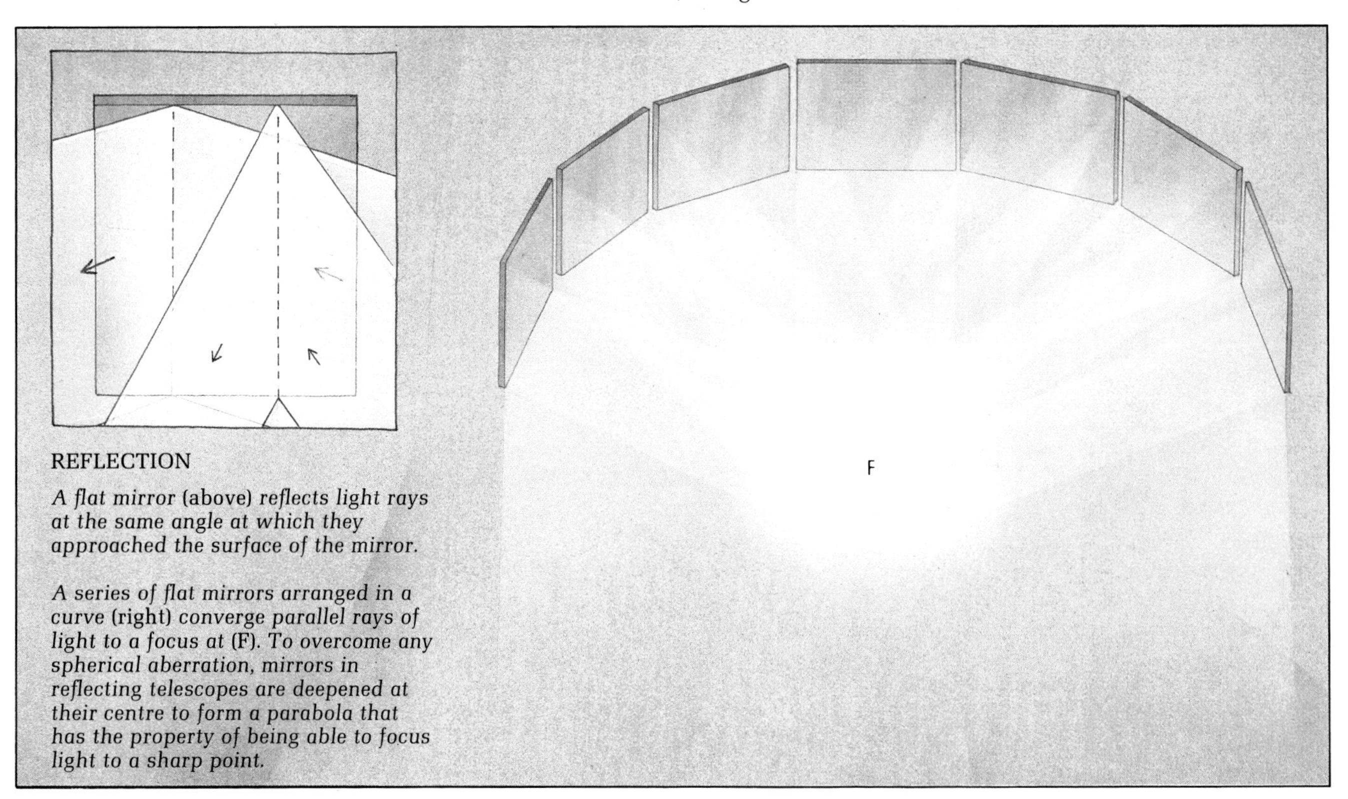

REFLECTION

A flat mirror (above) reflects light rays at the same angle at which they approached the surface of the mirror.

A series of flat mirrors arranged in a curve (right) converge parallel rays of light to a focus at (F). To overcome any spherical aberration, mirrors in reflecting telescopes are deepened at their centre to form a parabola that has the property of being able to focus light to a sharp point.

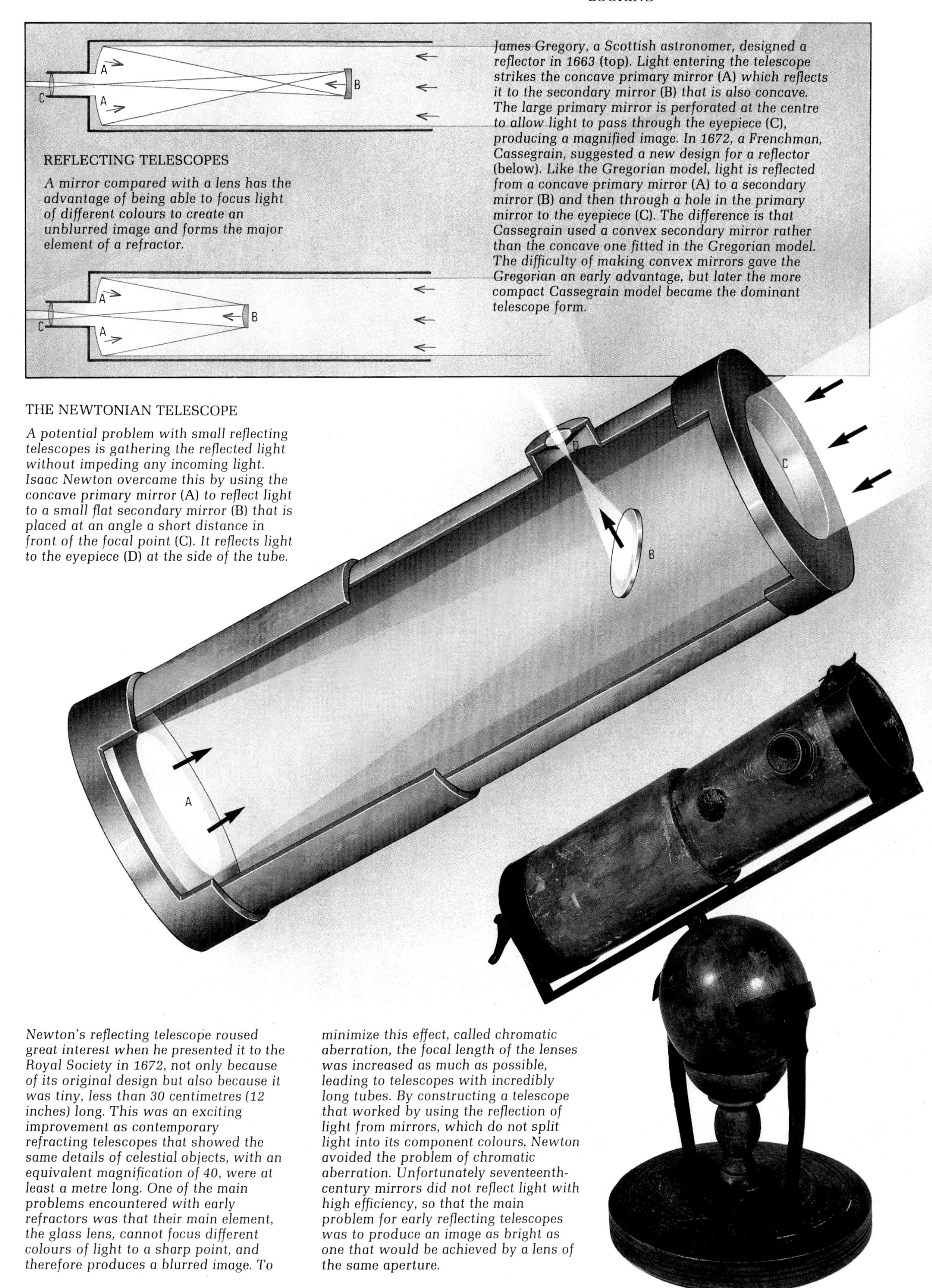

James Gregory, a Scottish astronomer, designed a reflector in 1663 (top). Light entering the telescope strikes the concave primary mirror (A) which reflects it to the secondary mirror (B) that is also concave. The large primary mirror is perforated at the centre to allow light to pass through the eyepiece (C), producing a magnified image. In 1672, a Frenchman, Cassegrain, suggested a new design for a reflector (below). Like the Gregorian model, light is reflected from a concave primary mirror (A) to a secondary mirror (B) and then through a hole in the primary mirror to the eyepiece (C). The difference is that Cassegrain used a convex secondary mirror rather than the concave one fitted in the Gregorian model. The difficulty of making convex mirrors gave the Gregorian an early advantage, but later the more compact Cassegrain model became the dominant telescope form.

REFLECTING TELESCOPES

A mirror compared with a lens has the advantage of being able to focus light of different colours to create an unblurred image and forms the major element of a refractor.

THE NEWTONIAN TELESCOPE

A potential problem with small reflecting telescopes is gathering the reflected light without impeding any incoming light. Isaac Newton overcame this by using the concave primary mirror (A) to reflect light to a small flat secondary mirror (B) that is placed at an angle a short distance in front of the focal point (C). It reflects light to the eyepiece (D) at the side of the tube.

Newton's reflecting telescope roused great interest when he presented it to the Royal Society in 1672, not only because of its original design but also because it was tiny, less than 30 centimetres (12 inches) long. This was an exciting improvement as contemporary refracting telescopes that showed the same details of celestial objects, with an equivalent magnification of 40, were at least a metre long. One of the main problems encountered with early refractors was that their main element, the glass lens, cannot focus different colours of light to a sharp point, and therefore produces a blurred image. To minimize this effect, called chromatic aberration, the focal length of the lenses was increased as much as possible, leading to telescopes with incredibly long tubes. By constructing a telescope that worked by using the reflection of light from mirrors, which do not split light into its component colours, Newton avoided the problem of chromatic aberration. Unfortunately seventeenth-century mirrors did not reflect light with high efficiency, so that the main problem for early reflecting telescopes was to produce an image as bright as one that would be achieved by a lens of the same aperture.

CHAPTER TWO

DATES AND DESTINIES

Pre-telescopic astronomy from Plato to Kepler via Arabia 500 BC–1630

"To find the golden number or prime add one to the year of our Lord and then divide by nineteen; the remainder, if any, is the golden number; but if nothing remaineth, then nineteen is the Golden Number."

Rules for the moveable and immoveable feasts, in the Book of Common Prayer.

Astronomy did not start with the first astronomical use of the telescope. The emergence of astronomy as a science occurred in the hundred years between 450 and 350 BC. At that time, as a result of Persian encouragement of the Greek inter-city wars, there was close contact between the Hellenic and Mesopotamian civilizations. One consequence of this contact was a mixing of two separate approaches towards astronomy. The Assyrians and Persians had been content to observe the motions of Sun, Moon, stars and planets, but had not indulged in any speculation on the underlying causes of the phenomena. In contrast, the Greeks had done a great deal of speculation and very little observation.

The third of the civilizations of classical antiquity, the Egyptian, contributed virtually nothing. Early on they established a calendar and picked out a set of thirty-six stars, the Decans, for timekeeping at night. When it became clear that the Decans were not very good timekeepers and that the calendar was wrong, the Egyptian hierarchy, inhibited perhaps by their appallingly inadequate arithmetic, did nothing at all to improve the system. The level of interest in the heavens over the Nile valley is perhaps best shown by the Egyptian eclipse records. About fifty total eclipses of the Sun were visible in Egypt between 1000 BC and the Arab conquest in AD 640, but the only record is for the last, in AD 601.

The combination of Greek and Persian traditions, supported by Greek geometry and Babylonian arithmetic, attacked two problems: first, the motion of the Sun and Moon, and second, the motion of the other "planets". The former was of great practical importance, as detailed knowledge of the motions of the Sun and Moon is necessary to establish a good calendar.

So long as man was a farmer, and taxation was based on a percentage of the harvest, a precise calendar was unnecessary. The month from new Moon to new Moon was a convenient interval, and the year was at first taken to be twelve lunar months long. This is only 354 days, so the error quickly accumulates; a spring month advances to the summer season in only a decade. The solution to this problem was to add an occasional thirteenth month to the year. The Egyptians added an extra month when observation of Sirius, the brightest star, showed it to be necessary, and they later set up a second, more arbitrary, calendar to run in parallel with the by then sacred lunar system. The Assyrian and Babylonian hierarchies added months without any obvious rule or pattern. The Jews were more pragmatic and added an extra month whenever it appeared that the harvest festival might arrive before the harvest. The Greeks, organized in city states rather than as a nation, added an extra month when the local town council saw fit.

This system sufficed for the needs of ancient, agricultural man, and continued until modern, or bureaucratic, man appeared on the scene. The first fairly accurate calendar was developed soon after the widespread adoption of the new invention of money. Clearly, if one is offering a year's pay for a year's work, or remitting tax on an annual basis, it is necessary to know whether the year is twelve or thirteen months long. To be a month late with the tribute money could be a fatal error.

The evolution of the calendar is an outstanding example of the aphorism that the deeper one digs the muddier one gets. Everyone knows what the day, the month and the year are. Or do they ? The *Explanatory Supplement to the Astronomical Ephemeris*, the standard source for numerical data, gives precise figures for three different kinds of day, five different kinds of months and four of the seven different sorts of year. The simple definition of a day as the time from noon to noon gives an interval that varies in length through the year (due to the Earth's tilt and its varying speed as it travels round the Sun). Similarly, the month measured from new Moon to new Moon can vary by more than half a day. Even after definitions have been agreed and the intervals accurately measured, the day, month and year are not related in any convenient manner. The "mean synodic month" – the average interval between new Moons – is 29·530589 days, the "tropical year" – the year defined by the seasons – for 1980 is 365·242193 days, and the ratio between these, 12·368266, the number of months in the year, is no nearer a helpful simple number.

The first calendar which related day, month and year in a manner that did not steadily accumulate sizeable and ever more obvious discrepancies was proposed by Meton in Athens around 432 BC. He established the equation 19 years = 235 months = 6940 days. The ratio of month to year, 235/19 = 12·36842, is out by only 0·00016, though both the year and the month calculated from this equation are a little too long – the month by a couple of minutes and the year by half an hour. Although Meton's system was never adopted by the Greeks, who developed their practice of reckoning by counting Olympiads, it is the pattern commemorated in the division by nineteen in the rules for finding Easter. The Persians adopted the nineteen-year Metonic cycle in 383 BC, fifty years before Alexander the Great brought the rest of Greek culture to

One of the oldest surviving observatories of the pre-telescopic era is the tower at Palenque in Mexico (left). *It was used by Mayan astronomers for observations in the sixth century AD.*

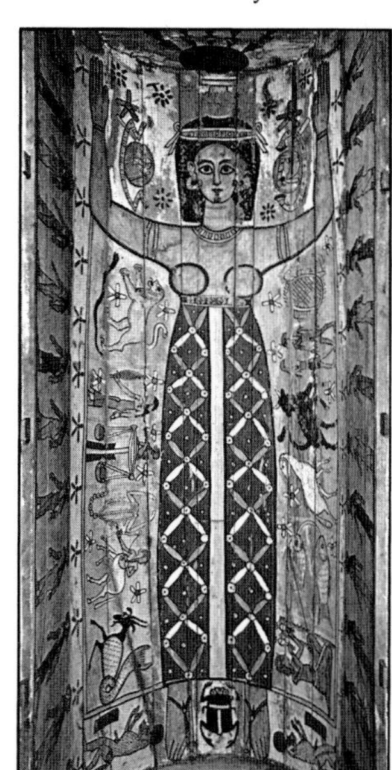

The Egyptian goddess Nut (right), *whose body represented the starry sky, framed by the signs of the zodiac.*

The remains of a Babylonian globe (below) *recording part of their elaborate calendar, which successfully accommodated both the lunar month and the solar year. Babylonian astronomers employed an ingenious system that fitted 235 lunar months, each beginning with a new Moon, into a repeating pattern of 19 solar years.*

Mesopotamia. The Jews had already acquired a Babylonian calendar during the captivity and so in later years added the new Babylonian nineteen-year cycle. This led to rules to fix the Jewish festival of the Passover, and finally the Council of Nicaea in AD 325 incorporated the cycle in the rules for fixing Easter.

The calendar is uninfluenced by the motion of five of the seven heavenly bodies that are not "fixed" stars, but the study of all seven "planets" was much influenced, and indeed driven, by astrology. Two critical developments occurred at the beginning of this period. The first was the idea of personal astrology, and the second the beginnings of predictive astrology. At the time of the Assyrian Empire, astrology was concerned with the interpretation of omens such as the first reappearance of a planet after it had been in line with, and thus obscured by, the Sun. These omens affected only the monarch or the realm and were current events. Since the year could be twelve or thirteen months and there was no clear rule allowing a decision in advance, prediction was severely limited.

The importance of astrology was enormous, and so was its influence on astronomers and others. Now that astrology has become a branch of the entertainment industry, it is hard to see why many of the greatest intellects of antiquity accepted and often furthered the art. The cynic might observe that a monarch can always use a cast iron excuse for not doing something, but if we are to accept the influence of astrology in classical times, we must suspend our disbelief.

The rise of astrology starts from the recognition that two of the seven wandering stars, the Sun and the Moon, have direct influence on the human race. The Sun clearly affects the life of the whole community, deciding such fundamentals as when one works and when one sleeps. The Moon appeared to have an equally obvious influence – Moon, month and menstruation are all words derived from the same root. Since the brightest

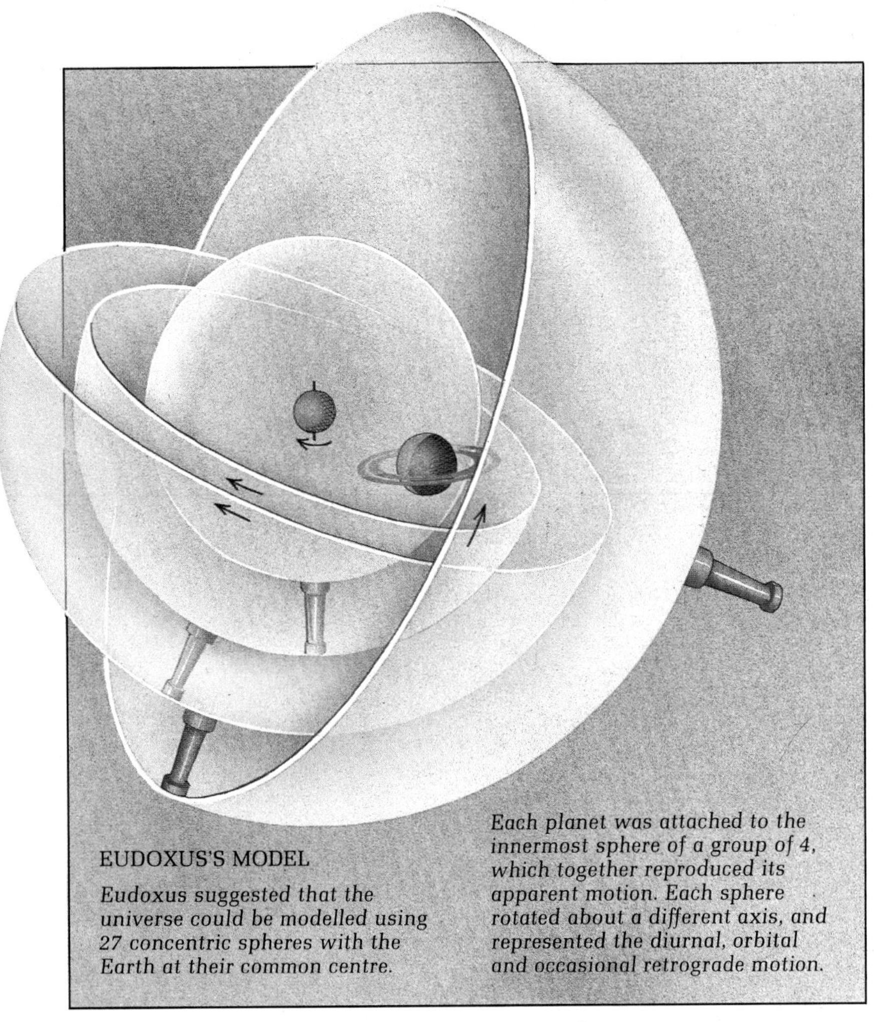

EUDOXUS'S MODEL

Eudoxus suggested that the universe could be modelled using 27 concentric spheres with the Earth at their common centre.

Each planet was attached to the innermost sphere of a group of 4, which together reproduced its apparent motion. Each sphere rotated about a different axis, and represented the diurnal, orbital and occasional retrograde motion.

In antiquity prominent groups of stars were given names – the constellations. Arab astronomers formalized the system and extended it to name the brighter stars. Thus Denebola, "the lion's tail", is still the name of the prominent star at the left of the Arabian painting of the constellation Leo (right).

A long-exposure photograph (below), *taken in a planetarium, demonstrates the movements of the planets. From Earth, each planet occasionally appears to stop its forward motion, travel in a backward loop (retrograde motion), and then resume its original east-to-west direction. Retrograde motion is the consequence of looking at the motion of a planet, from another planet (the Earth), which is itself moving but at a different speed.*

"planets" have such immediate effect, it is reasonable to suppose that the fainter planets have more subtle influences. A list of plausible parallels can be drawn up. Saturn, the slowest planet, is associated with laziness. Jupiter, bright and high in the night sky, is tied in with ambition. Mars, red, if not blood red, is correlated with war, and Venus, seen when going to bed or getting up, but never in the middle of the night, is perhaps understandably linked with love and lust. Last of the seven, Mercury, is hard to see and fast moving, so an assignment to the god of thieves is reasonable enough. The employment of the planets for astrology also gives an answer to the question that most sharply separates the astronomer from the astrologer – what are the planets for? Rational man holds that if the planets were divinely created there must have been a reason.

If the position of the planets is of significance in personal or national fortune, it becomes necessary to be able to say not only where the planets are now but also to calculate where they were and where they will be; an astrologer may have to prepare a horoscope for an hour when observations were not made or recorded or for a future time. Eudoxus, a pupil of Plato, whose life spanned the first half of the third century BC, was the first to make an attempt to get to grips with the underlying problems of planetary motion. As seen from the Earth, the motion of the planets is complicated and irregular, but they do have important common characteristics, and these were listed by the Assyrians. Fundamental to the whole system is the fact that the stars are to be found in all directions, and they are still present during daylight; if it were not for the brightness of the daytime sky they would be always visible. The Sun's position can be described with reference to these stars and it travels once a year round a well-defined path, the ecliptic. Finally, the Moon and the other planets all follow paths that never get far from this ecliptic, covering a narrow band called the zodiac.

The details of the planets' motions are much less straightforward. The most striking feature is that at regular intervals a planet appears to stop its steady clockwise motion round the zodiac, go backwards for a while, stop again and then resume its forward progress. The intervals between these episodes, and the duration of the backward, or retrograde, motion vary from planet to planet. The complete trajectories of the planets, unlike that of the Sun, do not repeat from year to year, one loop of retrograde motion taking place at a different point in the zodiac from the preceding or following loops.

In order to explain these observations, Eudoxus invented a new device, the mathematical model. This was an enormous step forward: Eudoxus's predecessors from Pythagoras to Plato had all described philosophical pictures of the universe, but these were not designed for calculation. The model Eudoxus devised was based on these Pythagorean and Platonic philosophies and leaned heavily on the desirability and mathematical convenience of uniform circular motion. In his imaginary system each planet was fixed to an individual spherical shell and each such shell was supported on two or three concentric shells. In turn, each of these was pivoted so that it could rotate within the next one out, but each shell had its axis in a different direction. With this model Eudoxus could re-create the main features of planetary motion. Twenty-seven imaginary spheres were necessary to represent the whole universe, and all were firmly centred on a spherical Earth.

As a pioneering effort Eudoxus's spheres were a brilliant invention and provided the first demonstration that a complex motion could be broken up into the sum of several simple components. This method of analysis, the resolution of each planet's variable motion into a combination of simpler uniform motions, was so nearly successful that for the next 2000 years all the many attempts to explain planetary motion used a similar approach.

The subsequent step was mainly retrograde, as Aristotle changed Eudoxus's mathematical model of the universe into a physical one. He replaced the imaginary spheres by real ones and attempted to find the source of motive power for the system. This, he assumed, had to come from the outermost sphere, the *primum mobile*, and extra shells were required to transfer the drive from one concentric sphere to the next, so that he finished with fifty-five no longer imaginary but "crystalline" concentric layers.

The first phase of early astronomy ended at the turn of the third century BC, when war swept through Greece again, and the Greek school of Socrates, Plato and Aristotle fell into decline, never to recover its pre-eminence. The next developments took place in the new Hellenic city of Alexandria in Egypt, where Alexander's general, Ptolemy, founded a university and library which were to grow rapidly to form the chief intellectual centre of the Mediterranean world. One of the earliest products of this new centre was the development of the first astronomical instruments that went beyond the primitive sundial, plumb-line and water clock. Chief among these were the armillary rings and their descendants, the astrolabe, the armillary sphere and the quadrant.

An armilla is a bracelet, a metal ring several centimetres in diameter, and two instruments were based on such a ring – the solstitial armillary, a measuring instrument, and the equinoctial armillary, an indicator. The latter is the simpler of the two, and comprises a metal ring set up at the correct angle so that the plane of the ring is parallel to the plane of the Earth's equator. Twice a year, once in the spring and once in the autumn, the shadow of the upper half of the ring will fall precisely on the lower half. These two times mark the equinoxes. Although simple, this device was not a trivial development. It is a common belief that early man, free from city lights and industrial pollution, must have enjoyed a crystal-clear sky for his astronomical measurements. This is not the case, particularly in Alexandria and Babylon. Both these cities are close to vast arid dust bowls and in both places the quality of the sky, especially near the horizon, is very poor. If conditions are bad, as they were for the 1970 eclipse expeditions to the southern edge of the Sahara, the dust in the air resembles light cloud cover and the daytime sky is nearer white than blue. It must have been very difficult to identify the equinox by looking for days when the Sun rose due east and set due west, and the equinoctial armillary was a major step forward.

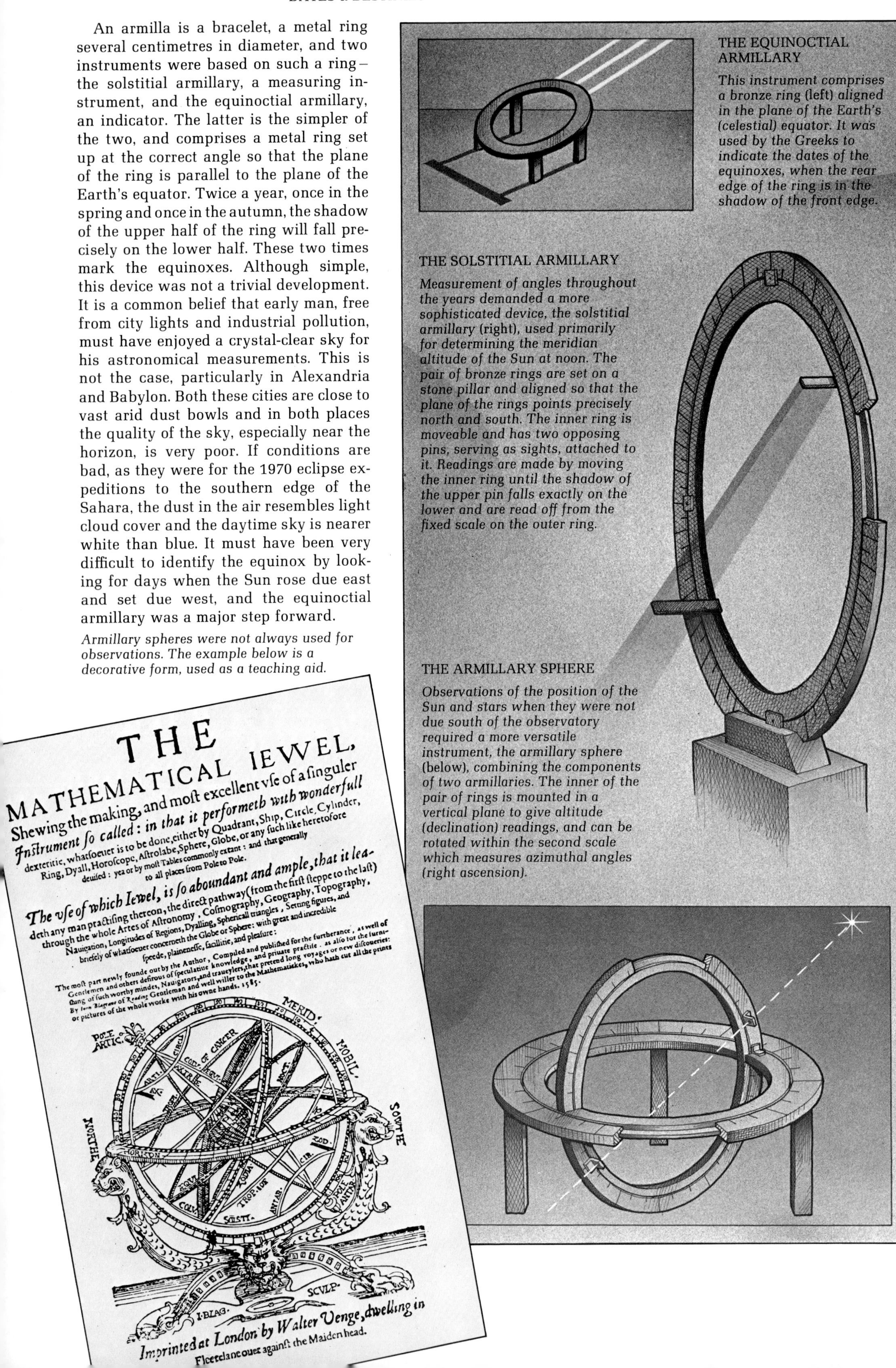

THE
MATHEMATICAL IEWEL,
Shewing the making, and moſt excellent vſe of a ſinguler
Inſtrument ſo called: in that it performeth with wonderfull
dexteritie, whatſoeuer is to be done, either by Quadrant, Ship, Circle, Cylinder,
Ring, Dyall, Horoſcope, Aſtrolabe, Sphere, Globe, or any ſuch like heretofore
deuiſed: yea or by moſt Tables commonly extant: and that generally
to all places from Pole to Pole.

The vſe of which Iewel, is ſo aboundant and ample, that it lea-
deth any man practiſing thereon, the direct pathway (from the firſt ſteppe to the laſt)
through the whole Artes of Aſtronomy, Coſmography, Geography, Topography,
Nauigation, Longitudes of Regions, Dyalling, Sphericall triangles, Setting figures, and
briefely of whatſoeuer concerneth the Globe or Sphere: with great and incredible
ſpeede, plaineneſſe, facillitie, and pleaſure:

The moſt part newly founde out by the Author, Compiled and publiſhed for the furtherance, as well of
Gentlemen and others deſirous of ſpeculatiue knowledge, and priuate practiſe. as alſo for the furni-
ſhing of ſuch worthy mindes, Nauigators, and traueylers, that pretend long voyages or new diſcoueries:
By Iohn Blagraue of Reading Gentleman and well willer to the Mathematickes, who hath cut all the prints
or pictures of the whole worke with his owne hands. 1585.

Imprinted at London by Walter Venge, dwelling in
Fleetelane ouer againſt the Maiden head.

Armillary spheres were not always used for observations. The example below is a decorative form, used as a teaching aid.

THE EQUINOCTIAL ARMILLARY

This instrument comprises a bronze ring (left) aligned in the plane of the Earth's (celestial) equator. It was used by the Greeks to indicate the dates of the equinoxes, when the rear edge of the ring is in the shadow of the front edge.

THE SOLSTITIAL ARMILLARY

Measurement of angles throughout the years demanded a more sophisticated device, the solstitial armillary (right), used primarily for determining the meridian altitude of the Sun at noon. The pair of bronze rings are set on a stone pillar and aligned so that the plane of the rings points precisely north and south. The inner ring is moveable and has two opposing pins, serving as sights, attached to it. Readings are made by moving the inner ring until the shadow of the upper pin falls exactly on the lower and are read off from the fixed scale on the outer ring.

THE ARMILLARY SPHERE

Observations of the position of the Sun and stars when they were not due south of the observatory required a more versatile instrument, the armillary sphere (below), combining the components of two armillaries. The inner of the pair of rings is mounted in a vertical plane to give altitude (declination) readings, and can be rotated within the second scale which measures azimuthal angles (right ascension).

The solstitial armillary is a more important device. This is a pair of rings, one of which revolves inside the other. The outer ring is graduated and held in a vertical plane with the horizontal diameter pointing south. The inner ring carries two indicators at opposite ends of a diameter. In use the inner ring is turned until the shadow of one indicator falls on the other and the position read from the fixed scale on the outer ring. With such an instrument it is possible to demonstrate the inequality of the intervals between winter solstice, spring equinox, summer solstice, autumn equinox and the next winter solstice.

The solstitial armillary has one major limitation in that it is designed to measure the altitude of the Sun at noon only, that is to work in the plane of the meridian. This is the best method to use when observing the Sun as errors in the alignment of the instrument have least effect. Unfortunately, some astronomical phenomena do not occur on the meridian, and an astronomer determining a star's position needs two measurements to fix its location in the sky. If he measures its altitude on the meridian to give the celestial equivalent of latitude, he has no measure of celestial "longitude". In recent centuries, astronomers have used clocks to convert the apparent motion of stars across the sky as the Earth steadily spins to this second co-ordinate, but in the absence of a reliable clock the Alexandrians had to develop a device to measure both co-ordinates simultaneously.

The new invention was the armillary sphere, an instrument that combines the components of two solstitial armillaries and replaces the shadow-measuring technique of solar astronomy by a pair of sights along which the astronomer peers. The two pairs of rings are set at right angles and mounted so that one of the rings is in the plane of the equator, a setting determined with the simple solstitial armillary. To use the armillary sphere, the innermost ring, with the sights on it, is moved so that the star or planet of interest can be seen. The two scales then give a pair of angles. One of these is the angle north or south of the celestial equator. This is the declination, the stellar equivalent of latitude, and is constant for a star and changes only very slowly for a planet. The other angle changes steadily as the Earth rotates, but with two armillary spheres and two observers it is possible to measure the constant differences in "right ascension", the celestial equivalent of longitude, between two stars. As with the Earth, the sky needs a zero for longitude, the equivalent of the Greenwich Meridian. This is defined as one of the two points where the Sun's path, the ecliptic, crosses the celestial equator, that is has zero declination. The moments of these crossings are more precise definitions of the equinoxes than merely looking for days when day-time and night-time are equal, and of the two crossing points the spring equinox is defined to have zero right ascension.

With instruments such as these the Alexandrian astronomers Aristyllus and Timocharis started, in about 290 BC, a programme of precise astronomical measurement and compiled a catalogue of the resulting star positions. At the same time a geometer, Aristarchus, described how to measure the distance to the Sun by making observations at solar eclipse, lunar eclipse and precisely at quarter Moon. He illustrated the method by taking arbitrary figures. He took the apparent diameter of the Sun to be 1°; the ratio of the radius of the Earth's shadow on the Moon at the lunar eclipse to the radius of the Moon to be 2; and the angle between the Earth, Sun and Moon to be 3°. His rough figures showed the Sun to be bigger than the Earth and he favoured a solar system that was centred on the Sun not the Earth. This idea was criticized as impious and was not pursued. However, Archimedes, when writing on the invention of a notation for big numbers, took up Aristarchus's method for the solar

His development of a counting system that could cope with large numbers led Archimedes, 287–212 BC, to make one of the earliest calculations of the immense size of the universe.

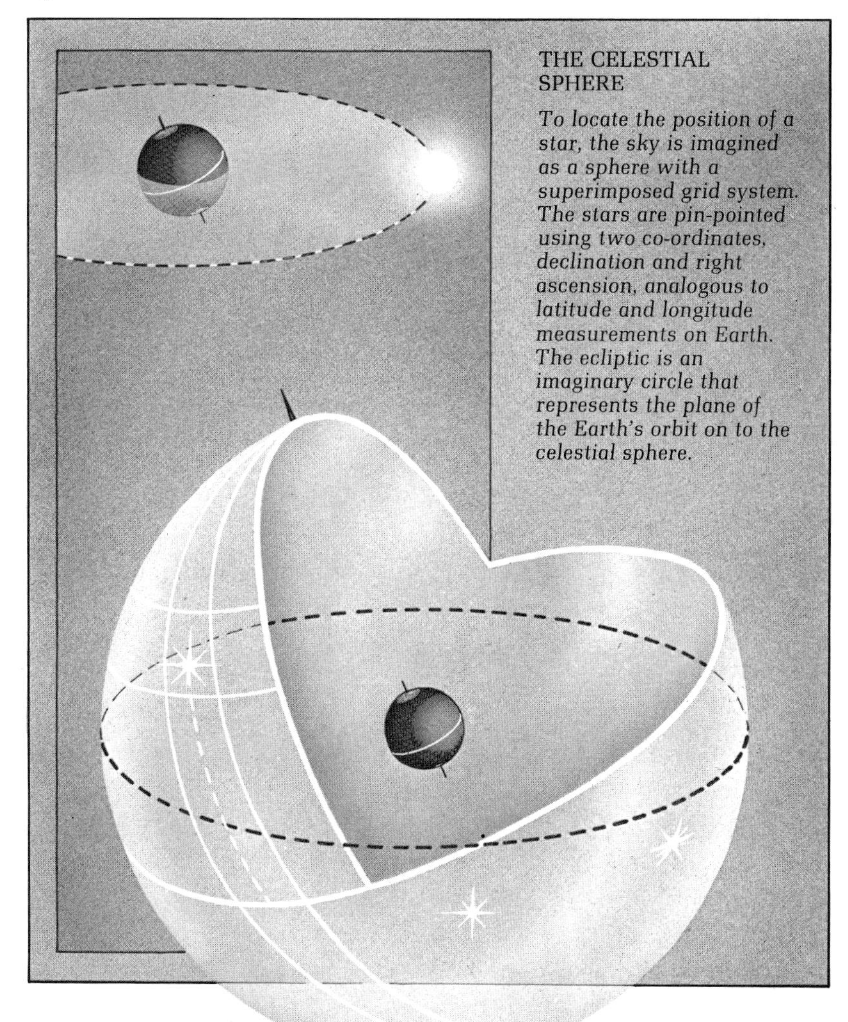

THE CELESTIAL SPHERE

To locate the position of a star, the sky is imagined as a sphere with a superimposed grid system. The stars are pin-pointed using two co-ordinates, declination and right ascension, analogous to latitude and longitude measurements on Earth. The ecliptic is an imaginary circle that represents the plane of the Earth's orbit on to the celestial sphere.

distance and adjusted the parameters at all opportunities to make the universe as big as possible, ending with a calculation of the number of grains of sand that would fit into this universe. The importance of this was in the new ability to calculate with enormous numbers, rather than in the astronomical guesswork. One device which did emerge from this was an instrument to measure the angular diameter of Sun and Moon. Known as "Archimedes' staff", it consists of a cylinder and a stick, and in use the cylinder is moved along the stick until it just covers the disc of the rising or setting Sun or Moon.

One of the greatest of Greek geometers, Apollonius of Perga, born c. 230 BC (below), developed the epicycle and deferent as a simple model to account for the motions of the planets.

The work of Aristyllus and Timocharis was extended by Eratosthenes in the latter half of the third century BC, with the compilation of a second catalogue giving the positions of nearly seven hundred stars. Eratosthenes was an all-rounder and was so nicknamed "pentathlos" in Alexandria, where he was director of the library. He compiled his star catalogue and measured the circumference of the Earth with impressive accuracy. The first conclusion of the new style of measurement was that Eudoxus's set of concentric spheres could not accurately describe the planetary motions. Fortunately Alexandria had the intellectual resources to do something about the failure and one of Eratosthenes' colleagues, the "great geometer" Apollonius of Perga, developed a new theoretical approach. This was based on a pair of uniform circular motions along arcs of different radii, the epicycle and the deferent. It is a more flexible approach than Eudoxus's, as half the components of the motion do not have to be centred on the Earth. About 225 BC Apollonius showed how to calculate the radii and rates of rotation of epicycle and deferent for each planet and the new theory very quickly gained acceptance.

EPICYCLE AND DEFERENT

Early astronomers found the complex motion of the planets difficult to explain in terms of their Earth-centred view of the universe. To overcome the problem, they invented a system using a combination of two circular motions. The planet moved at the end of an imaginary arm, the epicycle, that itself moved round at the end of a longer arm, the deferent.

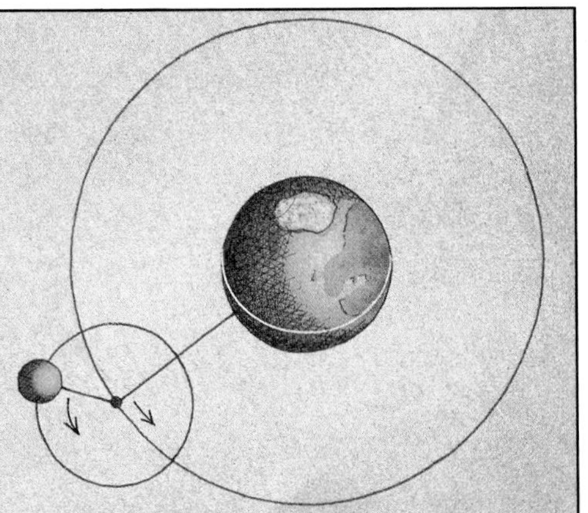

Historically, we have now reached the time when the Great Wall of China was built, so it is opportune to pause and look at Oriental astronomy. Chinese interaction with ancient Middle Eastern civilization seems to have been limited to a brief contact around 1400 BC, from which the Chinese acquired bronze and writing. In isolation, with or without a wall, the Chinese and Koreans developed their own astronomy, again with strong astrological overtones and using armillary spheres for measurement. Chinese astronomy declined in the Middle Ages, so that in the sixteenth century the subject had to be learned again from visiting Europeans, mostly Jesuit missionaries. However, astronomical records compiled by the ancient Chinese and their Korean neighbours are still of great interest, as these include dated observations of the first appearance of many "new" stars – novae and supernovae – including the supernova of AD 1054, which has left the famous Crab Nebula as its debris.

Meanwhile, in Alexandria the situation was deteriorating and the next major astronomer of the Alexandrian school, Hipparchus, left Egypt to work in Rhodes. Hipparchus was, without a doubt, the greatest practical astronomer of antiquity and he improved or originated almost all classical astronomical measurements. He attempted by three different methods to determine the distance from the Earth to the Sun. In practice Aristarchus's method does not work, because one must be able to identify the instant of quarter Moon (half dark, half bright) to within ten minutes if the measurement is to be of any use. Two of Hipparchus's new methods were based on observation of the solar eclipse in March 129 BC. This eclipse was total near Byzantium, but only eighty per cent of the Sun was covered in Alexandria. From this it was possible to deduce that the distance to the Sun is 490 Earth radii. An alternative approach based on the argument that if the Sun were close there would be an observable effect on its position with respect to the stars, gave an estimate of a similar order of magnitude. This answer, 3 million kilometres ($1\frac{4}{5}$ million miles), is about a fiftieth of the correct value, but here was firm evidence that astronomical distances were of a totally different magnitude from everyday experience. Hipparchus's next achievement was to show that even Apollonius's improved theory of the planets' motions did not fit his new and more precise measurements. He did not, however, propose a new theory but concentrated instead on making an extensive and accurate series of measurements of planetary positions.

Hipparchus's greatest discovery was catalysed by the appearance of a new star in 134 BC. It is hard to be sure that such a star has newly appeared and was not there last night, especially as a new star visible to the naked eye is a very rare event. In case another star should appear unexpectedly, Hipparchus compiled a catalogue of the positions of all the stars he could measure, including an estimate of their brightness ranging from "first magnitude" for the brightest to "sixth magnitude" for the faintest. He then compared his results with the catalogues of Eratosthenes and of Aristyllus and Timocharis. The right ascension measurements in the 70-year-old catalogue of Eratosthenes were a degree different from Hipparchus's measurements, while the 150-year-old measurements of the earlier catalogue were out by two degrees. Hipparchus realized that this difference was too large to be ascribed to errors in the measurements and must therefore come from the movement of the zero point on the right ascension scale. The Sun's position in the sky at the spring equinox is not constant from year to year, but is slowly and steadily changing. This slow apparent movement of the Sun's position against the stars is in the same direction as the general solar and planetary motions, so the phenomenon that Hipparchus recognized is called the precession of the equinoxes.

Only in the last 300 years has precise astronomical measurement been recognized as having any military significance. Prior to that it was a gentle luxury that could not be afforded in times of strife. Such a time followed Hipparchus, as the inexorable Roman military machine smashed one Hellenic kingdom after another. In Egypt a series of increasingly inept Ptolemies aided the decline. The only event of astronomical interest, apart from the Star of Bethlehem, was forced

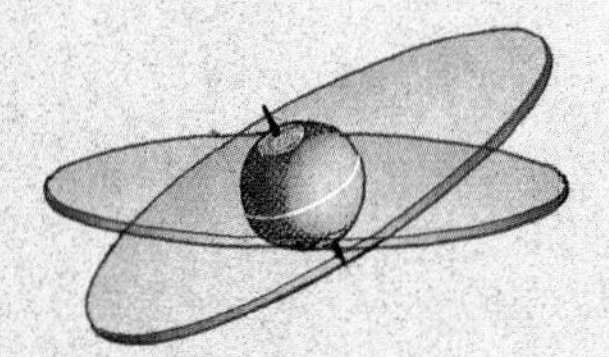

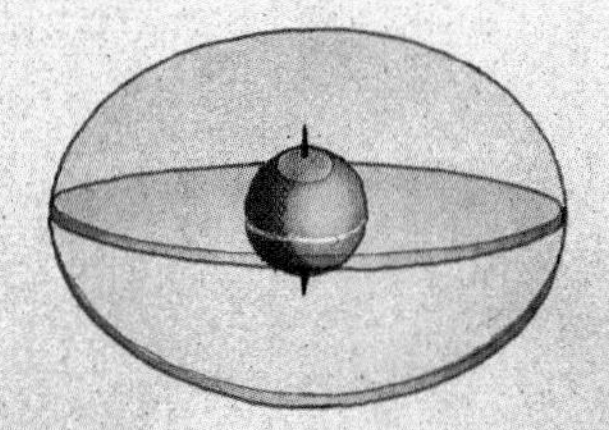

PRECESSION

The Sun and Moon both exert gravitational forces on the Earth that combine to alter the tilt of the Earth's axis in a pattern which repeats every 26,000 years. In consequence, the point of intersection between the celestial equator and the ecliptic at the vernal equinox – the zero point of the celestial co-ordinate system – moves through the signs of the zodiac. This slow motion affects the measured positions of all the stars. At present the Earth's axis points towards the Pole Star and the vernal equinox is in the constellation of Pisces, slowly approaching Aquarius.

The Crab Nebula (above) is an expanding cloud of gas, the remains of a supernova explosion that was seen and recorded by Chinese astronomers as a new bright star in AD 1054. However, the original explosion occurred in 3000 BC, and the light has taken more than 4000 years to reach the Earth.

Observations that were made by early Chinese astronomers, such as the mythological Chitasei Goyo (left), have proved to be a vital resource for modern astronomers. Ancient Oriental records and star charts are used to identify past supernovae.

through by that energetic codifier, lawmaker and general administrator, Julius Caesar. In 46 BC, Caesar reformed the Roman calendar.

This calendar was by then badly in need of reform. Rome was still using an approach similar to that used centuries before by the Assyrians and the Medes, where a month was occasionally added on the instructions of the High Priest. In the last years of the republic, the High Priest had not proved as upright as he should, and the extra month had become a political weapon for curtailing or extending the period of office of the consuls. By 46 BC the calendar had slipped, or been pushed, about three months out of step. Caesar's action in putting it right and setting up the new pattern of arbitrary months, unrelated to the Moon, plus a leap year every fourth year, took place in what was called, quite understandably, the year of confusion. The modern equivalent would be 29 February recurring for eighty-one consecutive days in a year that postponed Christmas to next Easter!

The last great astronomer of antiquity was Claudius Ptolemy, who worked in Alexandria in the first half of the second century AD. He was not related to the extinct Ptolemaic dynasty which had ended in a blaze of glory in 30 BC as Cleopatra committed suicide. He was, however, a great astronomer. He compiled the astronomical knowledge of his day into a single text, the *Mathematical Collection* (*Mathematike Syntaxis*), soon renamed the *Great Collection* (*Megale Syntaxis*). Ptolemy took Hipparchus's planetary measurements and developed an extension of Apollonius's theory that was a very good fit to the observations. He described two new instruments for more precise observation. Finally, as an astronomical observer he told lies – lies of such consistency and reasonableness that they stood as truth for nearly 2000 years. The net effect of these efforts was to establish a mathematical model of the universe, the Ptolemaic system, and to present it to the world in convincing and authoritative manner.

The main improvement Ptolemy added to Apollonius's system of epicycles and deferents was the addition of the equant, a mathematical manoeuvre to get away from Earth-centred motions. He allowed the inner end of the deferent to be placed a short distance away from the Earth, and required the uniform circular motion to take place not about the pivot point of the deferent's motion but about a new and arbitrary point, the equant, on an extension of the line joining the centre of the Earth to this pivot point. Ptolemy was quite clear that this procedure was a matter of mathematical convenience and not a physical picture, and was justified only because it worked. The computation required to get from observation to theory is forbidding: for each planet it is necessary to find the values of ten parameters – four lengths, four angles and two angular velocities.

The instruments Ptolemy described were improvements on the solstitial armillary. They were a quadrant, a graduated quarter-circle where the use of a smaller arc allowed a larger radius and so greater precision, and Ptolemy's Rules, an alternative to the quadrant which used a straight graduated bar instead of a circular arc. The Rules, or Triquetrum, measures the straight line joining the ends of an arc rather than the angle described by the arc

Julius Caesar, 100–44 BC, forced through a long overdue reform of the Roman calendar in 46 BC. He imposed the new Julian calendar, devised with the help of the Alexandrian astronomer Sosigenes.

and much of Ptolemy's book uses the modified trigonometry needed to exploit such measurements.

The most important of Ptolemy's achievements was that the book he wrote was a complete summary of all he knew, including Hipparchus's catalogue and the Babylonian eclipse records, which by then extended for almost a thousand years. The Babylonian Empire set up by the senior of Alexander's generals, Seleucus, had maintained a continuous but rather arid astronomy through the years. Its astronomers' observational approach was to determine the precise number of years after which a given planet returned to the same part of its motion in the same part of the sky. Thus for Mars forty-two revolutions take seventy-nine years, to within a few days, while for Venus eight revolutions is very close to five years. Given these numbers, prediction is much eased. To foretell the position of Mars you look up the records of where it was seventy-nine years ago and assume the pattern will repeat. The Babylonians seem to have stuck firmly to a purely arithmetical approach to astronomy; there is no equivalent of the Greek and Alexandrian models. Ptolemy found the Babylonian records useful and employed their eclipse results to improve Hipparchus's value of the mean distance to the Sun. His final value was 1210 Earth radii or nearly 8 million kilometres (5 million miles). This very inaccurate number happens to put the Sun outside the largest of the Aristotelian crystal spheres needed to carry the Moon, Mercury and Venus and was taken throughout the Middle Ages as a demonstration that Aristotle was right.

Ptolemy's book was the definitive text for ancient astronomy, so much so that scarcely any other text was felt to be worth copying during the interval that followed. This makes the history of astronomy hard to unravel and, of course, adds much weight to the influence of the Alexandrian school. Considerable labour was involved in calculating the position of a planet from the theory given in Ptolemy's book. There were no logarithm tables and no decimal notation to aid the human computer, who needed to be pretty good at sexagesimal long division, the sky being divided into 360 degrees and each degree into 60 minutes. To aid the working astrologer, Ptolemy wrote a sequel, known as the *Handy Tables*. This gave details of the position and motion of the Sun, Moon and planets at regular intervals for a "Sothic cycle" (the interval between instances when a 365-day calendar gives the same date for the rising of Sirius) of 1461 years. As the calculation ran from 323 BC to AD 1151 the tables were copied and re-copied: even more manuscripts of the *Handy Tables* are known than of the book on which it is based.

The last aspect of Ptolemy's work, his observations, are a hilarious contrast to all his other efforts. He could understand the instruments and their purposes, he could reduce the data and extend the interpretation of the results. What, it appears, he could not do was to use the apparatus he describes. What he did instead was to follow the route well known to all student astronomers when things go awry. He looked up the previous measurement (usually by Hipparchus) or he used his own theories to work out what result he ought to have obtained. He then presented this as if measured. The weakness of this approach is that you still agree with Hipparchus's measurements even when they are wrong. Most of the time this did not matter as the measurements were being used to illustrate methods of calculation and were particular to the latitude of Alexandria and often to a particular day. The ill-effect of Ptolemy's cheating was that his book gained an impressive authority, as theory and observation agreed so well.

After Ptolemy's death in AD168 nothing happened in the study of astronomy for a very long time. In the years immediately after his death there was little that could be done. Theory and observation agreed and with the *Great Collection* published, research interest in the subject was minimal. By the time interesting discrepancies had accumulated, the peoples of the eastern Mediterranean had developed a new occupation – literally beating the hell out of one another in pursuit of an accurate definition of God. While Vandal, Alan, Hun and Goth pounded the Western Roman Empire to pieces, the Eastern Empire's factions were sectarian rather than racial with Arians, Sabellians, Monophysites and Nestorians matching frightfulness with frightfulness. In AD 640 an outsider joined in the game, and the Mohammedan Arabs took Alexandria. The new fanatics completed the destruction of the library, already much damaged by the Christians.

Ironically, it was the descendants of these destroying Arabs who were to preserve and extend Ptolemy's work. The first stimulus to Arab astronomy seems to have come from India, where much classical astronomy had been learned, and this knowledge augmented by further additions carried south by Hunnish invasions. Ptolemaic astronomy reached the newly founded Baghdad during the reign of Haroun al Rashid, and by AD 800 the Greek texts were being translated into Arabic. The first effects were the compilation by Abu Ma'shar of an encyclopedia of astrology, a book of vast influence, and the writing of a commentary on the newly translated *Megale Syntaxis*, now renamed the *Al-majasti*, by Anu al-Abbas al Farghani. With these two new works Arab astronomers began to look at the discrepancies that had built up between Ptolemy's predictions and their own new observations. By the end of the ninth century, al Khwarizmi and Habash, "the computer", had calculated a new set of tables, and Thabut ibn Qurra had developed a theory that the precession of the equinox was not a steady motion but an oscillation.

The tenth and eleventh centuries saw the first flowering of this new school of astronomy. Abu Abdullah al Battani reworked the theory to include ibn Qurra's added complexity and compiled a new and better star catalogue. Like Ptolemy he invented a new instrument, the mural quadrant, but unlike Ptolemy he was a skilled observer. His master work, the *Opus Astronomicon*, includes for the first time observations of sufficient subtlety to show that the closest approach of Earth and Sun is not a fixed time after the equinox but is also affected by precession. At the beginning of the tenth century Abd al Rahman ibn Umar, nicknamed as-Sufi, the wise, compiled another list of the fixed stars, for the first time grouping them into constellations. Except for minor changes the boundary lines of the constellations of today are still where as-Sufi placed them.

At about the same time a second school of Arab astronomy had been founded in Toledo, Spain, at the other end of the Moslem world. Computation was more to the taste of the Western school than observation, and its main efforts were devoted to improvements in the theory and to calculation. The latter led to the Toledan tables of AD 970, tables that were revised and extended by al Zarquali in 1060. Both Moslem schools were disrupted by invasion: the Turks entered Baghdad in 1055 and the Christian king of León took Toledo in 1085. The Spanish school was the first to recover, three men of world class emerging in the second half of the twelfth century. The chief astronomer at Toledo was al Bitruji, who reinvented Eudoxus's set of concentric spheres, though in a far more elaborate form, and his two influential contemporaries were ibn Ruashd, who wrote a long and detailed commentary to the works of Aristotle, and Maimonides, who was expelled from Spain and became adviser to the Turkish leader Saladin.

The astrolabe (far right) was used both for observation and for computation. The face shown was used for calculation, while observations were made from the reverse side, which carried sights. The device could be used for telling the time, finding the position of the stars, determining latitude and even measuring heights. Its fundamental components are the metal fretwork chart of the brighter stars, with the celestial pole represented at its centre, and an engraved set of position lines on the underlying plate. In use, the sights are aligned on a specific star so the latitude or the time could be deduced by using the metal star map and the scale beneath as a pocket calculator.

The torquetum (right), from Peter Apian's Astronomicum Caesareum *of 1540, was devised so that observations could be made directly in the celestial co-ordinates. It consists of a pair of sights attached to the uppermost pair of scales. Beneath this are additional adjustments that allow the instrument to be set up so that both the latitude angle of the observer and the 23½ degree tilt of the Earth's axis can be offset. When correctly set, the uppermost scales will read declination and right ascension, thus eliminating a great deal of the tedious arithmetic needed to convert altitude and azimuth readings into celestial angles.*

In the east, the revival in astronomy was brief and essentially spanned the life of one man, Nasir ed Din al Tusi. He started his career as an astronomer in Baghdad and later became a prisoner and compulsory astrological adviser to the Shaikh al Jebel, the "Old Man of the Mountains", who was the leader of the murderous Moslem fanatics, the Assassins. Al Tusi was freed from the Shaikh's mountain fortress by Hulagu, the grandson of Genghis Khan, and became adviser to his new protector. After a rather sinister involvement with the Mongol capture of Baghdad, he retired to run a new observatory near Tabriz. During this unusually eventful life al Tusi found time to develop two new astronomical instruments, the azimuth quadrant and the torquetum, the last new devices for observational astronomy before the birth of the Renaissance in northern Europe a couple of centuries later.

The technical evolution of the Arabian centuries is steady and logical. Ptolemy had pointed out that the quadrant was more accurate than the armillary ring because a larger radius could be used. Extending the argument, big quadrants were better than little ones, provided that they were rigid. Rigidity was achieved by mounting the device on a wall, and so the mural quadrant was born. The same argument that leads from solstitial armillary to armillary sphere, lack of versatility, leads to the azimuth quadrant, which is a quadrant mounted on a rotating table. A quadrant of this type measures angles with reference to the local scene, the altitude above the horizon and the angle east or west of due south. Every measurement in this local system must then be converted to the celestial angles, declination and right ascension, by repeated mathematical calculation, before any further work is possible. The torquetum avoids this conversion by having four rotary motions and associated scales. The lower pair of scales is used in setting up the instrument to align the upper pair of scales in the correct orientation to read celestial angles. The complicated torquetum was also of importance as the device needed to check the accuracy of the most famous of the Arab astronomical instruments, the astrolabe.

The origin of the astrolabe can be traced back before the Moslem era, certainly to Theon of Alexandria in the fourth century AD and possibly as far back as Hipparchus. None the less, it was the Arab astronomers and craftsmen who developed and popularized its use. The astrolabe is two instruments in one: one face of the device is for making observations, the other is a pocket calculator. The observing face is related to the solstitial armillary and carries sights and a scale.

The first person to argue forcefully against the traditional Earth-centred picture of the universe was Nicolas Copernicus (right). His alternative model, with the Sun at the centre, accounted for the motions of the planets and was published in 1543, the year of his death. Nearly a century later it was placed on the forbidden book list by the Church in Rome.

From antiquity people were aware that the planets always lie within a few degrees of the ecliptic, the plane of the Earth's orbit around the Sun, and the stars within this narrow band were separated into 12 groups, the signs of the zodiac. They are shown (below) in an illustration from Regiomontanus's Epitomy of Ptolemy, *published in 1496.*

In use the astrolabe is held up by the top ring, so that it is hanging vertically. For stellar observations the observer looks along the sights, carried on a rotating arm, and sets them in alignment with the star observed. The resulting angle shown on the scale is the zenith angle of that star for the particular time and latitude. The importance of the observation was that if you knew two out of the three – star position, time and latitude – you could calculate the third. An astronomer, knowing the latitude of his observatory and the time (usually midnight) could determine the position of the star observed. Much more important, the educated astrolabe owner, knowing the star positions, could determine the time if he was at home, or the latitude when voyaging. The same measurement could be made with an azimuth quadrant, but a good deal of tedious calculation was then needed to get the required answer. This is where the second face of the astrolabe comes in. This face is a very elaborate and ingenious computer – a two-dimensional trigonometric slide-rule a millennium older than the modern straight logarithmic device, which was invented by William Oughtred in about 1630. The astrolabe has also enjoyed a better fate than other pocket timekeepers, the nocturnal and the pocket sundial. The instrument is not only useful but in the hands of skilled Arab craftsmen it was also elegant, even beautiful. As such, expensive presentation astrolabes were prestige gifts and have been preserved when their more humble relations were thrown away.

The next stage of the development of astronomy was again the result of military confrontation. Christians fought Arabs at both ends of the Mediterranean; in the west this was for possession of Spain, while in the east the Crusades brought the leaders of western Europe up against Saladin, their almost infinitely more civilized opponent. In both cases western Europe learned a great deal, and this knowledge helped to trigger the Renaissance. The first step was to translate the new knowledge into Latin, a process that not only changed the title of Ptolemy's work once more, from the *Almajasti* into the *Almagest*, but also mangled most of the names of the Arab astronomers involved in the transmission and extension of Ptolemy's works. Abu Ma'shar became Albumazar; al Battani, Albategnius; Thabut ibn Qurra, Tobit; al Zarquali, Arzachael, and ibn Ruashd, Averrhoes. After translation the next priority was to incorporate the new knowledge into a Christian background. This process was dominated by two men, Albert of Cologne and Thomas Aquinas. These two accomplished the synthesis of Aristotelian philosophy and Christian doctrine, though not without some controversy.

The first independent western European effort in astronomy took place in Spain, where King Alfonso X of León and Castile sponsored the calculation of a new set of tables in 1252. The "computers" faced a formidable task, and took about a quarter of a century to complete the work. The reason for this was that they adopted the Moslem theory of the planetary motions. This had extended Ptolemy's epicycle system by accounting for all discrepancies between theory and observation by adding more and more epicycles, each needing to have its radius, rotation rate, plane of motion and starting position determined. Alfonso, who was paying for all this, commented that, if during the creation the Almighty had asked his opinion, he would have recommended a much less convoluted universe.

It was a good deal later, the middle of the fifteenth century, before European observational astronomy achieved sufficient accuracy to rival the old data. Georg Purbach, at the University of Vienna, began to build and use precise instruments, and his pupil, Johannes Müller, better known as Regiomontanus, furthered the art. The fifteenth century also saw the last of the great Moslem observatories, built by Ulugh Beg in Samarkand. This new school took the idea of a big quadrant to its limit, building an arc 18 metres (60 feet) in radius as part of a six-storey building.

Regiomontanus not only observed the sky, he also computed new tables and was the first astronomer to benefit from the new invention of printing. His patron, in addition to supporting both observation and calculation, also bought the necessary printing press. The users of the tables, the astrologers, were still in great form. Even religious books such as the famous *Très Riches Heures* of the Duc de Berry contain astrological information. In Florence, Agostino Chigi had his villa decorated by Raphael with a mural designed to stress the outstanding good fortune that his horoscope promised and which he had achieved. However, not all the astrologers were too concerned about accuracy. Martin Luther was born in 1483, but the planets were more favourably placed in 1484, so his birth date was quietly adjusted to the latter. The great astrological fiasco came in 1524, when virtually all the planets and the Sun were in Pisces. This rare event, occurring once in every 1600 years, foretold a major upheavel and, as Pisces is a water sign, there was clearly going to be a great flood. There was not.

The main astronomical problem of the sixteenth century was, once again, the calendar. The Julian calendar is based on a $365\frac{1}{4}$-day year. This is slightly longer than the tropical year, and the calendar gets out of step by a whole day every 138 years. Since the first ecumenical Council of Nicaea in AD 325 the error had built up to about ten days and once again the date of Easter was in doubt. In 1475 the Pope, Sixtus IV, invited prominent astronomers to Rome to review the problem. Regiomontanus accepted the summons, went to Rome, caught the plague and died. When Pope Sixtus died a few years later the problem of calendar reform was left to simmer throughout the next century.

By the time of Regiomontanus's death Aristotle's theories had been fully integrated into Renaissance thought, with translation direct from the Greek now available in place of the Arabic intermediaries. Regiomontanus himself had published his teacher Purbach's book the *Theoriae Planetorum*, which worked exclusively with a model based, like Aristotle's, on solid crystal spheres. But all this was questioned by a fresh approach developed by Nicolas Copernicus, a Pole who had learned mathematics in Cracow, astronomy in Bologna, medicine in Padua and law in Ferrara. While still a student at Cracow, Copernicus began astronomical work on the reform of the calendar. As the error in the Julian calendar is only eleven minutes a year, roughly half the size of the correction resulting from the precession of the equinoxes, Copernicus was soon involved in the conceptually difficult problem of what is moving and what fixed. This he thought on in surprising isolation; neither the *Almagest* nor Purbach's new book was available in Cracow. The result of his thoughts, which took him a lifetime to develop and present, was a new pattern for the universe, with the Sun in the centre and an Earth which both revolved round the Sun once a year and rotates on its axis once a day. This came to be called the Copernican system.

Copernicus' new theory had very little immediate impact. In Verona, Fracostoro, an astronomer ten years younger than Copernicus, continued to refine the crystal sphere model of Aristotle and al Bitruji. He finished with eighty spheres, in contrast to Copernicus's thirty-four Sun-centred epicycles. The latter, heliocentric, theory was clearly the easier model for calculation, and Erasmus Reinhold, professor at Wittenberg, soon produced new tables, the *Tabulae Prutenicae*, based on Copernicus's work. These replaced the updated Alfonsine tables recalculated by Zaaut, tables that had been used by Columbus on his voyages.

By the beginning of the sixteenth century observational astronomy had advanced very little beyond the standards set by the medieval Arabs. Copernicus had made his own observations, but had preferred Ptolemy's values for some critical measurements. The first major worker to build on the foundations laid by Regiomontanus was William IV, Landgrave of Hesse Cassel. William was able to bring to his estate at Cassel three outstanding technicians, Andreas Schoener, the chief observer, Jost Burgi, a brilliant clockmaker, and Eberhard Baldewein, an ex-tailor who became head of the castle workshop. These three built the first recognizably modern observatory – the first observatory with a rotating dome, the first with a reasonably accurate clock. The clocks were sufficiently accurate to demonstrate at last that the sky did indeed rotate smoothly (or that the Earth did, if Copernicus was right).

The other important innovation at Cassel was the development of the astronomical sextant. This instrument, the last new device in pre-telescopic astronomy, was designed to measure the angular separation between pairs of stars, unlike the other devices available, all of which measured the position of one star with reference to the local horizon and meridian. If a mural quadrant, for instance, is badly calibrated or poorly aligned, the errors are hard to detect. In contrast, the sextant can be used to measure angular distances from star to star right round the sky, finishing with the star from which measurement started. If this is done with inaccurate apparatus the error is immediately apparent as the sum of all the measured differences in right ascension should add to exactly a full circle. The sexant needed two observers, and William IV and Schoener began a programme requiring the careful observation of the positions of a thousand stars in order to provide a precise network of reference points for all subsequent astronomy. This programme was however not completed. In 1575 a Danish visitor to Cassel, Tycho Brahe, so impressed William that he wrote to the Danish King Frederick II adding his authority as both ruler and astronomer to the argument that Brahe's unique talents were a credit to Denmark and the King should therefore set up a new observatory for his subject.

Tycho Brahe was twenty-nine at the time of his visit to Cassel and already had a considerable reputation as an astronomer. As a law student at Leipzig he had proved more interested in the errors of the planetary tables than in his legal studies. In 1568 he had visited Augsburg, and had worked on the design and construction of a giant azimuth quadrant for the Burgermeister, Peter Hainzel, as well as buying a 1·5-metre (5-foot) sextant and a globe for his own use. The new sextant, back in Denmark, had been used to show that the new star that blazed out in 1572, since called Tycho's nova, was a star and not a comet. The next year Tycho published a short book on the nova, and in 1574, after royal assurance that lecturing was not beneath the haughty aristocrats' dignity, he gave a course on astronomy at Copenhagen University. At the time of William's recommendation he had already

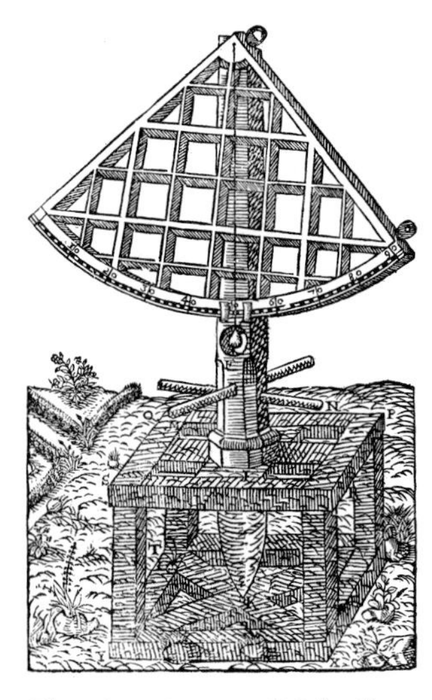

The giant 6-metre (19-foot) quadrant (above) was designed by Tycho when he was only 22, for his friend Peter Hainzel.

Tycho Brahe 1546–1601, made a series of astronomical measurements of unprecedented accuracy, using pre-telescopic instruments such as giant mural quadrants and sextants.

The sextant (above), a more precise instrument than the quadrant, was designed and used by Tycho himself to measure the altitude of stars. It was calibrated to a fraction of a degree.

made plans to leave Denmark and settle once more in Augsburg.

Frederick II moved rapidly and gave to Tycho a life tenancy of Ven, an island 5 kilometres (3 miles) long and 3 kilometres (2 miles) across, a short distance from the harbour of Elsinore. This grant was supplemented by the income from an estate in Norway and of a sinecure canonry at Roskilde, together with a promise of funds to build an observatory on the island.

The result was magnificent: not one but two observatories, Uraniborg and Stjerneborg, equipped with no fewer than twenty-eight instruments – mural and azimuth quadrants, sextants, triquetra, armillary spheres – two of everything and three of most. Space was provided for Tycho and his family, for ten or a dozen students and for twenty staff and assistants. One result of this expenditure (perhaps £10 million in modern terms) was that Brahe was able to pursue his passion for accuracy with ruthless diligence. A hard taskmaster, he even organized the observing programme so that students had to submit their observations as they were made, carefully excluding opportunities for them to compare results with their fellow students. The main innovation in the work at Ven, greatly aided by the vast array of equipment, was the study of instrumental accuracy. Different methods of calibration and different types of sight were tried while the final results were, for the first time, derived by averaging the observations of several observers and several instruments.

Modern hindsight shows that Tycho's sextant achieved a precision of a twentieth of a degree. His later work was so good that he was the first astronomer who had to apply a correction to his observation to allow for the refraction of starlight by the Earth's atmosphere, the minimum correction he applied being one-sixtieth of a degree. Following the pattern set by William of Hesse Cassel, Tycho established a set of twenty-one standard stars and then a network of an additional 756, all with precisely measured positions. Against this background he measured the positions of the Sun, Moon, planets and several comets, using observations of the Moon and Venus, both visible in daylight, to link the Sun with the stellar standards.

All this came to a bitter end in 1597. Frederick II had died and the new king, Christian IV, and his regency council, hard up because of the expenses of running a war, were less sympathetic to their expensive subject. The haughty and arro-

Tycho Brahe housed his superb array of instruments at Uraniborg (below), on the island of Ven, Denmark. With these, he managed to make measurements that had a precision of the order of one minute of arc. In particular, he meticulously recorded the position of Mars and the appearance of a new star in 1572. His dying wish was that his assistant, Johannes Kepler, should confirm Tycho's erroneous model of the solar system.

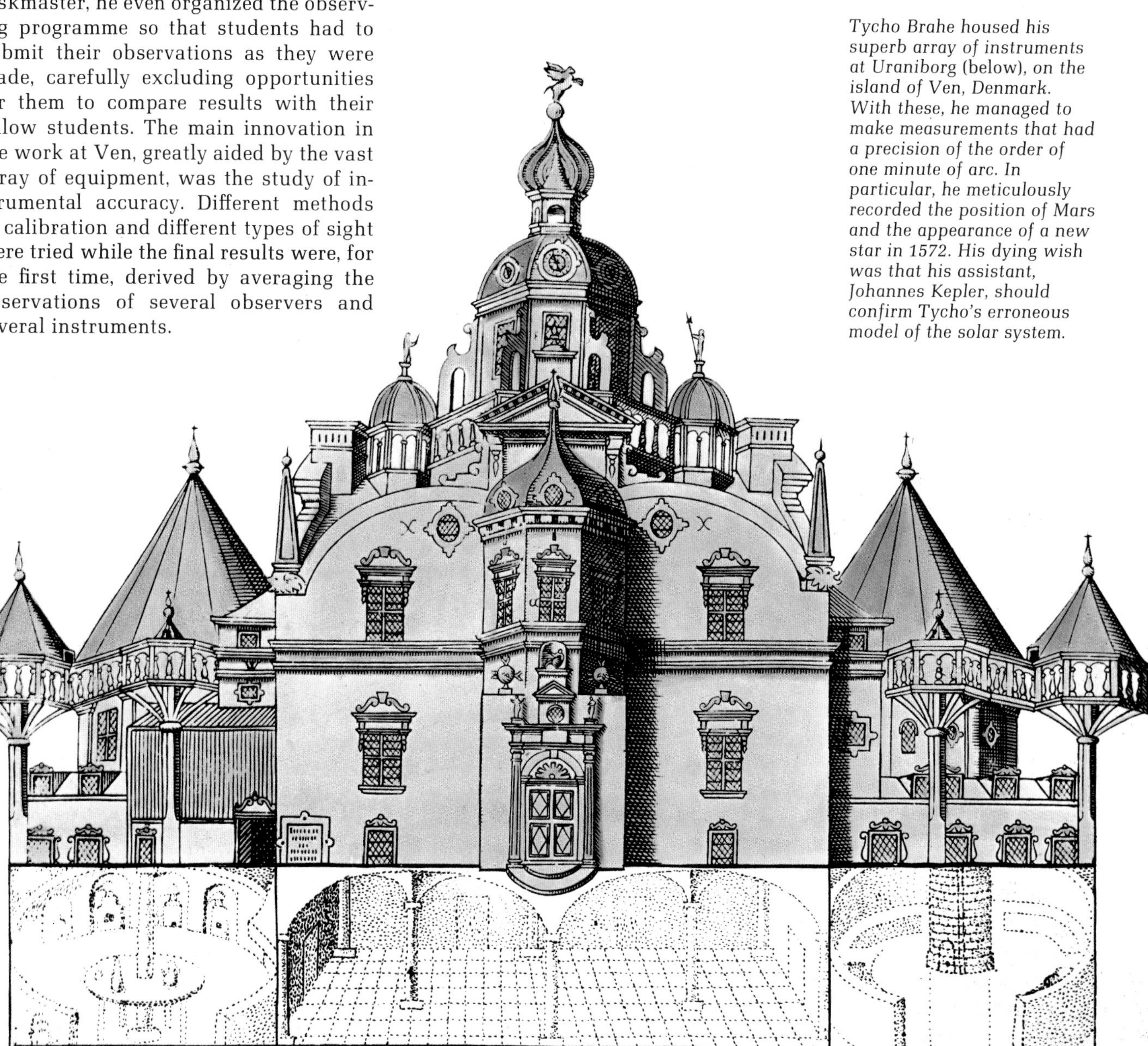

Johannes Kepler, 1571–1630, having laboriously analysed Tycho's data on planetary positions, found that they could be fitted to a set of empirical rules, two of which he presented (below) *in 1609.*

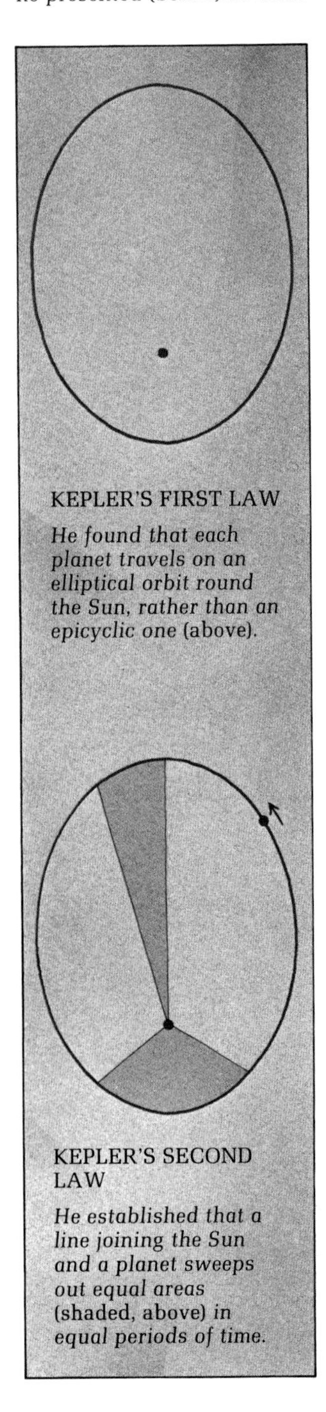

KEPLER'S FIRST LAW

He found that each planet travels on an elliptical orbit round the Sun, rather than an epicyclic one (above).

KEPLER'S SECOND LAW

He established that a line joining the Sun and a planet sweeps out equal areas (shaded, above) *in equal periods of time.*

gant Brahe was as bad a landlord as he was a brilliant astronomer. The tenants on Ven made frequent complaints to the King of ill treatment; the cathedral chapter at Roskilde was incensed at the way in which Tycho ignored his obligations to them, and the Danish seamen complained that the Ven lighthouse was left unlit. Christian IV withdrew his support and took back all the estates granted to Tycho except the lordship of Ven, a life appointment. When the astronomer brought his instruments to Copenhagen, he was forbidden to set up a new observatory. Tycho left, taking all the instruments and his observations with him, going first to Hamburg and then, in 1599, to Prague, where he became court astronomer to Rudolf II. Two years later he died, and soon after Christian IV gave Ven to his mistress and the observatories fell into ruin.

The last years in Prague, however bitter to the still irascible Tycho, were of enormous importance as they brought him into contact with Johannes Kepler, the man who was to turn the volumes of observation into a new world order, making of them the very foundation stones of the dawning age of reason.

Like Tycho Brahe, Kepler was not in Prague of his own choice. After being a scholarship student at Tübingen, he had been appointed mathematics teacher in Graz in 1594 and had already done some work in optics and on planetary orbits, assisted by his old professor, Michael Mastlin. He was a Lutheran, and so was frequently caught up in the religious troubles of his age. Expelled from Graz with the other Lutheran teachers in 1598, he spent a year in Prague, returned to Graz and was again expelled in 1600. He arrived back in Prague two months before Tycho Brahe's death. Kepler had already made a bet that, given Tycho's data, he could find an accurate planetary orbit in a week. He found, when Tycho bequeathed him all the data, that it took rather longer. In fact it was five years before Kepler obtained his first accurate orbit, that of Mars. He was however able to show fairly quickly that even after calculating seventy trial orbits, the epicycle-equant theory did not fit even the simplest of the observations, those taken when the Sun, the Earth and Mars were in a straight line. An orbit derived from four such oppositions failed to fit the fifth opposition by an eighth of a degree, a discrepancy that was significant only as a result of the great accuracy of the Ven observations.

By 1605 Kepler had found two laws of planetary motion. He first discovered what we now know as his second law – the equal area law – which states that the line joining planet to Sun sweeps out equal areas in equal times even though the planet's speed as it moves about the Sun is not constant. Kepler's "first law" stated that the orbit of Mars was not an epicycle, as proposed by both Ptolemy and Copernicus, but a simple oval, or ellipse. About this time attention was diverted from the planets by another new star, "Kepler's nova", by observation of a bright comet, by the mistaken identification of a sunspot as a transit of Mercury across the face of the Sun, and by endless legal wrangles with Tycho's heirs about the use and ownership of all the data. The two laws were published in 1609 in a book as unreadable as it is important, Kepler's *New Astronomy*. Like Copernicus's book, this new thesis had virtually no impact at all, and Kepler turned his attention to optics. In 1611 he published his second book on optics, the *Dioptrice*, and his twin interests led to the jargon of one field being used in the other, so that even now we speak of the Sun at one focus of the Earth's elliptical Keplerian orbit. Otherwise 1611 was a bad year. Kepler's wife and sons died and in 1612 the Lutherans were thrown out of Prague and Kepler went to Linz. There he continued his work and devised his third law, which relates the planetary periods to their distances from the Sun.

Kepler was able to publish this law and a seven-book summary of a lifetime's theoretical astronomy before his Lutheran background caused yet another expulsion. This time he went to Ulm, where in 1627 he completed the Rudolphine tables, still named after Rudolf of Prague, who had died fifteen years before. These, the first modern astronomical tables, included the latest aid to computation, the logarithm, as well as Tycho's star catalogue.

Tycho Brahe's measurements and Kepler's new astronomy stand at the divide between two eras. On the one hand they were the culmination of two millennia of observations with the unaided eye, and on the other they provided the basic data on which Newton and others could build in the years ahead. The marks of the old astronomy are still with us: counting systems in units of sixties, as for hours and minutes, go back to Babylon and beyond. We have already seen that Easter is fixed using rules from classical Greece. The astrologers' technical vocabulary has been incorporated into our own: words such as saturnine, jovial, martial and mercurial are still in use, though venereal has rather come down in the world. The constellations are still much as mapped out long ago, and many of them still bear Arab names. Perhaps the best known of all the old astronomers is neither Ptolemy nor Copernicus, but Abu Ma'shar, whose name lives on, a little mangled, as the wicked uncle in children's pantomime.

CHAPTER THREE

MEASURING

The application of the telescope to positional astronomy, 1630–1763

"I have either found out, or stumbled on . . . a most certain and easy way, whereby the distance between the least stars, visible only by a perspective glass, may be readily given, I suppose to a second [of arc]; affording the diminutions and augmentations of the planets strangely precise."

Letter from William Gascoigne, 1640.

Although the telescope produced a rash of astronomical discoveries, it was not immediately applied to astronomical measurement in the tradition of William IV of Hesse-Cassel and Tycho Brahe. The immediate technical reason, advanced most strongly by Hevelius, was that optics were not to be trusted. A great deal of effort had gone into the development of the astronomical sextant, which, in 1620, was the most precise instrument that had ever been built. In contrast, a simple lens can produce a very badly distorted image, as can be seen by examining a piece of squared paper through a magnifying glass. Such distortion is totally unacceptable in a device whose entire purpose is to make accurate measurements. The other, underlying, reason for the delay was both political and cultural. The countries involved descended dramatically from a civilized peak at the beginning of the seventeenth century to the misery of the Thirty Years' War, in which the whole of northern and central Europe was scourged by the most barbaric of the many wars the continent has known. The twin centres of technological excellence, Nuremburg and Augsburg, were both hit hard, and while in Silesia Kepler was swept into the train of the condottiere Albrecht Wallenstein, the spread of the war into northern Italy strengthened the resolve of the Inquisition, which was soon to start proceedings against Galileo. The main result was that neither Kepler nor Galileo was able to lead, instruct, inspire or equip a new generation of astronomers.

The two countries in which there was still freedom to pursue astronomical studies were England and France. The former was defended by its isolation, the latter by the diplomatic manoeuvres of Cardinal Richelieu. In both countries the first task was to develop the skills needed to build apparatus as good as the instruments no longer obtainable from Augsburg and Nuremburg.

The first steps forward were taken in France, where two separate groups emerged. One was a group of provincial amateurs whose main interest was in observing the planets. The best known of them, the wealthy Nicolas de Peiresc, owned some forty telescopes and used them to unravel the orbits of Jupiter's four satellites, a problem that had defeated Galileo. The other group were Jesuit teachers, and in 1631 one of this second group, Pierre Vernier, developed a graduated scale of angles that could be read accurately. Instead of using a pointer to indicate the position on a scale, Vernier used a short scale marked in slightly different intervals from the main scale. On an astronomical instrument, it was now possible to read an angle to an accuracy of a sixtieth of a degree when the main scale was only marked every half degree.

The first combination of telescope and measuring instrument was also a French achievement, due to Jean Baptiste Morin in 1634. Morin was Regius Professor of Mathematics in Paris and a firm believer that the Earth was the centre of the universe, being sure that Copernicus and Kepler were wrong. While his own observations did not prove this, as he devoutly wished, he must have been much encouraged by the theorists. The French mathematician René Descartes had persuasively presented a theory of the origin of the solar system in terms of spinning vortices, and this carried more weight than Kepler's empirical laws of planetary motion. Both Ismael Bouilleau in France and Seth Ward in England showed that Kepler's starting assumption, that an epicycle would not fit the orbit of Mars, was wrong. Fortunately, the telescope observations of the phases of Mercury and evidence that the Moon's orbit was an ellipse reinforced the Sun-centred view, and the resurgence of an Earth-centred universe was only limited.

Across the Channel in England the first precision astronomical equipment was developed by Robert Norwood, "a sailor and reader in mathematics". In 1633 Norwood built a 1·5-metre (5-foot) sextant, still without lenses, and used it to measure the difference in latitude of London and York. He then measured the distance from one to the other with great care and so

THE VERNIER SCALE

The ability of Pierre Vernier's invention to measure small fractions without requiring a finely divided scale is based on the ease with which the eye can detect when two engraved marks are precisely aligned. In essence, a smaller, moving scale has divisions only nine-tenths as long as those on a longer fixed scale (A). The units are obtained by reading back from the zero mark on the moving scale, i.e. 5. The decimal figure is 8. On the instrument scale (B) the first figure is 19. On the Vernier, the only mark that is directly opposite a mark on the main scale is 7, making the reading shown to be 19·7. This little device enormously reduced the time required to construct and calibrate precision instruments.

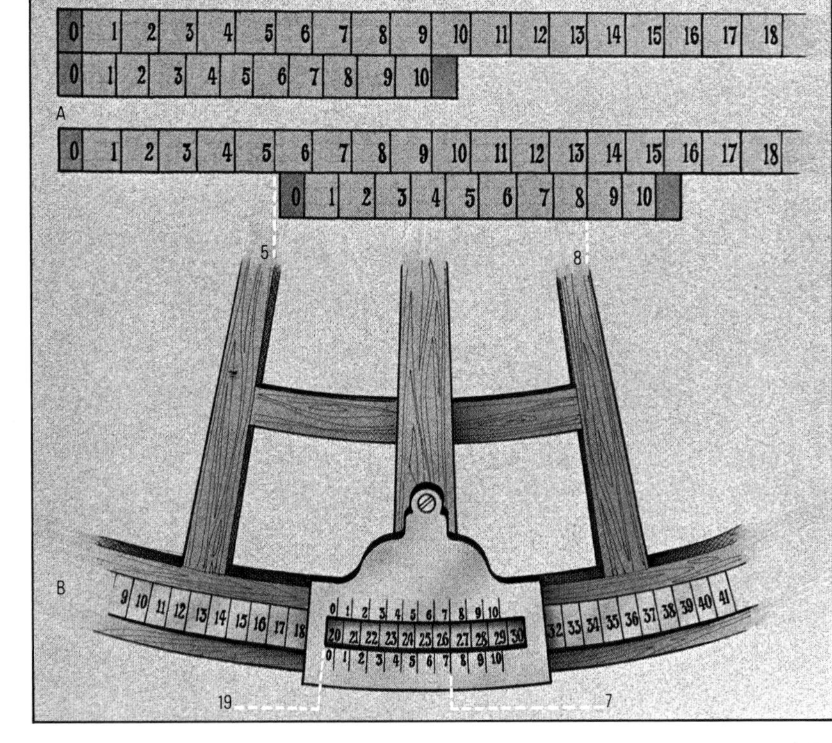

determined the size of the Earth. He found one degree of latitude equal to 111,915 metres (367,176 feet), a figure which we now know was in error by only a half per cent. At about the same time a group of young Englishmen purchased or built small telescopes and used them first for observation, and then for measurement. Jeremiah Horrocks, then only nineteen, employed his observations to improve Kepler's Rudolphine tables of the planets and was able to predict that Venus would be silhouetted against the Sun in the afternoon of Sunday, 24 November, 1639. He and his friend William Crabtree prepared to observe this rare happening and Horrocks later wrote an account of their observations. As a very junior curate, Horrocks was required to attend to his duties, and in those puritan times church service could not be missed, no matter how curious the reason. Several long services were unfortunately a necessary part of Sunday's proceedings. The second service of the day ended only thirty-five minutes before sunset and Horrocks hurried to his telescope. "I then beheld a most agreeable sight, a spot, which had been the object of my most sanguine wishes . . . just wholly entered on the Sun's disc on the left side. I . . . immediately applied myself with the utmost care to prosecute my observations." His colleague Crabtree reacted rather differently. Crabtree was a draper in Manchester, a town noted for the greyness of its weather, and 24 November was cloudy. When at last the Sun broke through, Crabtree was so surprised and delighted at his first sight of Venus clearly visible against the Sun's disc that before he could recover his equilibrium and make measurements the clouds closed in again.

It was the youngest of the group, William Gascoigne, who had the luck to discover how to make an accurate measuring device of Kepler's telescope. As he recalled in his letters, a spider, quite by chance, spun its web within his telescope, exactly at the common focus of the two lenses. This cross-hair was in the same place as the focus of the telescope objective, where image of star or planet was formed; and it was magnified by the eyepiece in exactly the same way as this image. Four enormous gains resulted from this happy accident. First, the objective lens and the cross-hair precisely define the direction in which the telescope is pointed, so a telescope could be used to measure position in the sky. Second, the inaccuracies due to optical distortion are avoided, as the distortion is negligible if the objective lens is set accurately perpendicular to the newly defined direction. Third, as the cross-hair and the star image are both in the same place, the eye does not have to focus alternately on distant star and nearby sights, as was the case with pretelescopic instruments. Last, the magnification by the eyepiece of both cross-hair and image enables the observer to measure far smaller angles than the naked eye can even resolve.

Given this invention by chance, Gascoigne quickly improved on the spider's

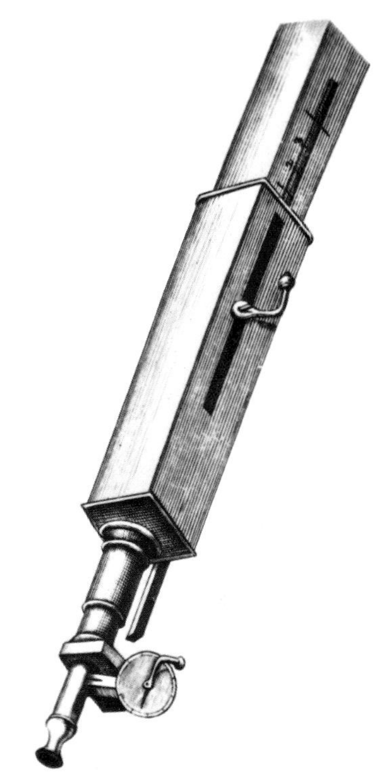

Gascoigne developed the first micrometer (above) and attached it to his Keplerian telescope as early as 1639. It was used to find diameters of the Moon and planets.

Engraving of a comet (below) *that was seen in the Nuremburg skies in December 1680. Accurate measurement of the positions of comets in the seventeenth century proved them to be members of the solar system in orbit round the Sun.*

Painting (left) by Ford Madox Brown (more than 200 years after the event) of Crabtree observing the rare sight of the planet Venus crossing the disc of the Sun during its transit on 24 November, 1639. Crabtree was too stunned by the sight to make any readings.

In the sixteenth and seventeenth centuries navigators used the cross-staff, above, to measure latitude from observation of the Pole Star or the Sun. Longitude could be found from the ship's course and speed.

An allegorical engraving (below) *of a meeting of the Académie des Sciences, that was founded in 1666 by Le Roi Soleil, Louis XIV. It attracted such international scientists as Cassini, Huygens and Roemer.*

work. He added a candle to his telescope, so that he could illuminate the cross-hair and see it in the dark, and he developed the first eyepiece micrometer. This device used two indicators mounted in place of the cross-hair, whose separation could be altered by turning a calibrated screw. With this he could measure the changes in the apparent diameters of each of the planets as they moved nearer and further from the Earth.

All this, alas, came to nothing. Horrocks, after a brilliant start including the demonstration that the Moon's orbit was an ellipse, died in 1641, aged 22, and in 1644 Gascoigne, then 24, was killed fighting for the Royalists at Marston Moor, the decisive battle of the English Civil War.

Although Morin, Vernier and Gascoigne had shown the way towards a new precision in astronomy, the next major figure was neither English nor French – nor did he take advantage of the new developments. Johannes Hevelius, the son of a wealthy Danzig brewer, had been a keen student of astronomy both at school and at the University of Leyden. When he was recalled to Danzig to work in the family business he abandoned astronomy for several years, but in 1639 his interest was revived, in part because of encouragement by his old tutor and in part by a solar eclipse. Within two years he had built up an observatory that finally extended across the roofs of three buildings. Although Hevelius used telescopes for observation with great success, he was adamant that accurate measurements required instruments firmly in the tradition of Tycho Brahe. In 1647, with a growing range of sextants and quadrants, he began work on a revision of Brahe's catalogue of precisely known star positions.

Hevelius had two advantages over Brahe. First, he could use his long-focus observing telescopes to see whether a star was single or consisted of several stars close together, and second, he could benefit from Descartes' publication of Snell's law of the refraction of light. Armed with this knowledge, Hevelius could use his sextants to measure angles between well-defined single stars and then make a much-improved correction to these angles to remove the distorting effects of refraction by the Earth's atmosphere. Although Hevelius refused to add a telescope to his measuring instruments he did introduce a number of improvements. He invested much effort in the improvement of non-telescopic sights, including the development of a screw-driven fine adjustment, and in the counterbalancing of his instruments. With all these improvements – none spectacular in itself, but together adding up to a significant gain – Hevelius was able to make observations several times more accurate than Tycho Brahe. For the first time positions could be measured to a precision of one arc minute (1/60 degree). Sadly, the improvement was not as great as Hevelius had hoped, and, indeed, less than he claimed for his instruments. He was involved in an acrimonious argument with Robert Hooke on the advantages and disadvantages of telescopic measurement, and just as he got over the bitterness of the row, disaster struck – his entire observatory was burnt to ashes. Although Hevelius made brave efforts to start anew, the old enthusiasm had gone and the new observatory was but a shadow of the old.

By the middle of the sixteenth century two new truths were gaining acceptance with the establishment. The technology that resulted from science's new experimental approach was useful. Among the recent inventions were not only the new telescope and its close relative the microscope, but also the thermometer, the barometer, better clocks and faster calculating methods such as Briggs's logarithms and Oughtred's slide-rule. It was also becoming clear that this stream of inventions would flow faster if the scientists concerned met one another in person to discuss their problems and ideas. The result was a burst of patronage and institution-forming. The first of the new societies was however short-lived. The Florentine Academy, formed in 1657, was disbanded in 1667, with one of its members, Antonio Oliva, committing suicide while in prison to avoid further interrogation by the Inquisition. In England, an informal group of scientists was given royal patronage and became the Royal Society in 1662, and four years later Louis XIV of France formed the Académie des Sciences.

There were three great challenges awaiting the astronomically-minded members of these new assemblies. Two were matters of fundamental importance – the size of the solar system and the distance to the stars – and the other was of more terrestrial application – the determination of longitude, especially during a long sea voyage. The British, with a growing navy, colonies in America and a healthy share of the Far Eastern trade through the East India Company, turned to the longitude problem. The French, with less commercial incentive and a rather more academic research group, set out to determine the size of the solar system.

The French efforts were directed to making a set of very accurate observations of Mars from which the distance from Earth to Mars could be deduced. Kepler's third law of planetary motion showed that once one distance in the solar system was known, then the others could be calculated. The first step was, as it had been for Tycho Brahe, to establish a network of known star positions, properly corrected for refraction. This work was undertaken by a small group of the Académie staff, particularly Jean Picard and Adrien Auzout. Unaware of Gascoigne's work, Auzout reinvented the cross-hair and the micrometer eyepiece and by 1667 he and Picard were making observations with a 2·7-metre (9-foot) quadrant and a 1·8-metre (6-foot) sextant, both equipped with telescopes. The first results included a much improved correction for atmospheric refraction and a recognition that the garden behind the Académie's meeting place was not the best observing site, as the surrounding buildings blocked too much sky. This, added to the rather more exciting discoveries of the other astronomer academicians, particularly Huygens, persuaded the King to have a royal observatory built. This building was started in 1667 and finished in 1672. It suffered badly from the contrast between the King's architect and his chief astronomer. Claude Perrault, the architect, was an urbane courtier, well used to the ritual of Louis' administration, while Giovanni Cassini, the astronomer, was newly arrived from Italy and spoke very bad French with a thick accent. Since the King was the sole authority and Cassini lost every argument, the new observatory was built with virtually no regard for the problems of astronomical observation.

Picard's approach to determine the Earth–Mars distance was to measure Mars' parallax. Parallax is the name given to the perspective effect whereby an observer sees a relatively near object against different parts of the distant background as he moves his own position. If the distant background is assumed to be very much farther away than the object under study, as is certainly true of Mars and the background stars, and the amount that the observer moves is also known, then the distance to the object can be calculated. The basic aim was to measure the position of Mars with respect to the background stars from two places on Earth as far apart as practicable. Picard's first step was to measure the size of the Earth, so that he knew the length of the baseline between the two observatories accurately. Like Robert Norwood, Picard carefully measured the length of a north–south line, in this case from Amiens to Montdidier, near Paris, and then determined the latitude at each location. From this he derived a better value of the Earth's radius and was set for the actual measurement. The Académie sent Picard's colleague Richer with a set of astronomical instruments and a clock to Cayenne in French Guiana, a few degrees from the equator

KEPLER'S THIRD LAW

This relates distance of a planet from the Sun (S) to the time taken for it to complete one orbit. If this distance is defined to be one Astronomical Unit (AU) for the Earth (E), then for any other planet such as Mars (M), shown below, the square of its orbital period, in years, is equal to the cube of its distance from the Sun in AUs. One AU is roughly 150 million kilometres (93 million miles).

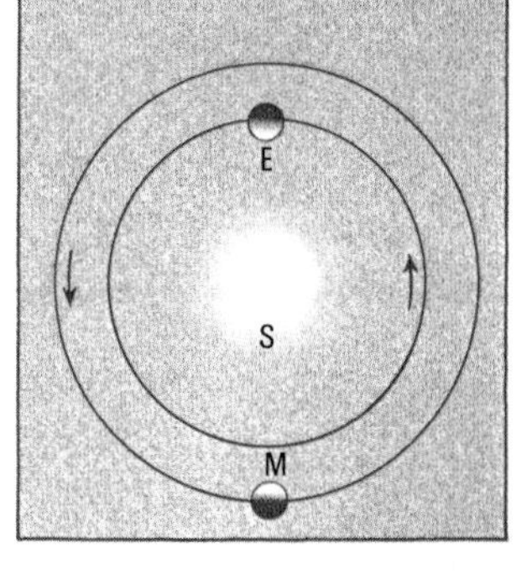

STELLAR PARALLAX

During the year the Earth moves from one side of its orbit (A), at which point a parallax measurement is taken, to the other side (B), when another measurement is made. Comparison of the two shows that the angle between the "nearby" star (C) and a more distant star (D) changes, and the distance to the former can be found from this.

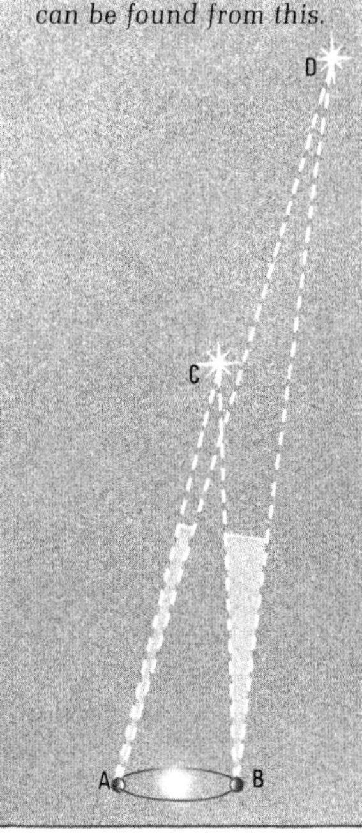

THE SOLAR SYSTEM

Above, the nine principal planets are drawn to a scale of 1 mm = 2400 km and are shown in their natural order from the Sun. On the right, the distances of the planetary orbits from the centre of the Sun are shown. For a sense of scale, they are superimposed over the Earth, with the centre of the Sun lying on the North Pole and the outermost planetary orbit, which is at present Neptune's, lying on the equator. Normally Pluto is the outermost planet, but due to its highly eccentric orbit for a time it travels within Neptune's orbit. (The dotted line is Pluto's mean orbit.) The motions of the planets, except Venus, are counter-clockwise.

THE SUN
It contains 99·86 per cent of the mass of the solar system and is entirely gaseous, being mostly hydrogen and helium. The mass, constitution and surface temperature are similar to those of millions of other stars.

MERCURY (1)
Close to the Sun, with a highly eccentric orbit, Mercury is a small, rocky planet without an atmosphere and with a cratered appearance. The planet rotates once every 58·7 days.

VENUS (2)
The brightest of the planets, due to the high reflectance of the white clouds that contain sulphuric acid and carbon dioxide and completely cover the surface. The planet rotates clockwise once every 243 days.

EARTH (3)
The most massive of the inner planets, spinning rapidly on its own axis once in 23 hours 56 minutes. In contrast, its moon rotates once a month, so that the same face is always towards the Earth.

MARS (4)
Mars orbits the Sun at a distance of 230 million kilometres (140 million miles) once every 687 days. It has two very small moons, the inner of which orbits the planet in less than one Martian day. Its atmosphere is composed mostly of carbon dioxide.

THE MINOR PLANETS
The number of minor planets, or asteroids, is more than half a million. One of them, Vesta, is occasionally bright enough to be seen with the naked eye. Most of the minor planets have orbits in the gap between Mars and Jupiter.

JUPITER (5)
The largest planet, a tenth the diameter of the Sun and, like the Sun, composed chiefly of hydrogen. A good pair of binoculars will show the four brightest of Jupiter's many moons. The giant planet rotates in less than 10 hours and centrifugal distortion makes the equatorial diameter 6 per cent longer than the distance from pole to pole.

SATURN (6)
Recent observations from the Voyager spacecraft have shown Saturn's rings to be much more complex than was apparent from Earth, the outermost ring being twisted and some of the inner rings eccentric. Like Jupiter, the planet spins round in roughly 10 hours and moves round the Sun every 29½ years. Titan, the largest of Saturn's moons, is massive enough to retain an atmosphere.

URANUS AND NEPTUNE (7 & 8)
These two planets are very similar, Neptune being slightly heavier and Uranus slightly the larger. Neptune, 4·5 billion kilometres (2·8 billion miles) from the Sun in an orbit that takes 165 years, is the colder of the two, with a surface temperature of −230°C. Uranus is 2·9 billion kilometres (1·8 billion miles) from the Sun and about 10°C warmer. Both planets have atmospheres of methane and hydrogen and each has five satellites. Uranus's axis is almost in the plane of its orbit.

PLUTO (9)
Pluto is not at present the planet farthest from the Sun. It travels round a highly inclined, eccentric orbit once every 248 years.

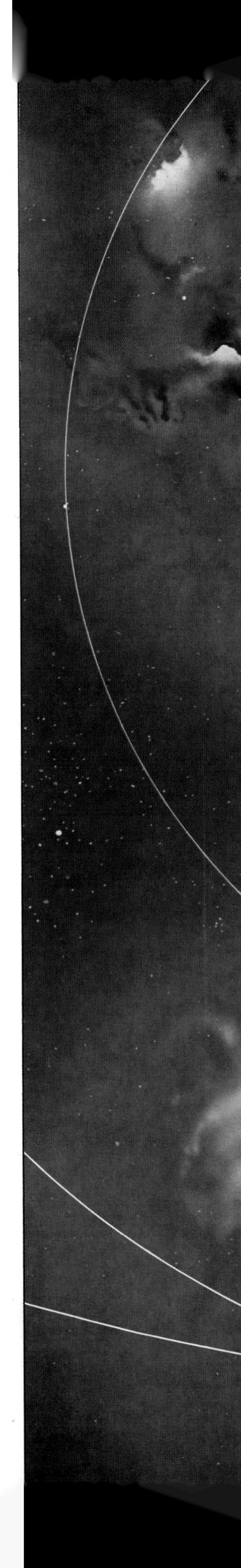

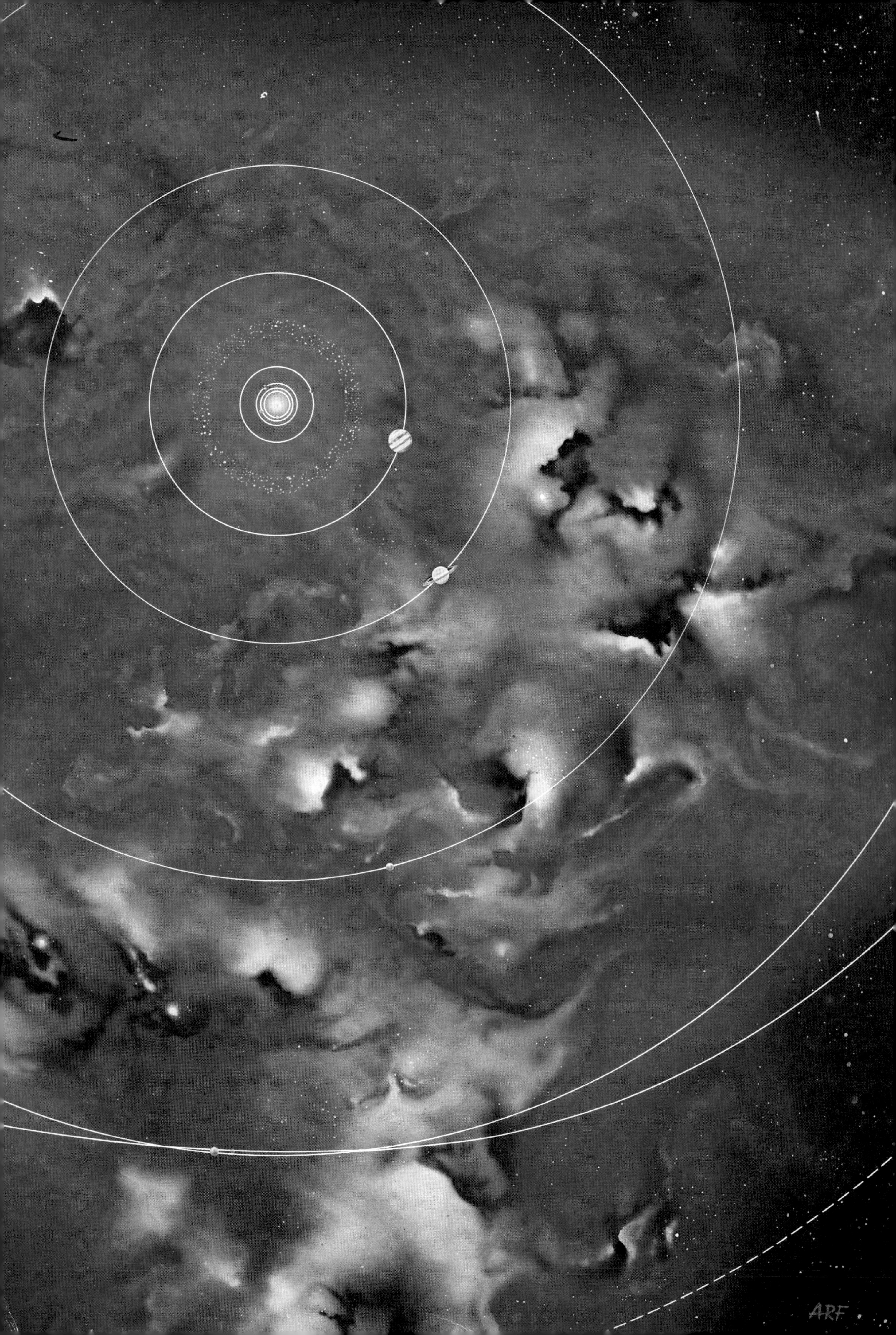
ARF

John Flamsteed was appointed as the first Astronomer Royal in 1675 specifically to make precise astronomical observations that would help in solving the longitude problem. Although he had to use his own money to purchase instruments, Flamsteed managed to compile the first accurate star catalogue, which was not published until after his death in 1719.

and the southernmost of the French colonies at the time. The position of Mars was carefully measured in Cayenne and Paris. Richer returned home, only to find that the expedition had failed. The differences between the two sets of measurements were smaller than had been hoped, too small in fact to be detected. Almost immediately, Cassini demonstrated that to determine a planet's position, rather than send the astronomer round the Earth, it was better to use the Earth to take the astronomer round. By using the Paris instruments to measure Mars' position against the stars in early morning and then in the late evening, he was able to measure the small angular shift, about a quarter of an arc minute. This was the first measurement which clearly showed an effect significantly smaller than an arc minute and completely vindicated the use of the telescope as a measuring instrument.

While Picard and Richer's work had demonstrated that Ptolemy's estimate of the distance from the Earth to the Sun was wrong, Cassini's result showed just how wrong. To describe the size of the solar system called for two new orders of magnitude. The angles measured were so small that seconds of arc, the old Babylonian sixtieth of a sixtieth of a degree, joined the language of experimental science. At the other extreme the distances were far larger than hitherto imagined: the mean distance of the Earth from the Sun turned out to be 140 million kilometres (88 million miles). Kepler's third law then put the most distant known planet, Saturn, almost ten times farther out, nearly 1500 million kilometres (930 million miles) away. Since the Picard–Cassini method requires that the background stars be much farther away than the planets, the universe leapt in size, stressing man's insignificance just as the new "Age of Reason" began.

The British efforts on the longitude problem were slow to start, the critical step resulting from "friends at court" rather than the direct action of the Royal Society committee on longitude. The crucial problem in measuring the longitude difference between two places is the determination of time. For the navigator, it is not the easy problem of finding the time at the place at which he is making his observation but the much more difficult problem of knowing what time it is in the port he left many months earlier. If, for instance, the navigator observes the Sun to be at its maximum altitude, then it is 12 noon local time, and if in addition he knows that it is 6 am at Greenwich, then he must be exactly a quarter of the way round the globe. The time difference is a quarter of a day, so the longitude difference between Greenwich and the observer is a quarter of 360 degrees, making his position 90° east or west of Greenwich: he probably knows whether he is in the Pacific or the Indian Ocean.

A navigator of 1670 had no reliable clock. Robert Hooke had tried out Christian Huygens's new pendulum clock on one of the Royal yachts and rapidly found that a swinging pendulum was not much use on a rolling ship. If terrestrial timekeepers were no use, the celestial alternatives must be tried. The time of total solar eclipses can be accurately calculated in advance but are too infrequent to be of use to a navigator at sea (though they can be used to fix the position of ports). The moons of Jupiter go round the planet almost as regularly as clockwork, and their much more frequent eclipses would be a solution, except that the telescopes needed to see them were far too clumsy to use on a ship. The only possible timekeeper was the Moon. So argued the Sieur de St Pierre, a Frenchman resident in London. Just as the clock hand moves round in front of the dial, so the Moon moves in front of the stars, and St Pierre claimed that given suitable measurements of the Moon's position, he could work out a longitude. St Pierre's method was sound in principle. He did not know that it would take almost a hundred years of careful measurements and calculation before it was accurate enough to be of general use. He did, however, know the King's mistress, Louise de Kerouaille, Duchess of Portsmouth. On 15 December 1674 a royal warrant required the Royal Society to supply the data St Pierre had asked for. Two months later the nine members of the Society to whom the task was delegated took on a 28-year-old assistant, John Flamsteed of Derby. Flamsteed was quite certain that St Pierre's method was useless, but he agreed to supply the figures that the King had commanded.

Charles II was very surprised to learn not only that the data needed to satisfy St

ANGULAR MEASUREMENT

The Babylonian system of counting in sixties is used for minutes and seconds of angle as well as time. The two angular units are a sixtieth of a degree, defined as an arc minute, and a sixtieth of a sixtieth of a degree, defined as an arc second. An arc second is roughly equivalent to the angular size of a full stop at a distance of 100 metres (110 yards)!

Pierre's request were not easily available but also that, in Flamsteed's view, the then-available data were too inaccurate to be of any use. On 4 March 1675 he issued another warrant. This appointed Flamsteed Astronomer Royal with a specific brief to make measurements of relevance to the longitude problem. St Pierre meanwhile grumbled at the delays until his persistent requests for more and better figures finally irritated the King into requiring that the method be tested forthwith. At this Flamsteed explained what St Pierre's undisclosed method would be, and even suggested that St Pierre had borrowed the whole scheme from Jean Morin. He then put in the figures and produced St Pierre's answer for him, showing, as he had predicted, that the method was a disaster. The measurements had been made by Hevelius, so their quality was unimpeachable. None the less, the longitude of Danzig came out at 8°26′ east of Greenwich, instead of the true figure of 19°15′. The error of 650 kilometres (400 miles) was enough to move the port from the Baltic into the North Sea.

It was explained to Charles II that to make the method work would require a great deal of careful work. Not only the observation of the Moon's position, but the new telescopic instruments also would need to be applied to remeasure Tycho Brahe's catalogue of background stars with greater accuracy. The high precision needed is a consequence of the Moon's slow motion. Although it is fast by astronomical standards, the Moon travels round the Earth much slower than the hour hand goes round a clock – once round every month, not twice a day. In addition, the motion of the Moon is affected by the Sun, and the elliptical orbit determined by Horrocks is only an approximation. Flamsteed realized its true motion was more complex, and Kepler's laws alone were not enough to predict its future path. So in June 1675 came a third royal warrant. This required the Master General of Ordnance to sell some old gunpowder and use the money to build a small observatory in the Royal Park at Greenwich, provided the whole enterprise cost less than £500.

Flamsteed was luckier in his architect than Cassini had been a few years earlier. Christopher Wren had already been a Professor of Astronomy. Like Auzout and Picard, Wren had, with Hooke, reinvented the cross-hair and the eyepiece micrometer. The following summer Flamsteed moved into his new observatory built "for astronomy and a little for pompe", and on 19 September 1676 he began observation.

Flamsteed's policy was first to establish the set of reference stars against which the Moon's path could be measured. He expected to do significantly better than Hevelius by using telescopic sights and very good clocks. The difficulty was that there were no more royal warrants. Flamsteed had an observatory but no instruments and no money for instruments. The situation was however eased by the generosity of Sir Jonas Moore, Surveyor-General of Ordnance. Moore persuaded Robert Hooke to make a 3-metre (10-foot) radius mural quadrant at very low cost and paid for the Master Smith at the Tower of London, Edward Sylvester, to make a 2·1-metre (7-foot) iron sextant. Moore also bought two clocks for the observatory, made by the most famous of all English clockmakers, Thomas Tompion. The sextant proved to be a highly reliable instrument, in contrast to Hooke's mural quadrant, which demonstrated the eternal truth that an instrument can be cheap or accurate, but very rarely cheap and accurate. With these tools Flamsteed started a

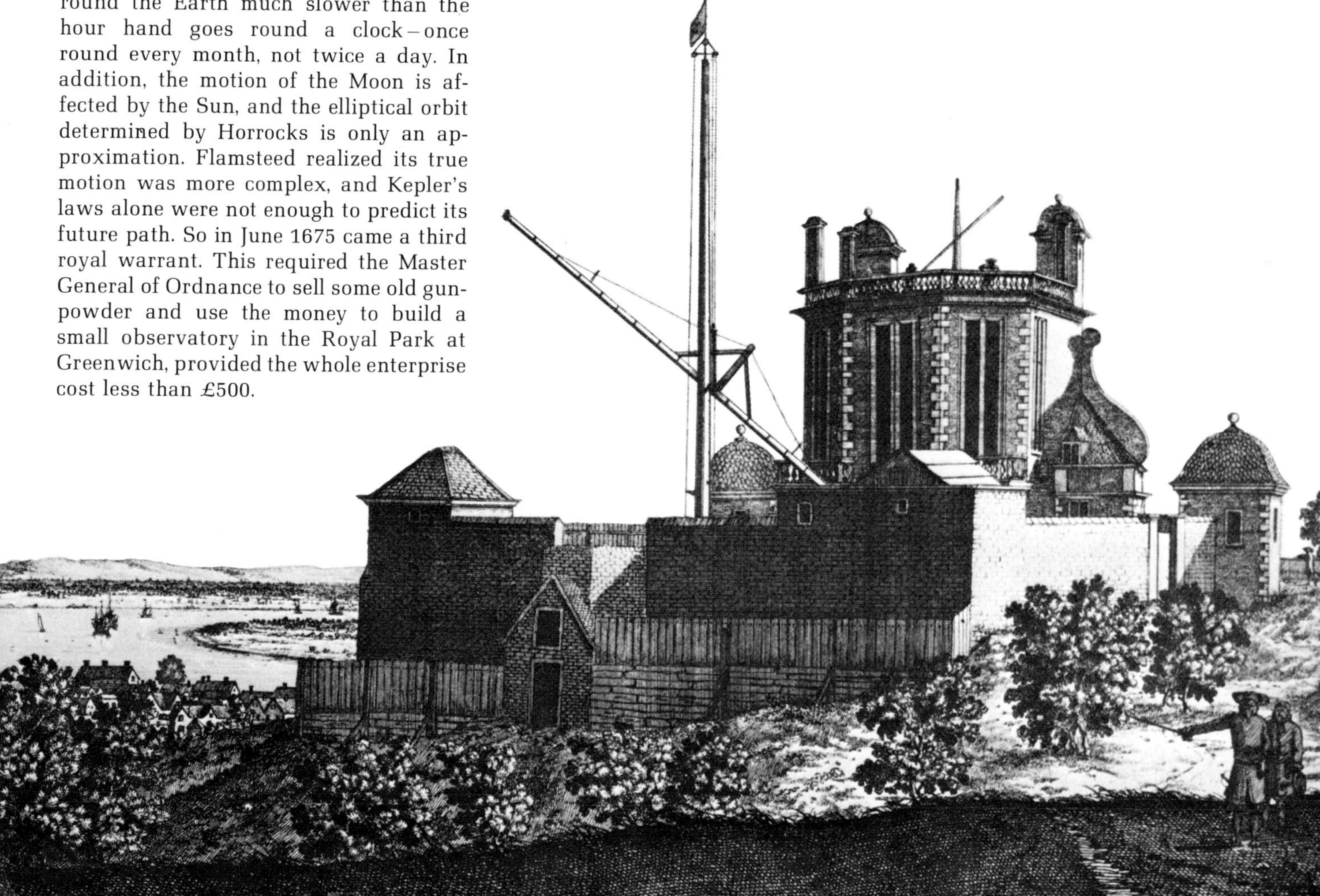

Flamsteed's observatory and house were built at Greenwich using materials from a gatehouse demolished at the Tower of London, from the old fort at Tilbury and from the prison that the building replaced. The nearness of the site to the River Thames aided the close links that gradually grew between the observatory and the Admiralty.

comprehensive programme of measurement that was still in progress at his death forty-three years later.

In view of Flamsteed's limited budget – it was twelve years before he acquired a good mural arc, and he had to pay for it himself – it is not surprising that the next developments took place elsewhere.

In 1672 Picard had visited Ven to check the latitude of Tycho Brahe's observatory, which was already in ruins. He did find some small discrepancies, but, more important, he brought a young Danish astronomer, Ole Römer, back to Paris. Römer had also independently invented an eyepiece micrometer, but his was a better model than Auzout's and soon replaced it in the Paris observatory. His observational skill was demonstrated by his careful study of the eclipses of Jupiter's innermost moon, Io. These events did not occur at regular intervals, but showed a systematic variation which was related to the distance from the Earth to Jupiter. Römer correctly interpreted the variation in ellipse times as being due to the time taken for light to travel from Jupiter to the Earth, and from his observations he deduced a value for the velocity of light. This was not a tactful discovery as Cassini, the head of the observatory, was certain that the velocity of light was infinite.

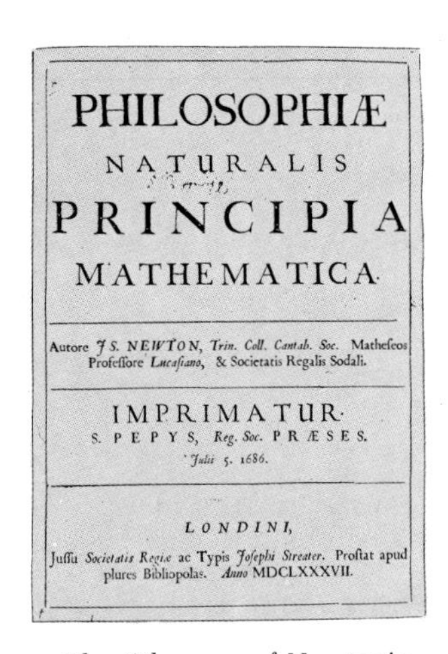
PHILOSOPHIÆ
NATURALIS
PRINCIPIA
MATHEMATICA.

Autore *J S. NEWTON*, *Trin. Coll. Cantab. Soc.* Matheseos Professore *Lucasiano*, & Societatis Regalis Sodali.

IMPRIMATUR.
S. PEPYS, *Reg. Soc.* PRÆSES.
Julii 5. 1686.

LONDINI,

Jussu *Societatis Regiæ* ac Typis *Josephi Streater*. Prostat apud plures Bibliopolas. *Anno* MDCLXXXVII.

The title page of Newton's Principia Mathematica, *which was first published in 1687. It laid down new foundations for astronomy, explaining the motions of the planets.*

Not long after this Römer returned to Denmark to become Professor of Astronomy. To continue his observations, he had to design the necessary instruments and have them constructed. The new equipment demonstrated Römer's genius. Unlike Picard, Hooke and others, Römer did not just add a telescope to the classical quadrant or sextant, but built two completely new devices. The first of these was the transit telescope, an instrument designed to do the same job as the mural quadrant, namely observe stars or planets as they crossed the north–south line. The second instrument, commissioned a few years later, was the equatorially mounted telescope. In both these machines the telescope was pivoted at its mid-point, not right up at the objective, as was the case with the telescopic quadrant and sextant. The gain in rigidity so produced was further enhanced by Römer's use of a new light but strong structural component, the hollow conical tube, in place of the more flexible cylindrical telescope tube.

During the 1680s progress in astronomy in England was almost unbelievably great, though not as a result of Flamsteed's work. Edmond Halley, Flamsteed's erstwhile assistant and then secretary to the Royal Society, persuaded Isaac Newton to write up his researches, and then paid for their publication. The result, the *Philosophiae Naturalis Principia Mathematica*, is a work of towering genius. Among many other things it put forward the law of universal gravitation, and showed that Kepler's laws of planetary motion were a natural consequence of gravitation. Few of Newton's contemporaries could follow or accept Newton's arguments. Almost inevitably, Robert Hooke said he thought of it first, and in this case Hooke was right: the difference was that while Hooke thought that Kepler's laws were a consequence of universal gravitation, Newton could prove it.

The last decade of the seventeenth century and the first of the eighteenth were uneventful. After Picard's death and Huygens's dismissal, morale at the Paris observatory was at a low ebb, while Sir Jonas Moore's death left Flamsteed without his chief sponsor. What stimulated a revival in interest was the loss of Vice-Admiral Sir Cloudesley Shovel and his fleet off the Scilly Isles in 1707. This naval disaster was attributed to the defects of contemporary navigation, especially the lack of a good method of determining longitude. The first action was to try to get all Flamsteed's measurements published, so that there would be a catalogue of standard stars available to all who needed it. Flamsteed, like many scientists, wanted his work to be as near perfect as possible, and he refused to hurry into print with these "preliminary" data – he had by then almost 30,000 observations with his mural quadrant alone. In 1710, Queen Anne issued a warrant establishing a board to oversee the Royal Observatory and specifically requiring that each year's observations be published within six months of the end of the year in question. Slowly Flamsteed produced his catalogue, until at his death in 1719 only the translation of the preface into Latin and the engraving of the plates remained to be completed. The final work, the *Historia Coelestis Britannica*, was published in 1725. It contained the positions of almost 3000 stars and was a colossal anticlimax.

There were three reasons for the catalogue's minimal impact. The first of these was that Flamsteed had not made a sufficiently careful study of the Moon's position to help with the longitude problem. That task was taken up by Edmond Halley, who was appointed Astronomer Royal after Flamsteed's death in 1719. Although he was sixty-four at the time and knew that a complete set of lunar observations would take eighteen years, Halley lived to be eighty-six and to complete his programme. The second reason was that Halley had shown, in 1718, that some at least of the fixed stars were not, after all, fixed but were moving: Arcturus, Sirius and Aldebaran were no longer in the positions measured by Hipparchus. The third blow was the demonstration by

Samuel Molyneux and James Bradley that new and exciting phenomena could be observed with instruments only a little more precise than Flamsteed had used.

Molyneux and Bradley's achievement rested on the development of a new instrument, the zenith sector. This type of telescope, pioneered by Hooke, was designed to move in the vertical plane and to measure the positions of only those few stars that cross the meridian almost directly above the observatory. Such a telescope is particularly free from errors due to mechanical distortion and atmospheric refraction: the astronomers hoped to use it to measure star distances by the principle of parallax. The idea was that the apparent positions of nearby stars should change as the Earth moves from one side of its orbit to the other. Attempts to measure this effect, first sought by Copernicus himself, had failed because the nearest stars are so distant. Molyneux and Bradley applied the new zenith sector to try to detect the parallax of Gamma Draconis, a star that crosses the meridian very close to the zenith above Molyneux's home at Kew in West London. In 1725 they found that the position of the star had changed between December and March by some twenty arc seconds, but the change was in the opposite direction to any possible parallax effect. A new zenith sector, this time at Bradley's aunt's home, showed that this shift was common to all the stars near the zenith. Furthermore, the motion was consistent with small changes in the position of the Pole Star that Picard and Flamsteed had found but could not explain. Bradley realized that these small changes were due to a combination of the Earth's motion and the finite velocity of light. Just as a man in a hurry in a rainstorm must lean forward if his umbrella is to keep his feet dry, so the vertical telescope, hurried along by the motion of the Earth, must also lean forward if the starlight is to reach the cross-hair at its base. The new zenith sector proved sufficiently accurate for Bradley to discover a second, smaller, periodic variation in star positions. Again it was not the hoped-for parallax, but an effect called nutation, which results from the attraction of the Moon on the equatorial bulge of the Earth. In the end Bradley never did succeed in measuring any star's parallax. The best he could do was to show that if there were such an effect, then it was less than one arc second for the favourite star, Gamma Draconis. This implied that the star was at least 30 million million kilometres ($18\frac{1}{2}$ million million miles) away, a figure of unimaginable size and still only a lower limit for the distance to a bright, and presumably nearby, star. It was more than a century before anyone could do better.

The zenith sector, because it hangs vertically, avoids any error due to flexure. Provided that a star passing directly overhead is sufficiently bright, it is possible to make measurements during daylight and so follow small changes in position throughout the year.

AN EQUATORIALLY MOUNTED TELESCOPE

An equatorial telescope is mounted so that one axis, the polar axis, points directly at the north celestial pole. Motion around this axis compensates for rotation of the Earth and greatly eases observation. In use the telescope rotates about the polar axis (A) at a constant rate, while the other axis, the declination axis (B), can be fixed for the duration of the observation.

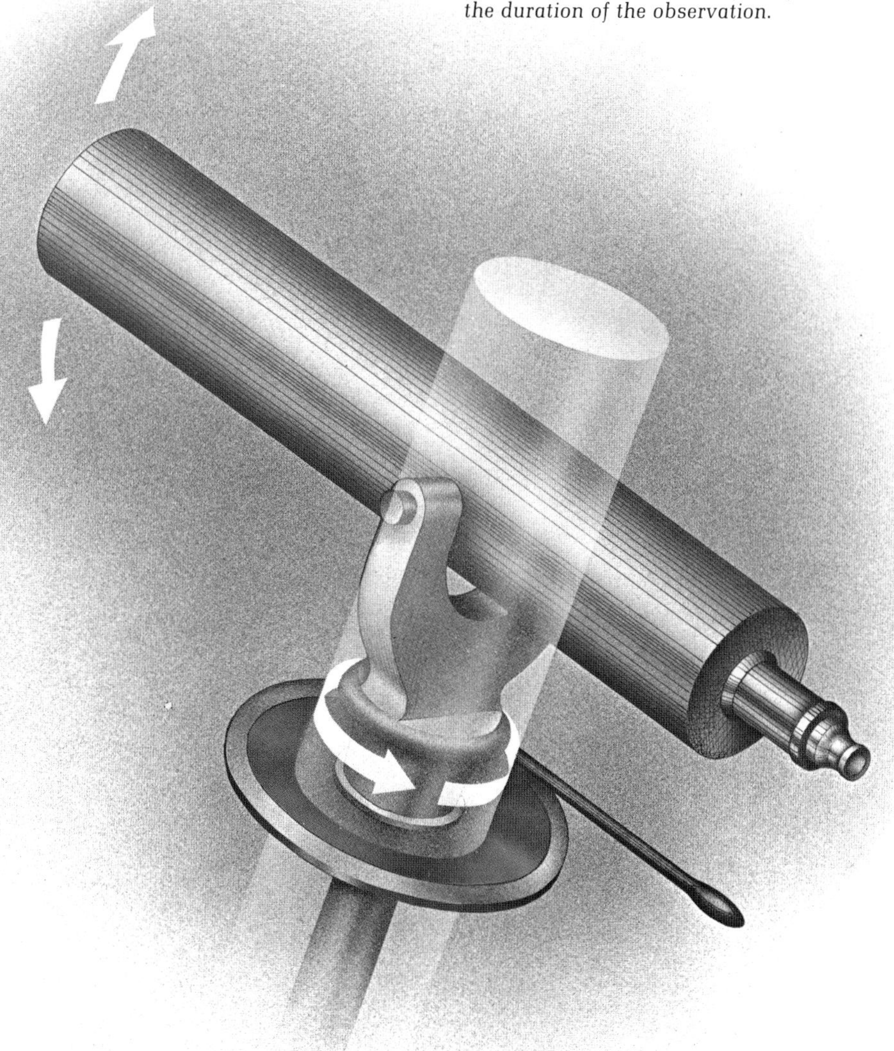

CHAPTER FOUR

BEYOND SATURN TO THE GALAXY

The rise of the reflector and William Herschel. 1721–1821

"The greatest improvement, which this invention has ever received, is indisputably and singly owing to Sir Isaac Newton: to whose extraordinary sagacity and very judicious experiments, the world first owes the discovery of the different refrangibility of the rays of light and the insuperable difficulties arising from thence in perfecting any refracting telescope."

Book 3, Chapter 2 of
A Compleat System of Optics
in four books, viz A Popular, a Mathematical, a Mechanical, and a Philosophical Treatise,
by Robert Smith, Professor of Astronomy at Cambridge and Master of Mechanicks to His Majesty, 1738.

When Edmund Halley took over as Astronomer Royal in 1720 he found the observatory at Greenwich stripped of instruments. Flamsteed's heirs had removed every one of the observatory's telescopes, taking the view that as Flamsteed had paid for them out of his own pocket, they were clearly his personal property. More important, they also removed the volumes of Flamsteed's observations, and it was nearly fifty years before the Royal Society recovered them, by which time the earliest data were over a hundred years old.

Halley was awarded a £500 grant for the purchase of instruments and with £400 of this he bought a transit telescope, a mural quadrant, the stone piers needed for their support and two new buildings to house them. The first of these instruments, the transit, was based on an existing 1·5-metre (5-foot) telescope of Hooke's, to which a cross axis and precision scale were added to make a telescope similar in function to Römer's pioneering design. The transit telescope was not an outstanding success, perhaps because it was made in a hurry to give Halley something with which to start his programme of lunar observations. The lattice-work of the axle was very susceptible to distortion due to small temperature changes, and once the second instrument, the mural quadrant, was complete, the transit telescope was scarcely ever used.

The new quadrant was built by George Graham, Tompion's successor as London's leading clockmaker, and it incorporated all the improvements developed in the preceding century. The telescope tube was braced to make it more rigid and counterweighted so that the load on the pivot remained unaltered as the telescope was moved in elevation. A screw drive and Vernier scale were used for setting and reading the position of the telescope on the quadrant. The new standard of positional measurement, which Bradley and Molyneux had pioneered with their zenith sectors, was now an accuracy of a few arc seconds. An arc second is a very small angle: with Halley's 2·5-metre (8-foot) radius quadrant a displacement along the scale of 0·01 centimetre (1/250 inch) (less than the thickness of this page) will give an error of eight arc seconds. The total error in the measurement of a position in the sky is compounded from many sources: distortion of the telescope due to flexure under its own weight, wear of the pivot, temperature effects, slow warping of the frame and inaccuracies in calibration and setting all contribute, and their sum must be controlled to a few parts in a million. The care needed is illustrated by the fate of Flamsteed's main quadrant, mounted on a necessarily heavy wall on made-up ground near the edge of the hill on which

Halley's mural quadrant (above) was the first of the new class of precise instruments used in the eighteenth-century.

Edmund Halley, 1656–1742, was Astronomer Royal from 1720 until his death.

the observatory stood. The whole device, wall, quadrant and all, had slowly tipped over a very small fraction of a degree, and this was enough to spoil the original careful alignment of the instrument. Halley made sure that the wall for his new quadrant was built on firmer ground farther from the edge.

Halley and Graham's quadrant became the pattern for a whole series of instruments, the majority built by Jonathan Sisson and John Bird, Graham's successors. Bird's quadrants in particular were built not only for Greenwich and Oxford but also for many European observatories, and they dominated precision astronomy for a long period. The main competition to Graham was from J. Langlois in Paris, whose quadrants and sextants were used by the French astronomers for a series of important observations. Chief among these were the measurement of the length of a degree of latitude near the equator by Pierre Bouguer and in the north of Lapland by Pierre de Maupertuis for comparison with a similar measurement near Paris by Cesare Cassini (Cassini III). The results, published in 1749, showed that one degree was a longer distance on the surface near the equator than at the Poles. This confirmed Isaac Newton's prediction that the Earth is a flattened sphere whose equator bulges outwards because of its daily rotation. In another French venture, Cassini's assistant, Nicolas de Lacaille, took one of Langlois's sextants to the Cape of Good Hope. In 1751, in a single-handed whirlwind campaign, he made observations of 10,000 southern stars and, in collaboration with Joseph Lalande in Berlin, measured the parallax of the Moon with much improved accuracy.

Although Molyneux and Bradley had obtained the first experimental evidence for the Earth moving round the Sun and Bouguer for its rotation, thus confirming the Copernican view long after its general acceptance, there was no doubt in the minds of the users that their instruments were far from perfect. The limitation lay in the performance of the telescope lenses, which suffered from the chromatic aberration which Newton had identified and described as incurable. Only lenses of very modest aperture could be employed in instruments compact enough to be used for accurate measurement: the telescope lens of Halley's mural quadrant was only 3·5 centimetres ($1\frac{1}{3}$ inches) in diameter and that of Lacaille's South African instrument less than 1·5 centimetres ($\frac{2}{3}$ inch). Even with so small a lens the image of a star was not a brilliant white dot but a small yellow-green blur about ten arc seconds across surrounded by a purple halo. Increasing the size of the lens in order to see fainter stars increased the size of the blurred image, and the accuracy of setting decreased to an unacceptable extent. The only possible solution appeared to be the reflecting telescope.

In the 1670s Newton and then Robert Hooke had made reflecting telescopes and these had made clear the disadvantages of the design as well as its advantages. The reflecting telescope was hard to make, inefficient and of limited life due to tarnishing of the mirror surface. The difficulty of manufacture results from a fundamental requirement that the surface of the mirror must be made several times more accurately than that of a lens for the same image quality. If a 10-centimetre (4-inch) aperture reflecting telescope is to give a star image only an arc second across, then the sum of the errors in polishing the two mirrors must not exceed an eighth of the wavelength of visible light; that is, no point on either of the mirrors may depart from the ideal concave or flat surface by more than thirty or forty millionths of a millimetre (about a millionth of an inch). Only very hard materials can be polished to this accuracy, and unfortunately the highly reflecting metals silver and tin are not among them. Furthermore, the technique used in the seventeenth and eighteenth centuries for "silvering" domestic mirrors was not of any help. This process used a small amount of mercury to make a layer of tin foil adhere to a polished glass plate. The thickness of the resulting layer cannot be controlled with precision, and therefore makes an uneven reflector when applied to the front of the main mirror. Coating the back of a polished glass mirror is of no use either: the result combines the disadvantages of the high precision needed for the mirror with the colour difficulties that arise from having glass in the path of the light.

The possible hard metallic alternatives to glass fall into two groups, steels and bronzes. Even highly polished steel makes a very poor reflector – there was no stainless steel in 1720 – and the polished surface rusts rapidly. The alternative group, the copper-tin alloys, are more promising and include a range of mixtures of long ancestry, known as speculum metals, that are hard enough to polish and also reflect light fairly well. "Fairly well" is still not very good: a newly polished speculum mirror reflects only about fifty per cent of the incident starlight, while a lens transmits more than ninety per cent. The copper in the mirror alloy leads to the slow build-up of a blackish-brown tarnish so that an old mirror is even less efficient. A telescope made with two mirrors each with a reflectance of thirty per cent will focus less than a tenth of the light incident on the first mirror into the final image, and in consequence a reflecting telescope must have a large aperture if it is to collect enough light to give a bright image. An instrument with a speculum primary mirror anything less than about 15 centimetres (6 inches) in diameter would not be worth building.

The challenge of constructing a reflecting telescope of sufficient aperture and precision to give a high-quality image was taken up by an English provincial gentleman, John Hadley. Hadley was the son of the High Sheriff of Hertfordshire and a man of both inventive mind and skilled hands. Before he started work on a reflecting telescope he had already been elected a Fellow of the Royal Society on the strength of other inventions. In 1721 John Hadley, with the help of his brothers, George and Henry, built a reflecting telescope which he demonstrated to the Royal Society. The telescope was a Newtonian, based on a concave speculum mirror 15 centimetres (6 inches) in diameter and 1·5 metres (5 feet) in focal length. The Society was impressed not only by the telescope but also by the ingenuity of its stand, which allowed for simple adjustment in elevation and azimuth. Hadley gave the telescope to the Society, who arranged for a comparison of its performance with that of a 20-centimetre (8-inch) aperture, 37-metre (120-foot) focal length telescope. This was based on a lens made by Huygens, which James Pound, Bradley's uncle, had borrowed from the Society some years before. Hadley, Bradley and Pound did the testing and concluded that the new telescope gave as sharp an image as the refractor, though less bright, and that it was much easier to manage. Pound also reported that the enclosed tube of the reflector meant that it could be used in twilight, when the aerial telescope was useless. This was a significant gain, not just giving a few more hours' work but extending the range of possible observations. It allowed astronomers to make detailed studies of Mercury, for instance, and search for surface markings and possible satellites near a small planet not often seen except in twilight skies.

Hadley's success and enthusiasm persuaded Molyneux and Bradley to turn their attention to the reflector, and Molyneux directed a long series of tests for a better speculum metal alloy. The best alloy must strike a balance between having too little copper, and thus being so brittle that it cannot be cast, and having too much, when the polished surface tarnishes rapidly. Other metals can be added to improve the reflectivity. Newton had suggested adding arsenic, but this boils at a temperature lower than the melting point of copper, so the alloying process is hard to control and the results uneven. A mixture of five metals – copper, tin, zinc, silver and antimony – was rejected as too expensive, and Molyneux finally settled on a copper–tin–zinc alloy, essentially a mixture of bronze and brass. Unlike many of their contemporaries, to whom secrecy was second nature, Hadley and Molyneux went to some trouble to publicize their methods, both contributing sections to Smith's monumental *Opticks*, which was published in 1738.

John Hadley went on to make several more telescopes, both of Newtonian and Gregorian design, but he is better remembered as the more famous of the two inventors of what is now the sextant – not the astronomical but the navigator's instrument – the device that has been used ever since for determining latitude. The size of arc varied: Hadley used only 45°, so his original instrument was called an octant. A colonial American, Thomas Godfrey, had invented a similar navigating device a year before Hadley using a 90° arc, calling his instrument a sea quadrant. Hadley's invention, announced in the city that held the headquarters of both the British Navy and the East India Company, naturally proved the more influential. Later instrument-makers finally agreed on a 60° arc and the device settled down into its present configuration as a sextant. While the new invention did little to further astronomy directly, it did both aid the growing market for precision scientific instruments and make possible precise seaborne measurement of the position of Moon and stars. The latter would be required if Flamsteed and Halley's work was to be used for the determination of longitude.

The first commercial instrument-maker to take up the challenge of building a reflecting telescope was Francis Hawksbee of Crane Court in London. Molyneux comments: "He deserves very well to be encouraged, being the first person who hath attempted it without the assistance of a fortune which could well bear the disappointment." Other makers were soon to follow. Mann, Scarlet and Hearne in London, Short in Edinburgh and Paris and Passemant in France all tried their skill, and Short soon emerged as the man with both the skill required to make precision telescopes and the contacts to market them successfully.

James Short was the orphaned son of an Edinburgh joiner. He went to Edinburgh University, where, after getting his Master of Arts degree, he stayed on to train as a preacher in the Church of Scotland. While a student he came under the influence of the mathematician Colin Maclaurin, and soon after qualifying in theology in 1731 Short turned Maclaurin's college rooms into an optical workshop. His work on mirrors was interrupted in 1736, when he was appointed mathematics tutor to the young Duke of Cumberland and moved to London. Whatever the Duke learned from his teacher it was certainly not an enduring respect for things Scottish. Ten years later he was to be the "Butcher of Culloden". James Short was soon involved in telescope manufacture in London, especially of telescopes for the nobility. In the middle of the eighteenth century the telescope enjoyed a fashionable vogue as a prestigious after-dinner diversion and many of Short's telescopes were built for this market, one that was much less critical than were the professional astronomers.

Short's main contributions to telescope developments were improvements in a polishing technique and in mounting. If a reflecting telescope is to have both a large aperture and a conveniently short focal length, then the primary mirror must have a paraboloidal rather than a spherical surface. This is very significantly more difficult to make and Short was the first to manufacture concave parabolic mirrors as regular production items. His work on the mounting of his telescopes stressed the need for the mirrors in a Gregorian telescope (his favourite style) to be accurately aligned, and he fitted some of his instruments with equatorial axes rather than Hadley's more clumsy alt-azimuth configuration. Short made a modest number of large-aperture telescopes and more than a thousand small ones. A 30-centimetre (12-inch) aperture Gregorian was sold to Lord Thomas Spencer, and another, of 45 centimetres (18 inches) aperture, made to the command of the King of Spain. Not all Short's telescopes were built for the use of wealthy dilettantes: a 24-centimetre ($9\frac{1}{2}$-inch) aperture, 1·9-metre (6-foot) focal length Newtonian supplied to Greenwich was highly successful and was used for the study of the eclipses of Jupiter's moons, a topic that demands both good definition of the image and very well polished mirrors.

The major astronomical events of the 1760s, when James Short was at the height of his profession, were the two transits of Venus. This time the scientific world was

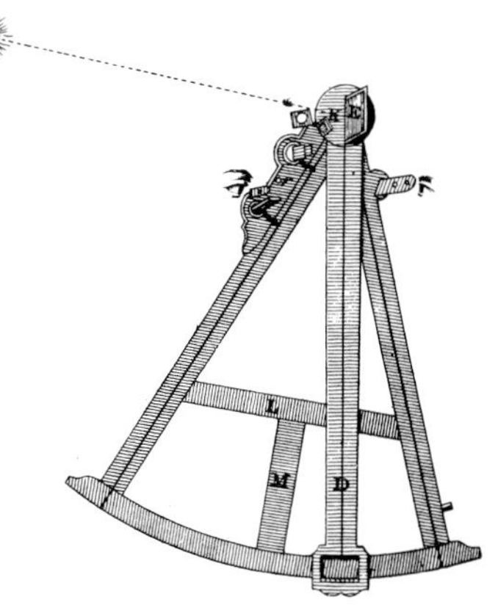

The octant (above), developed by John Hadley in 1731, was the immediate precursor of the navigator's sextant. The instrument made the determination of latitude simple and accurate.

The reflecting telescope (below), built by James Short, used Gregorian optics, largely because concave mirrors are more easily made than the convex ones needed for the more compact Cassegrain optical system. The "finder" telescope, mounted on top of the main tube, is used to point the instrument accurately in the right direction.

much better organized than in 1639 when Horrocks and Crabtree had observed the previous event. In 1716 Halley had drawn up a list of recommended sites from which to observe the two transits, and he stressed the great importance of the phenomena, which he knew he would not live to see. The size of the solar system had so far only been measured by the determination of planetary parallax, a technique that was near the limit of the best instruments available and so not very accurate. The advantage of the Venus transit is that the distance from the Earth to the Sun can be deduced by measuring the different times the planet takes to cross the face of the Sun when seen from different latitudes. The transit takes about five hours, and for the best results observations are needed from places as far apart in latitude as possible. A careful choice of observing sites gives a difference in transit time of roughly half an hour, and a host of expeditions were organized by the scientific societies of the day.

James Bradley, now the Astronomer Royal, arranged for his deputy Charles Mason to go to the East Indies, accompanied by Jeremiah Dixon as his assistant. Nevil Maskelyne, who was later to become the fifth Astronomer Royal, went to St Helena. The ship carrying Mason and Dixon was forced to return to port after a battle with a French man-o'-war, and in the end they only reached the Cape of Good Hope. Maskelyne found St Helena still as cloudy as when Halley had been there almost a hundred years before, and the chief outcome of his voyage was practical experience of navigation at sea, which was to stand him in good stead when years later he published the first Nautical Almanac. Bradley made observations at Greenwich while James Short, "by command of the Duke of York", took his instruments to Savile House, where a crowd of spectators did not assist in the making of accurate observations.

All this proved to be but a rehearsal for the second transit in 1769. At this event eight nations arranged some 150 observations, and expeditions were sent all over the globe. Mason and Dixon, just back from their famous survey of the Maryland–Pennsylvania border, went to Ireland and an island to the north of Norway respectively. Captain Cook took Charles Green to Tahiti in the *Endeavour*; Joseph Dymond and William Wales went to Hudson's Bay; and Maskelyne stayed in Greenwich, while his assistant Bayly was sent to the North Cape. Each group took a quadrant, a clock, two of Short's reflecting telescopes, a compass, a barometer and a thermometer; the Hudson's Bay group added a collapsible wooden observatory. George III made a grant of

THE TRANSIT OF VENUS, 1769

The orbit of Venus round the Sun is in a plane that is slightly tilted with respect to that of the Earth. As a result, the precise alignment of the Earth, Venus and the Sun, so that Venus can be seen in silhouette "in transit" across the face of the Sun, is a rare event. Five occurrences have been observed, in 1639, 1761, 1769, 1874 and 1882. The next is due in 2004. The transits of the eighteenth century offered a chance to establish accurately the size of the solar system. The 1769 transit was observed by 151 astronomers at seventy-seven different stations spread across the globe. The map (below) shows the parts of the world from which the transit was then visible.

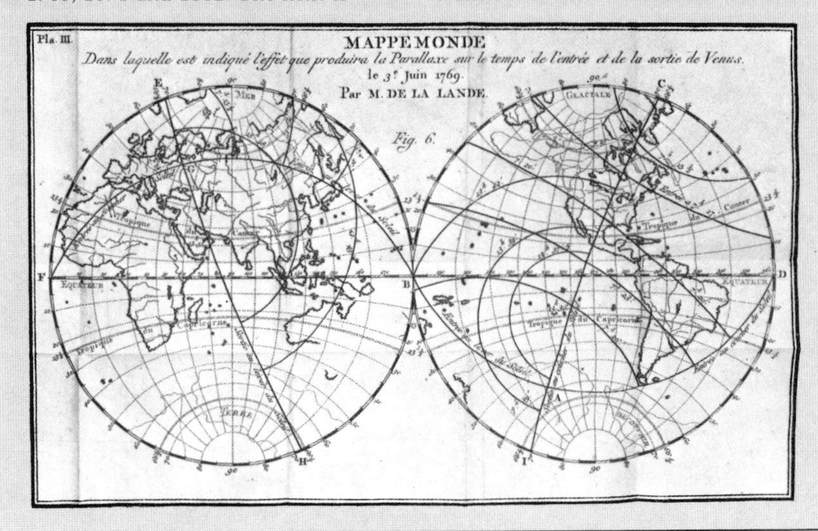

The Endeavour *leaving Whitby Harbour, England, in 1768* (left). *The ship was purchased by the Admiralty for carrying the Royal Society's expedition under the command of James Cook to observe the transit of Venus at Tahiti. In addition to a crew of five officers and eighty-eight men, she carried a scientific team of six.* (Below) *A view of their destination at Tahiti, named after the event as* Port Venus.

The small quadrant (below) made by John Bird, is typical of the instruments taken by Captain Cook to Tahiti in 1769. The quadrant was needed to find the latitude and to check the accuracy of the expedition's clocks.

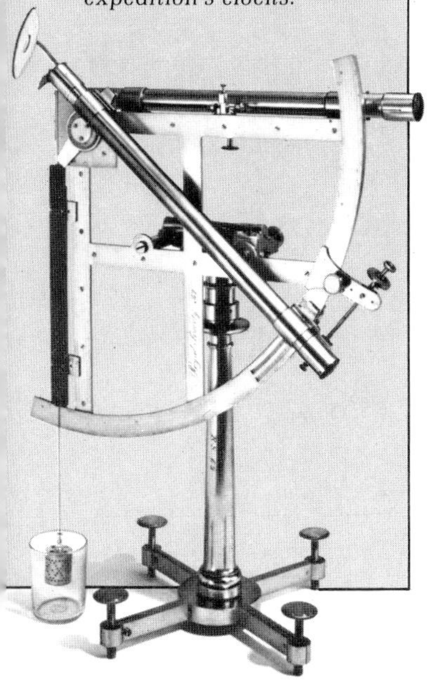

£4000, which was generous enough to leave a balance at the end of the day. This surplus was used by Maskelyne for a later expedition to Scotland to measure the gravitational attraction that a mountain exerts on a plumb line and thus indirectly to determine the mass of the Earth and the mathematical constant in Newton's law of universal gravitation.

Of the other nations observing the transit, the French seem to have been the least fortunate. Guillaume le Gentil had been sent to Pondicherry, near Madras, for the 1761 transit and had no luck because by the time he arrived the town had been taken by the British; he was still at sea *en route* for Mauritius on the day of the transit. Rather than make the long journey back to Paris with no data he waited eight years in the Far East for the second transit. This time he reached Pondicherry, but the crucial day was cloudy! He may have consoled himself with the fate of his colleague, Jean d'Auteroche, who had better weather and worse luck. In 1761 d'Auteroche went by sledge right across European Russia to observe successfully from Tobolsk, 500 kilometres (310 miles) east of the Urals. In 1769, he travelled west rather than east and, as the Spanish fleet with whom he was due to travel was unlikely to arrive in time for the transit observations, he crossed the Atlantic in a small boat. In contrast to Maskelyne, d'Auteroche's view of navigation by lunar observation was less than enthusiastic: "The tedious calculations which this method requires, with the accuracy and attention requisite in the observation itself, make it doubtful to me whether it will ever be fit for the use of trading vessels." When he reached Vera Cruz, d'Auteroche, with his party and their Spanish escort, trekked across Mexico, sailed to the southern tip of Baja California and set up the expedition's telescope and clock just in time to make observations. The journey had taken from 18 September, 1768, to 19 May, 1769. After the transit the party waited a fortnight to observe a lunar eclipse, needed to fix the longitude of their observatory. The wait was disastrous. The astronomers caught an "epidemical distemper" and three of the party of four died, d'Auteroche making the critical eclipse observation while fatally ill. Monsieur Pauly, second-in-command and "the King's Engineer and Geographer" caught the same disease as the others, but survived to carry the precious results back to Paris.

The results of all these expeditions were rather a disappointment, principally because the dense atmosphere of Venus distorted the image of the planet just at the crucial moments that the transit began and ended. The measurements obtained by the different observers of the second transit gave values that ranged from 143 to 150 million kilometres (89 to 93 million miles) for the "Astronomical Unit", the mean Earth–Sun distance. A careful analysis by Lexell gave 147 million kilometres (91 million miles) as the most probable distance. It represented a significant gain in precision over the earlier planetary parallax measurements, but was not nearly as good as the many expedition sponsors had hoped.

The interval between the two Venus transits marked a watershed in the development of the telescope. First, in the early 1760s, Peter Dollond developed a much improved refracting telescope. Then the following year an organist and music teacher in Halifax, England, took up astronomy as a hobby, and last, in 1768, James Short died taking his secret optical methods with him to the grave. No other professional could be found who was able to match Short's skill, but the organist William Herschel outstripped him and became not only a superb telescope maker but also one of the greatest astronomical observers of all time.

Herschel was born in Hanover in November 1738, the second son in a family of ten children of an oboeist in the band of the Hanoverian Guards. His father taught him to play the violin and oboe and also communicated some of his enthusiasm for astronomy. In 1753 William joined his father Isaac and his elder brother Jacob in the regimental band, an appointment that ended abruptly in 1757 when the Guards were thrashed by the French at the battle of Hastenbeck. Isaac Herschel became a prisoner of war and his two sons left for England, unable to return home because the French had occupied Hanover. After a series of musical appointments, mostly in Yorkshire, in 1766 William Herschel became organist at the Octagon Chapel in Bath, the fashionable spa in which William Pitt, newly created Earl of Chatham, was nursing the gout and the beginnings of a nervous breakdown.

The duties of the chapel organist were not onerous and Herschel was free to give music lessons and organize concerts, which provided a prosperous background for a steadily increasing application to astronomy. When William fetched his younger sister Caroline from Hanover in 1772, she recorded that they broke their journey in London and "in the evening, when all the shops were lighted up, we went to see all that was to be seen in that part of London; of which I remember the optician shops, for I do not think we stopt at any other".

Early the next year Herschel bought a quadrant and several books on astronomy. The summer was spent in making

Dr Maskelyne FRS
Astronomer Royal

Nevil Maskelyne, a London curate with a taste for astronomical calculation, became fifth Astronomer Royal (1765–1811) after visiting St Helena to observe (unsuccessfully) the 1761 transit of Venus and sailing to the West Indies in 1763 to test John Harrison's fourth marine chronometer.

longer and longer focal length refractors, using purchased lenses. First a 1·2-metre then a 3·6-, 4·5- and finally a 9-metre (4-, 12-, 15- and 30-foot) long instrument was built, the last "almost impossible to manage". In early September Herschel hired a 60-centimetre (2-foot) long Gregorian reflector which was so much more convenient that he resolved to build one for himself. By the end of the month he had bought the tools and some half finished mirrors from an acquaintance and was carefully following instructions by Hadley and Molyneux in Smith's books, now thirty-five years old, but far from out of date. The first telescope, a Gregorian, was not a success, but the second, a Newtonian, showed the rings of Saturn and the Orion Nebula, and the hobby exploded to occupy the whole house. In 1773 Caroline commented on "a cabinet maker making a tube and stands in the drawing room, Alex [William's younger brother] putting up a lathe in a bedroom for making patterns and turning eyepieces, while I was to amuse myself making tubes of pasteboard". A year later Herschel moved house, by which time he had built two Newtonian reflectors, one of 11 centimetres ($4\frac{1}{3}$ inches) aperture and 2 metres ($6\frac{1}{2}$ feet) focal length, the other 23 centimetres (9 inches) aperture and 3 metres (10 feet) focus.

The mixture of music, carpentry and optical polishing by day was complemented by observation at night. Herschel's telescopes produced images of unprecedented quality, and in 1777 he began a review of all the stars in Flamsteed's catalogue. The star images in his reflectors were perhaps a tenth of the size of the coloured blurs that Flamsteed and Halley had contended with, and Herschel could see that many of the stars were not single points of light but close pairs. The first review used the 11-centimetre ($4\frac{1}{3}$-inch) aperture telescope and covered all stars brighter than fourth magnitude. There were two reasons for undertaking the review. The first was to sort out the single stars, whose previously measured positions could still be assumed to be accurate, from the close pairs, where previous observation had recorded only a position at the centre of an unresolved blur. Herschel's second motive was to find close pairs of stars, one bright and so probably near the Earth, the other faint and presumably far away. These chance alignments should offer a good observational base for the search for the elusive stellar parallax and hence the distances to the stars. Atmospheric refraction, Bradley's aberration of starlight and errors in setting the telescope would affect both stars equally, so any small difference in the nearby star's apparent position due to the Earth's motion would be easier to find. One visitor to the Herschels' house around the time of his first review was pronounced by William to be "a devil of a fellow" – Nevil Maskelyne had got through several hours' conversation without once mentioning to his host that he was the Astronomer Royal!

In March 1781, just after moving house, Herschel's second survey reached the constellation of Gemini. On the thirteenth he saw in the field of view of his telescope a faint object that, unlike the stars nearby, showed as a clear disc. During the night the new discovery moved a perceptible amount and by the next night it had shifted its position by almost an arc minute, enough to prove that it was a member of the solar system. Herschel's first opinion was that the new discovery was a comet, and he described it as such in a letter to the two senior British establishment astronomers, Maskelyne, the Astronomer Royal, and Hornsby, the Professor of Astronomy at Oxford. The quite staggering superiority of Herschel's telescope was shown by the response. Hornsby could not even find the new object, and while Maskelyne found it, he could not measure how far it had moved from Herschel's reference stars. Maskelyne could see the stars concerned, but they were so faint in his telescope that they were only visible in total darkness: if he illuminated the cross-hair to make measurements, he could no longer see the stars he had to measure.

Surprise at the performance of Herschel's telescopes turned to astonishment and disbelief when his preliminary results were published by the Royal Society. The report contained a measurement of the diameter of the "comet" – five arc seconds – and made almost casual reference to the routine use of eyepieces of enormous magnification. If there was one thing that professional astronomers knew had to be explained to their amateur colleagues, it was that high-power eyepieces were no use. If a telescope gave a blurred image using an eyepiece that provided a magnification of 200 times, then an eyepiece magnification of 1000 times just made that blur five times bigger. No more detail could be seen, and the carefully collected starlight would be spread out so that the blur was not only bigger but fainter. Yet here was a provincial amateur not only using unheard-of magnifications but making new discoveries with them!

The discovery went from "new and interesting" to "completely unprecedented" as further measurements were made and the orbit of Herschel's "comet" determined. But it was no comet: for the first time in recorded history a new planet had been discovered. The new orbit was almost twice as far away from the Sun as Saturn, hitherto the most distant planet, and the new member of the solar system was so distant it would take eighty-four years to go once round the Sun. As soon as the period and distance were known, it was easy to calculate that its five-arc-second diameter meant that the new planet was a giant, the third biggest in the solar system (after Jupiter and Saturn) and four times the diameter of the Earth. In the same way that Galileo had named Jupiter's moons after the Medicean rulers of Florence, Herschel called his planet "Georgium sidus", the Georgian Star, in honour of King George III. And just as the international scientific community had rejected Galileo's choice, so Herschel's new name was soon replaced by that of Saturn's father, the mythological Uranus.

The immediate result of Herschel's spectacular discovery was that he was commanded to bring his telescope to court to show the new phenomenon to the King. Herschel first visited Maskelyne at Greenwich, where he compared his telescope with both those at the observatory and also with a telescope by Short which belonged to Alexander Aubert and was reputed to be the finest Short had made. Both Maskelyne and Aubert agreed on the superiority of Herschel's reflector and also that with such excellent images of stars it was indeed possible to use high magnification. Then, after several prestigious dinner parties and a week off with the 'flu, Herschel went to Windsor. The evening of 2 July was fine, and in Herschel's words, "My instrument gave a general satisfaction; the King has very good eyes and enjoys observations with the telescopes exceedingly." The following week, after some prompting from his friends, Herschel asked the King for support for his astronomical work and was appointed Royal astronomer, with a salary of £200 a year; the sole duty of the office was to allow the royal family to look through his telescopes whenever they had a fancy to do so. The salary was less than half Herschel's income in Bath, but he rented a new house near Windsor, moved his furniture and was setting up his telescopes by the end of the month. Before leaving Bath, Herschel even found time to cast a spare 30-centimetre (12-inch) aperture mirror for a newly completed 6-metre (20-foot) telescope, a fortunate precaution as the very cold weather that winter cracked the original. With these instruments, soon augmented by a 6-metre with a 48-centimetre (19-inch) diameter mirror, Herschel continued his review of the heavens. This time there were three motives. As well as extending his double star research, Herschel intended to inspect closely and catalogue all the nebulae, and also to make counts as accurate as possible of the number of stars to be seen in different directions.

The double star work was slowly moving in a new direction. Herschel had found there were far too many close pairs of stars for the pairing to be the result of chance alignment. This wrecked the basic assumption that a close pair of stars, one bright, one faint, would be at different distances and so usable for parallax studies; the stars must be actually associated with one another in space. In compensation there were hints that the components of some star pairs were rotating about one another. Herschel's work on nebulae was stimulated by a list of non-stellar objects produced by Charles Messier in Paris. Messier was an astronomer with a particular enthusiasm for discovering new comets. When first found, the distinguishing feature of a distant comet is that its image is less sharp than the neighbouring stars, and to avoid being misled Messier had made a list of all the diffuse objects he could see in the sky that were stationary. On investigating these objects Herschel found that his better telescopes could resolve most of them into clusters of stars. He was initially of the opinion that all nebulae were just assemblies of stars and would be seen as such, given sufficiently powerful precision telescopes.

Charles Messier, 1730–1817, whose colophon is shown above, compiled the first catalogue of astronomical objects that were not stars as an aid to his search for new comets. The catalogue number, prefixed by the letter M for Messier, is still used as a shorthand notation for about a hundred objects. These include M1, the Crab Nebula, and M31, the Andromeda galaxy.

Herschel's third motive, that of counting the variations in the number of stars he could see as he pointed his telescope in different directions, was a research requiring not only good instruments but also a profound intellectual insight. He made an enormous mental leap to form a mind's-eye view of the solar system and its environs as seen from outside. Unaware that Thomas Wright of Durham had made, and then withdrawn, similar suggestions some twenty-five years before, Herschel realized that the Sun might itself be within a star cluster. The appearance of the Milky Way could then be explained by assuming this assembly of stars was a flat disc, many stars being seen when one looks in the plane of the disc and relatively few when looking out of that plane. He assumed that he could see right through the disc in all directions and also that "taken one with another" all stars were equally bright. With this model Herschel was able to convert his star counts into distances – multiples of the distance to Sirius, the brightest star, rather than miles or kilometres – and to get the first clear impression of the shape of the galaxy.

These researches involved close collaboration between Herschel and his sister Caroline, who sat at the foot of the telescope making notes so that her brother

The portrait (below) of William Herschel shows him holding a diagram of "the Georgian planet and its satellites".

Caroline, William Herschel's sister and devoted assistant (above), was a highly capable astronomer in her own right, both as an observer and a computer.

would not spoil the dark-adapted sensitivity of his eyes by using a lamp to make notes himself. Research on the three main lines was not the only activity. The high definition of Herschel's telescope allowed him to define the axis of rotation of Mars and to refute claims of mountains on Venus. The superiority of his instruments was at times something of an embarrassment, and he wrote to Aubert asking him please to confirm that the Pole Star was double, as no one else could see this and he was widely disbelieved.

In addition to observation, Herschel augmented the salary granted by the King by making and selling reflectors, mostly either 2- or 3-metre ($6\frac{1}{2}$- or 10-foot) instruments, and his records show that he made about fifty telescopes in the period between 1782 and 1785. A 2-metre instrument cost £65; later on the inflationary effect of the Napoleonic wars pushed this up, and in 1799 Fanny Burney's diary records that the younger William Pitt had bought a telescope for £100. This particular instrument did not work on arrival at Pitt's home in Dover. The main cause seems to have been adherence to the unchanging rule never to read the instructions until all else has failed.

The activities of observing and telescope-making were maintained in spite of appalling weather. The winter of 1783 was unusually cold, the ink froze, mirrors broke. Caroline, necessarily moving round the snow-covered 6-metre (20-foot) telescope in pitch darkness, slipped and was badly hurt, and William resorted to rubbing himself all over with a raw onion in an unsuccessful attempt to keep off the ague. The damp situation of the Herschels' house did not help health or observation and in 1785 and again in 1786 the Herschel family moved, finally settling in Slough – in spite of its name a drier site than the house near Windsor had proved. Then, with a grant from the King, the family started on the construction of a 12-metre (40-foot) focal length, 1·2-metre (48-inch) aperture telescope. This was more than twice the size of Herschel's large aperture 6-metre (20-foot) telescope, itself the biggest in the world, and undertaken in spite of the failure of an earlier project to cast a 90-centimetre (36-inch) diameter mirror. On that occasion the mould broke and molten speculum metal spilled out on to the floor, the stones of which broke up explosively, filling the room with a mixed shrapnel of stone and red-hot metal. The first mirror for the new telescope was cast in 1785 by a professional London founder. The result weighed almost half a tonne, in spite of being thinner than required, and Herschel organized a team of workers to help with the grinding and polishing. The mirror

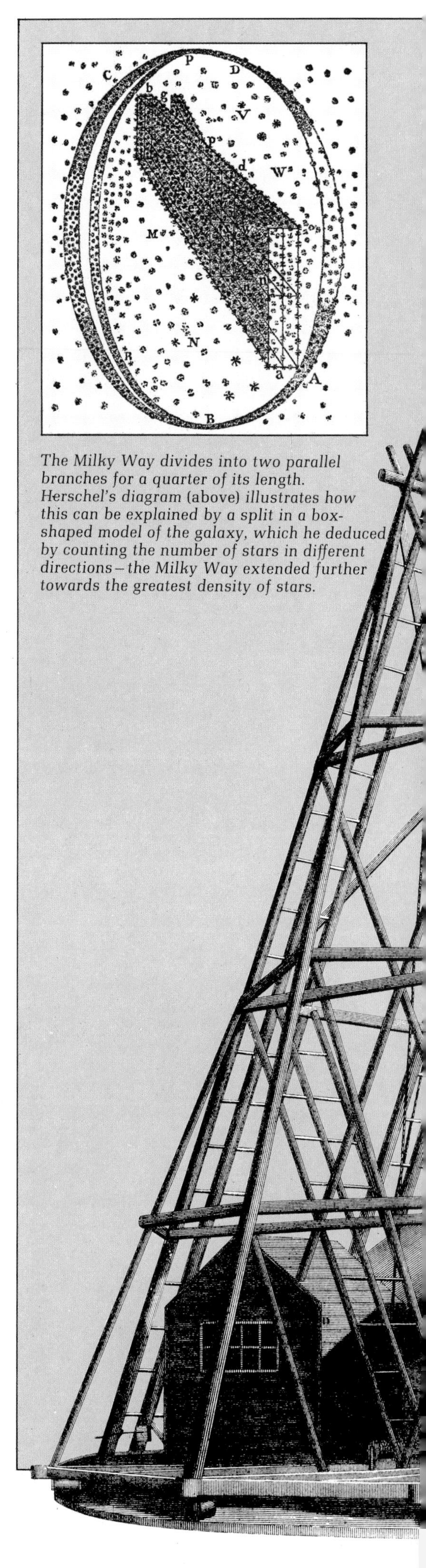

The Milky Way divides into two parallel branches for a quarter of its length. Herschel's diagram (above) illustrates how this can be explained by a split in a box-shaped model of the galaxy, which he deduced by counting the number of stars in different directions – the Milky Way extended further towards the greatest density of stars.

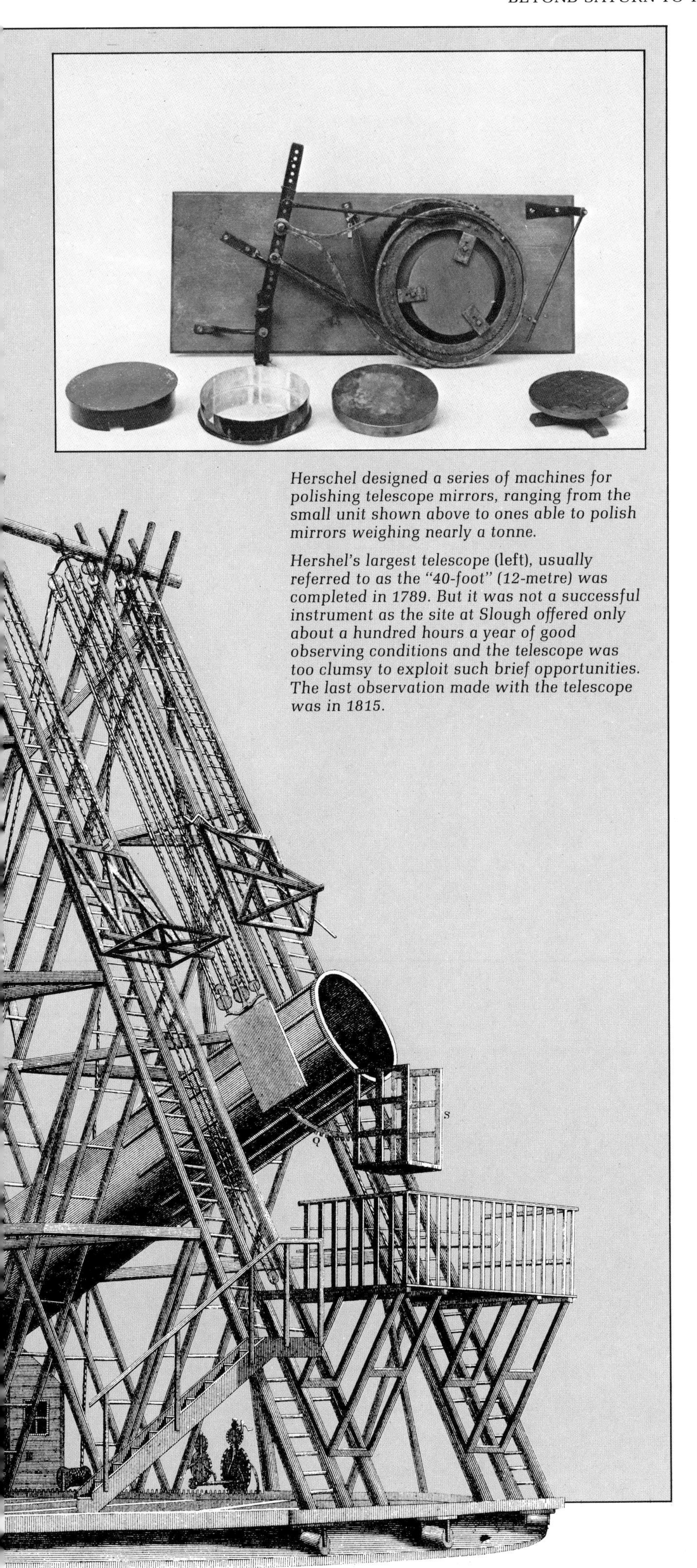

Herschel designed a series of machines for polishing telescope mirrors, ranging from the small unit shown above to ones able to polish mirrors weighing nearly a tonne.

Hershel's largest telescope (left), usually referred to as the "40-foot" (12-metre) was completed in 1789. But it was not a successful instrument as the site at Slough offered only about a hundred hours a year of good observing conditions and the telescope was too clumsy to exploit such brief opportunities. The last observation made with the telescope was in 1815.

was fitted with an iron ring with twelve numbered handles and moved on the polishing tool by twelve men in numbered overalls. The numbers enabled Herschel to call out instructions controlling the direction of the polishing strokes. The process was not very successful, and a second casting was made, which cracked, and then a third. This last was thicker and heavier than the earlier casting and it took twenty-four men to move it on the polisher. After much tedious shoving and hauling the mirror was still giving poor images and Herschel spent almost a year developing a polishing machine. This was more controllable than the clumsy team of bored labourers and on 27 August, 1789, the mirror was finally deemed good enough to put in the telescope that had been built to receive it. The new machine was an almost instant success. The second night Herschel discovered Enceladus, the sixth satellite of Saturn, and saw more clearly than ever some spots on the planet.

In modern times the completion of the world's largest telescope, followed immediately by a new discovery to demonstrate the success of the project, would lead to a flood of publications. Papers on the optics, the structure, the site, the new discovery and anything else that might be more or less fit to print would go forth, all stressing the skill of the project scientists and engineers and the sagacity and benevolence of their funding agency. Herschel's approach was rather more modest. He wrote to Sir Joseph Banks, the President of the Royal Society and Editor of the Society's *Philosophical Transactions*:

"Would you think it proper, as my paper on Nebulae is now printing, and I believe one of the last things in the volume, to add at the bottom of it? – P.S. Saturn has six satellites. 40 feet reflector. Wm Herschel."

Although the giant telescope got off to such a good start it was not used as much as the smaller instruments. It was unwieldy to operate, and because the mirror had a high copper content (to give a speculum alloy that could be cast in so large a piece) it tarnished rapidly. The majority of Herschel's research was still done with the 6-metre (20-foot) telescope and later, when Herschel was in his sixties, with a 3-metre (10-foot) instrument of unusually large aperture, namely 60 centimetres (24 inches). The quantity of original work between 1789 and 1822 was extraordinary: a second catalogue of nebulae in 1789 was followed by a third in 1802; studies of Saturn revealed a seventh moon and the rate of rotation of the planet; and work on "variable" stars (stars that fluctuate in brightness) was followed by the measurement of the brightness of stars. Other stellar researches included the first observation of stellar spectra, more studies of double stars and, in 1811, the recognition that some nebulae were not aggregates of stars but truly diffuse matter, while within the solar system he worked on the size of Venus, the rotation of Jupiter's moons, the comets of 1807 and 1811, and the structure of the Sun. In parallel with this Caroline was equally busy, not only helping her brother but working in her own right. She discovered six comets, and when William's results failed to agree with those of Flamsteed she reworked Flamsteed's results and corrected the many hundreds of measurement errors in his catalogue.

The two most important of Herschel's later researches were connected with the Sun. In 1800 he undertook a study of the heating effects of different colours in the solar spectrum, mainly to help in the design of filters which would prevent the Sun's rays from distorting his telescope mirrors. He found that while red light had a greater heating effect than green, there was beyond the red an "invisible light" that obeyed the usual laws of optics and carried much of the Sun's radiant heat. This, the discovery of the infra-red, would have sufficed alone to win Herschel an international renown had he done no astronomy at all. The other problem was more subtle. If, as Herschel had shown, the Sun was a star in a huge galaxy, then the system must be in motion – either inwards as the galaxy collapsed under gravity, or in balanced rotation like the solar system. There is a perspective effect associated with motion; as one travels towards a pair of objects, trees, lamp posts or whatever, they appear to move apart towards the sides of one's field of view, and the reverse, a moving together, occurs as one moves away. Herschel analysed the motion of all those stars whose displacements had been measured and concluded that the Sun was indeed moving and that the motion was towards a point in the constellation of Hercules.

The whole of Herschel's career as an astronomer – almost fifty years from his first telescope in Bath to his death at the age of nearly eighty-four – stressed the near inevitability with which the combination of vastly improved equipment and a clear insight lead to major discovery. The improvement in our detailed knowledge of the solar system was enormous, but the demonstration that the Sun was a moving star inside the structured island of our galaxy is one of the major steps in man's understanding of his universe. Ironically, in all this Herschel never did discover the stellar parallax he had so confidently set out to find at the beginning of his astronomical career.

CHAPTER FIVE

THE IMPOSSIBLE ACHIEVED

The achromat from Moor Hall to Bessel, 1730–1840

"There is no fact in the history of science more singular than that Newton should have believed that all bodies . . . separated the red and violet rays to equal distances when the refraction of the mean rays was the same. This opinion, unsupported by experiments and not even sanctioned by any theoretical views seems to have been impressed on his mind with all the force of an axiom. . . . When, under the influence of this blind conviction he pronounced the improvement of the refracting telescope to be desperate, he checked for a long time the progress of this branch of science, and furnished to future Philosophers a lesson which cannot be too deeply studied."

The Life of Sir Isaac Newton,
by David Brewster, 1831.

The image quality of seventeenth- and early eighteenth-century refracting telescopes was severely limited because the telescope lenses could not focus the different colours forming the white light emitted by a star into a single sharp image. This colour-associated defect, chromatic aberration, was inseparable from the focal properties of the lens used. It either restricted the astronomer to a refracting telescope with an overall length roughly a hundred times its aperture or drove him to use the more demanding and less efficient reflector. The possibility of a telescope of reasonable aperture that was both short and efficient was realized when, in 1758, a patent was taken out in London for a lens which largely overcame the bad effects of chromatic aberration.

The patentee of the first lens that would give an image free of coloured edges was John Dollond, then aged fifty-two and an established authority on optics. Only four years earlier Dollond had written two letters, one to the Swiss theorist Leonhard Euler and the other to the telescope-maker James Short, in which he had explained in the most lofty and patronizing tones that anybody who was anybody knew that chromatic aberration was unavoidable. When Dollond later explained the background to his highly profitable change of view, he implied that he had made his discovery unprompted and unaided. Even at the time there was evidence that this was not the case. Three men in particular – Leonhard Euler, Samuel Klingenstierna and Chester Moor Hall – had been prepared to question Newton's authority prior to Dollond's patent.

It is one of the ironies of experimental science that although Newton was in error, the line of argument that led to his refutation was itself wrong. Euler, Klingenstierna and Moor Hall "observed" that the lens system of the eye was free of chromatic aberration, and concluded that a corrected lens, an "achromat" was indeed possible. They were all wrong. The eye suffers from quite severe chromatic aberration, as Newton well knew. None the less, the "observation" was linked with the structure of the eye to suggest a new approach. The lens of the eye has optical properties similar to glass, but unlike the lenses of astronomical telescopes, this lens is immersed in a second refracting medium which has optical properties similar to water. It was argued that a combination of water and glass components might be achromatic and therefore worth study. Furthermore, if an achromatic lens existed in the eye, then Newton's rule that the dispersion of white light into its colours was proportional to the deviation in angle suffered by the beam must necessarily be wrong. Euler, a brilliant mathematician, assumed a different, but equally arbitrary, rule which showed that it was then possible to design a lens that might be colour-free by using two glass components to enclose a water-filled space. Klingenstierna was more subtle. He too was a mathematician and he pointed out that if Snell's law of refraction was correct, then Newton's rule could be true either for a pair of thin prisms for which the deviations were small or for two fat prisms for which deviations were large. It was quite impossible for Newton and Snell both to be right for both thin and thick prisms. Euler's theory was published in 1747. Klingenstierna sent his conclusions to Dollond in 1755 in a letter in which he strongly recommended that further experiments should be made.

The earliest precursor of Dollond, Chester Moor Hall, never published anything, but possibly had the greatest influence. Moor Hall, in contrast to Euler and Klingenstierna, was an experimenter and he found empirically that Newton's rule was incorrect. More important, he found that two different types of glass available to him differed so much in their relation of dispersion to deviation that he could design an all-solid achromat: water was unnecessary. The two glasses were known in the trade as "crown" and "flint" glass. Crown glass was the traditional window and spectacle glass. Flint glass

John Dollond, born in 1706, was a silk weaver and became an expert in optics. He was appointed Optician to the King in 1761.

was new, being a mixture developed for tableware and chandeliers in which the usual soda–lime–sand combination was modified by the addition of lead oxide. This recipe was first tried by George Ravenscroft in 1673, seven years after Newton's prism experiments, and during the next twenty-five years glass-makers added more and more lead oxide, making heavier and heavier glasses. (The most extreme of the "lead crystal" glasses make most unattractive tableware; they are oily in appearance and so soft that they are very easily scratched.) Technically, these were important as the optical properties were significantly different from crown glass. The new glasses had higher refractive indices than the traditional material so that a flint glass prism bent the light rays through a greater angle than a similar crown glass prism and, in addition, the dispersion of the flint glass was greater. As a result, the new prism produced a greater spread of the colours in the spectrum. The critical point was that while the deviation was increased by twenty per cent the dispersion was up not by twenty per cent, as Newton predicted, but by fifty per cent.

Chester Moor Hall was a most unusual man, a modest barrister who pursued optics as a hobby, not a profession. He proceeded cautiously to test his ideas by designing and ordering the components of a 6-centimetre ($2\frac{1}{2}$-inch) aperture achromat made from two glass lenses. In 1733 he asked two different opticians, James Mann and Edward Scarlett, each to make him a single lens, one a convex lens of crown glass and the other a concave lens of flint glass. By happy chance, both were busy and sub-contracted the work – and both to the same man, George Bass. Bass made the two lenses of the same diameter and specified by the same designer and, naturally curious, he put them together to see the effect. The first man to look through a working achromatic lens, Bass did not have the commercial acumen to follow up his unexpected knowledge. Moor Hall explained his new invention to two working opticians, Bird and Ayscough, but one was suffering from a surfeit of prosperity and the other from its absence. Bird was the leading instrument manufacturer, but was too busy making quadrants to need a new idea, while Ayscough was too worried about bankruptcy to start a new development.

Some time in the mid-1750s, twenty years after the event, Bass told John Dollond that he knew an achromatic lens was possible because he had made one. This information, coupled with Klingenstierna's letter, persuaded Dollond to go and do some experiments for himself instead of just accepting Newton's authority. Dollond's experiments were straightforward and effective. First he cemented together sheets of glass to make a wedge-shaped trough, filled this with water, and put into the wedge a glass prism. By trial and error he adjusted the wedge angle of the trough until a beam of light passing first through a prism of glass and then through a prism-shaped space full of water emerged undeviated from its original path. This combination, with the glass and water components giving equal and opposite deviations, still produced a clearly dispersed spectrum of colours. Newton was indeed wrong. The next step was to make the more important water-and-glass prism combination which gave deviation without dispersion. This done, it was clear to Dolland that not only were the established rules wrong but that the difference between glass and water was too great to make a practicable achromatic lens from these materials.

Dollond then carried out a similar series of experiments in which he compared

Peter Dollond, 1730–1820, whose business card is shown above, became Master of the Worshipful Company of Spectacle Makers in 1774, a post which he was to hold twice more.

crown and flint glasses, and finished up with a combination of three prisms, two of crown and one of flint glass, which deviated but did not disperse the incident light. These experiments were not sufficiently precise to allow him immediately to design an achromatic lens, but after a good deal of trial and error Dollond did manage to make his first successfully corrected lens. As George Bass had remembered, this was a combination of a convex lens of crown glass and a concave lens of flint glass.

The next step was taken by John Dollond's son, Peter, who was the first man to realize that the achromatic lens was not just an interesting diversion but a device that would yield a major commercial advantage to its inventor. Peter Dollond had set up an optical works of his own in 1750, using his own business skills to produce a range of optical components whose design relied heavily on his father's knowledge and reputation. In 1758, as soon as John Dollond's trial lenses were finished, Peter persuaded his father to patent the invention. There was a lot of jealous resentment by other London opticians of the resulting increase in the Dollonds' prosperity. This culminated in two unsuccessful law suits in which the patent was challenged, chiefly on the ground that Moor Hall had demonstrated an unarguable priority. In his summary of the evidence Lord Camden, the judge of the second case, agreed that Moor Hall was the inventor but "It was not the person who locked his invention in his scritoire that ought to profit by a patent, but he who brought it forth for the benefit of the public."

The patent gave the Dollonds a monopoly of the manufacture of achromatic lenses, but it did not solve two of the major difficulties. It was exceedingly difficult to get good discs of flint glass from which to make lenses and, after John Dollond's death in 1761, there was no one in the firm able to improve the design by calculation. Peter Dollond's production methods relied more and more heavily on "guess and try" backed by increasing technical skill, and by this method he developed a three-component lens, two crown glass lenses enclosing one of flint.

Meanwhile, the theory of achromatic lenses was not neglected. In 1761 the Royal Society published a Latin translation of Klingenstierna's work, much to Peter Dollond's annoyance, and in France Alexis Clairault repeated John Dollond's experiments and then designed a range of better lenses. As a young man Clairault had accompanied Pierre Maupertuis' expedition to measure the length of a degree of latitude in Lapland, and had played a major part in the interpretation of the results. He had later confirmed his high reputation by a theoretical analysis of the effects of Jupiter and Saturn on the orbit of Halley's comet, which improved by more than a factor of ten the accuracy of the predicted return date of the comet – a prediction which was borne out by observation when the comet returned in 1758. Clairault's analysis of chromatic aberration showed that three components were unnecessary. A two-component "doublet" lens was equally effective. In addition, the curvatures of both faces of the convex and the inner face of the concave components could all be made the same, greatly reducing the cost of production. Clairault widened his analysis to include the effects of spherical as well as chromatic aberration. As with Short and Herschel's mirrors, the image given by a simple lens is not perfect if the optical surfaces are parts of a sphere (spherical) rather than paraboloidal or hyperboloidal. A non-spherical lens is extremely difficult to manufacture. With a two-component lens, however, it is possible to make the errors of the two halves equal and opposite, so that they cancel out to give a good image. Clairault's designs were made up by several opticians in 1763. The first example, made by Passemant, was incorporated in a telescope given to Louis XV.

The application of the new achromatic lens to astronomical rather than terrestrial telescopes began at the 1761 transit of Venus, when Klingenstierna in Stockholm, Mayer in Germany and La Caille in France used the new lenses, presumably supplied by Peter Dollond, for their observations. The new telescopes gained little from this first use, news of their good performance being swamped by the general disappointment at the low quality of the overall results. Events took a turn for the better when Peter Dollond lent his first three-component achromat to the Astronomer Royal, Maskelyne. The lens was compared with the same long-focus telescope that had been used to check Hadley's reflector some forty years before: on 25 September, 1765, all existing refracting telescopes became obsolete and Peter Dollond's fortune was assured. Prior to the new development a 10-centimetre (4-inch) aperture lens would need to be mounted in a telescope 10 or even 20 metres (33 or 66 feet) long if close double stars were to be resolved. Now the same resolution could be achieved with a lens whose focal length was only 105 centimetres (42 inches), combining the convenience of manageable size with the high efficiency of lens optics.

The predicted arrival of Halley's comet (shown above at its return in 1910) encouraged popular interest in astronomy in the mid-eighteenth century. A spacecraft may study the comet at its next return in 1986.

The new type of lens was applied almost immediately to the instruments of positional astronomy – quadrant, sextant, mural arc and so on – as well as to "observing" telescopes. The first divided-scale instrument to benefit was French and was fitted with a Clairault design objective, made by an amateur, M. de l'Etang. The lens had an aperture of 3·8 centimetres ($1\frac{1}{2}$ inches) and a focal length of 73 centimetres (29 inches) – a modest combination but one that collected ten times as much light as La Caille's traditional quadrant telescope, whose lens was of similar focal length but only 1·2 centimetres ($\frac{1}{2}$ inch) aperture. By the time of the second Venus transit the achromat was firmly established and was used for twenty-seven of the 150 successful observations. Fifteen French astronomers had the new lenses, including the luckless d'Auteroche, and the British contingent

CHROMATIC ABERRATION

A simple lens, that is a lens made from a single piece of glass, has a different focal length for each colour in the spectrum. It is impossible for such a lens to give a sharp image in white light – the defect of chromatic aberration.

The image of a white object formed by a simple lens is shown above, left. It has coloured edges, usually orange or blue, due to the chromatic aberration effect of the lens. The extent of the coloured edges depends on the diameter of the lens and gets worse as the lens gets bigger. In comparison, the same object formed by an achromatic lens produces a sharp-edged white image (above right).

AN ACHROMATIC LENS

The correction of chromatic aberration requires the combination of two lenses of different types of glass. The net result is a "doublet" which still focuses the light but with the opposite colour defects of the two components balancing each other, as shown below.

would have been more numerous had the Royal Society been willing to pay Peter Dollond's very high prices.

While Maskelyne was organizing and equipping the British expeditions for the 1769 transit, his home base, the Greenwich Observatory, was in the midst of a quiet revolution. Two external developments had taken place which required the entire policy of the Observatory to be rethought. Suddenly there were two solutions to the longitude problem. First, a German professor of mathematics, Tobias Mayer, combined a theory built on the work of Euler, Clairault and d'Alembert with Bradley's measurements to come up with a set of accurate predictions of the Moon's motion. At last, the lunar method could be used to determine longitude. Then, while Maskelyne, using Mayer's method of calculation, was compiling his first *Nautical Almanac*, the instrument-maker John Harrison crowned a lifetime's work on chronometry by producing a clock which would keep time at sea to an unprecedented accuracy. His clock's precision was demonstrated when it was taken by ship on a trip to Jamaica and back: the longitude error was less than 30 kilometres (20 miles), even after months of pitching and rolling in a naval frigate. This was more accurate than Mayer's tables, and it also avoided all the calculations that the lunar distance method required. The only problem was that the chronometer was neither cheap nor capable of mass production. When Cook sailed to Tahiti on his second expedition he took with him the first copy of Harrison's chronometer and the device enjoyed much prestige. However, Cook's ship HMS *Resolution* cost £2750, the chronometer £500.

The Observatory rapidly found two major new roles in its service to navigation. Maskelyne established a team of human computers to work out the predictions of Mayer's theory for each coming year, for publication in the annual *Nautical Almanac*. In parallel the Observatory took on an assistant whose job was to test chronometers. The twin approach was essential as the *Almanac* was still necessary for navigators who could not afford the new chronometers. All this deflected the Observatory from its fundamental observation programme, and the new achromatic lenses were gradually incorporated into the telescopes at Greenwich. The slow progress followed from the difficulties accompanying the manufacture of flint glass, which meant that the vast majority of achromatic lenses were of fairly small aperture and the perpetual cry of astronomers for "more light" was not easily answered. While Herschel improved and enlarged the reflector, Peter

Dollond, joined by his brother John, concentrated on the exploitation of his lenses in less demanding applications. The Dollonds' work led to the traditional terrestrial telescope: an achromatic objective lens of a few centimetres aperture set in a collapsible mounting.

The majority of Dollond's few large lenses were made into telescopes by Peter Dollond's brother-in-law, "Honest Jesse" Ramsden. Born near Halifax in 1735, Ramsden was the great nephew of Abraham Sharp, who had built Flamsteed's successful quadrant in 1688. His first job was as a clerk in a cloth warehouse, but he left to be apprenticed to a London instrument-maker. He set up in business on his own in 1762 and soon afterwards married Peter Dollond's sister Sarah. Jesse Ramsden's first astronomical instruments were small achromatic refractors and were notable for the way in which a mounting used by James Short (a development of the fifteenth-century torquetum) was utilized and improved. His chief contribution to precision engineering in general, and astronomy in particular, was the development of a precise screw thread. This modest-sounding device is of fundamental importance. Nearly every screw found in the modern world is made by copying another screw, and it is extremely difficult to generate the accurate master screw thread that can be copied. It took Ramsden nearly fifteen years to perfect a screw that was a precise helix to within 0·001 millimetre (40 millionths of an inch).

The new screw was applied as the critical component of a dividing engine, a machine for engraving the accurate scales needed on all the angle-measuring devices required in astronomy and navigation. John Bird, the leading maker of quadrants in the mid-eighteenth century, had made his graduated scales by the methods of classical geometry, constructing 90-, 60- and 30-degree angles and then bisecting these angles repeatedly, often to produce scales with 96 divisions in a right angle. The first effect of Ramsden's new dividing engine was that the gain in precision led to an immediate reduction in the size of graduated scales, so that the navigators' sextant was cut in size from a cumbersome 50-centimetre (20-inch) radius to a convenient 25 centimetres (10 inches). Even smaller sextants, down to 10-centimetre (4-inch) radius, were built for the explorers and empire-builders who travelled by land, and so did not have a convenient ship to carry their luggage. The second effect of Ramsden's engine was a consequence of the new ability to graduate a complete circle. Several Vernier scales, reading at intervals round such a scale, give much greater precision than can be achieved with the single Vernier of a quadrant. Not only are several readings better than one, but when they are averaged the result is to eliminate any small errors in the concentricity of scale and pivot. Bird's mural arcs and quadrants were replaced by mural circles, combining the new scales with the reduction in focal length and increase in aperture due to the achromatic lens. This produced compact instruments able both to see faint stars and to measure their position with high precision.

Ramsden was not only a precision engineer. His name is still associated with the "Ramsden eyepiece" and upon occasion his firm made their own achromatic lenses. (Dollond had given Ramsden a share of the patent as a wedding present.) Ramsden's first love, however, was for precise machinery, a passion pursued with complete disregard for commercial considerations. This meant that if, halfway through a job, Ramsden saw a better

The Nautical Almanac and John Harrison's sea-going chronometer (below) provided two solutions to the navigator's problem of finding the longitude. The almanac used predictions of the position of the Moon based on the work of Tobias Mayer. Both Mayer and Harrison applied for awards from a prize fund administered by a reluctant British Admiralty Board of Longitude. Awards of £3000 to Mayer's widow and £18,750 to Harrison were made, each after a ten-year delay.

way, then everything stopped for the new idea to be tried. Many instruments were completed, though usually late, and a telescope for Dublin Observatory set something of a record, being delivered twenty-seven years after the date of the order. The start of the first national "Ordnance" survey was delayed for some years because Ramsden had not completed the theodolites. The Royal Observatory gave up and cancelled their order for a telescope after only six years.

The instruments that Ramsden did complete were superbly accurate. The Ordnance Survey was carried out independently from several baselines, including one on Salisbury Plain and one 600 kilometres (380 miles) away in Ireland. When complete it was possible to calculate how long the English baseline would be if its length had been deduced from measurements starting in Ireland. The error was 12·5 centimetres (5 inches), one five-millionth of the overall distance.

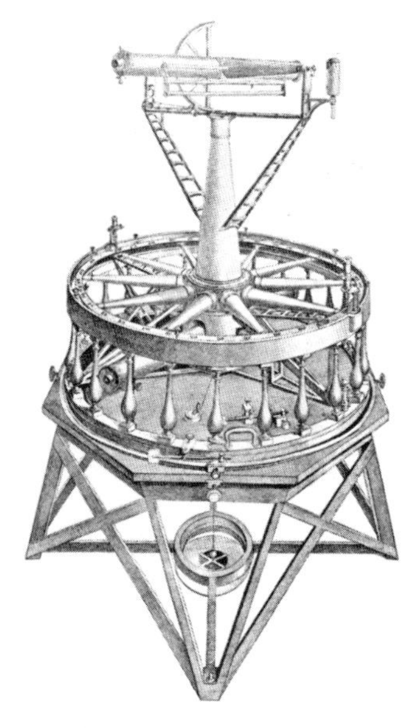

The precision of Ramsden's dividing engine and the use of light but stiff tubular structures were exploited in surveying as well as astronomy. Ramsden's large theodolite, shown above, was mounted on a scaffolding directly above Bradley's transit telescope at Greenwich, to link the national survey to a site of precisely known position.

The most famous of Ramsden's astronomical instruments was Giuseppi Piazzi's "Palermo circle", a 1·5-metre (5-foot) focal length achromatic refractor fitted in an alt-azimuth mounting. The motions of the telescope both in altitude and azimuth were measured on complete circular scales and the structure was designed to be as stiff as possible, using hollow conical and cylindrical tubes as the main structural members. Piazzi was originally Professor of Mathematics at Palermo; when in his forties he was sent abroad to learn astronomy at Paris and Greenwich as the first step in the foundation of the twin observatories at Naples and Palermo. He obtained delivery of his telescope by laying siege to Ramsden's workshop, visiting almost daily for progress reports and even working on the instrument himself. The programme was interrupted by a visit to Herschel's reflector, when Piazzi fell off the ladder in the dark and broke his arm, but eventually his new telescope was complete and he returned home to see it set up on the roof of the royal palace in Palermo. The instrument amply demonstrated the new order of accuracy, and Piazzi was able to show that motions of the "fixed" stars were the rule rather than the exception. The motions were, however, minute: the fastest moving objects he observed – a pair of stars in the constellation Cygnus, one of fifth and the other sixth magnitude – were moving across the sky at a velocity of five arc seconds a year. The average is much less; the mean velocity of the fifty brightest stars (the Pole Star is the faintest of these) is 0·29 arc seconds per year. It is impossible to measure such small quantities unless the instrument is almost unbelievably stable; a distortion of a tiny fraction of a millimetre in a decade is enough to destroy the whole programme.

Piazzi, like Herschel, steadily worked his way round the sky, measuring and remeasuring the positions of 7600 stars. Again, like Herschel, his methodical diligence was rewarded, when on the opening night of the nineteenth century he too discovered a new minor planet. The planet, which he called Ceres after the classical patron goddess of Sicily, was not, like Uranus, a far off giant, but a minute speck orbiting the Sun in an almost circular path between Mars and Jupiter. The timing of the discovery was remarkable. Johann Bode, the Director of Berlin Observatory and the man who had named Herschel's planet Uranus, had already written to Piazzi asking him to help search for a planet in the Mars–Jupiter gap; Piazzi found Ceres before the letter arrived. (Bode's reason for the search was an empirical and unexplained relation, now called the Titius–Bode law, which describes planetary distances from the Sun.).

The other dramatic gain from the discovery of Ceres was that it drew international attention to the invention of a theory of errors by a young genius, Carl Gauss. This work provided the theoretical tools needed to derive reliable results from a number of necessarily imperfect sets of observations. Prior to Gauss's work there had been no dispassionate method for the assessment of experimental precision, nor any way in which to combine many observations other than the crude evaluation of an average. Gauss used his theory to successfully predict the position of Ceres after a period of unobservability when it lay close to the Sun in the sky, on the basis of only a few prior observations by Piazzi. There were two surprises in store when Ceres was studied more carefully. Even Herschel was unable to see it as a disc rather than a point of light, so the planet was much smaller even than the Moon. Second, the little planet was not alone. In the next few years three more were found, all in orbits very similar to Ceres. Wilhelm Olbers, a Bremen doctor, found two, Pallas and Vesta, and his colleague Karl Harding discovered Juno.

Jesse Ramsden died just before Piazzi's discovery of Ceres, leaving a workshop full of half-completed instruments and a reputation for honesty and integrity. His pursuit of accuracy led him to the development of better instruments and, outside his workshop, to the preservation of historical as well as technical precision. In reply to his brother-in-law's hagiographical portrayal of John Dollond as the singlehanded inventor of the achromat, it was Ramsden who quietly published a more balanced account. The orders awaiting completion in Ramsden's workshop were taken over by his foreman, Matthew Berge, a policy which worked well as long as innovation was not necessary. Berge's sextants were fine instruments, but his attempt to complete a telescope for Sir George Shuckburgh was less successful. The Shuckburgh telescope, based on a 10-centimetre (4-inch) aperture Dollond achromat, was very similar in structure to Piazzi's Palermo circle. The critical change was that while Piazzi's instrument moved on vertical and horizontal axes (an alt-azimuth mount) the new telescope was pivoted to move on the more convenient polar and declination axes. Tipping the Palermo design over at an angle to make an equatorial mounting, however, put a skew load on the main frame and even with six columns where Piazzi's circle used four, the mounting was too flexible.

Sir George Shuckburgh bequeathed his telescope to the Royal Observatory, where it proved something of a Greek gift. The Astronomer Royal reported, fifty years later, that it would cost £100 to make serviceable, "And when this is done, we should have a small indifferent telescope, on a weak frame, in a position bad beyond anything that an astronomer ever imagined." It was not Berge's fault that the telescope was mounted in a converted summerhouse where the main Observatory building blocked out much of the sky, and it would almost certainly have had a stiffer mount had Ramsden lived to see its defects. The main point of the Shuckburgh equatorial was that it broke new ground. It was the first large equatorially mounted telescope, a logical consequence of the short overall length of the achromatic refractor. Its priority is still commemorated in that the tilted yoke Ramsden designed is still called an "English mounting".

Ramsden's death was not the only setback to telescope-making in England. The craft was also suffering severely from the effects of a whole series of governments with a consistent hostility to glass. The first step had been a tax on windows, imposed by William Pitt as a simple property tax that would adjust roughly with the opulence of the house. Caroline Herschel had been most upset to find that the house near Windsor that her brother had rented had thirty windows and would be uncomfortably expensive. Then cut glass tableware and chandeliers caught the attention of the exchequer and both were deemed taxable luxuries. The duty was, disastrously, set on the weight of glass used, not by item or finished value, so the optically vital but heavy lead glasses ceased to be economic and were dropped from production. Many years later, after the optical glass industry had been totally destroyed, the government made a series of grants for its re-establishment, no doubt claiming that the grants were part of its usual far-sighted policy.

"Honest" Jesse Ramsden, 1735–1800, seated beside his dividing engine. In the background is the "Palermo Circle", the achromatic telescope used by Piazzi in his studies of the motion of the stars.

Although early nineteenth-century British telescope-makers were forced to use imported glass, their skills were still formidable. Chief of Ramsden's successors were Edward Troughton and Charles Tulley. Troughton developed the mural circle, building in particular an instrument for Greenwich. This telescope, and one of two copies of it made by Thomas Jones, were used by the early nineteenth-century Astronomer Royal, John Pond, for the continuation of Bradley's work to establish the precise position of a small group of fundamental reference stars. Pond added to the accuracy of the circle by using a pool of mercury to give a perfectly horizontal mirror. This technique allowed a pair of observations to be taken of the same star using two circles, one looking up, the other down at the reflection in the mercury mirror. The mirror is more accurate than a plumb line, provided it is kept clean and free from draughts, and Pond spent a lot of time on finding the right design. The mural circle had, however, reached its limit. As telescopes increased in aperture the lopsided load on the pivot became more and more of an embarrassment. A pair of pivots, one either side of the telescope tube, was needed and the symmetrical transit circle came to the fore. Troughton built one such instrument, for Stephen Groombridge. This instrument, very clearly a descendant of Ramsden's Palermo design, was to study the motions of the "fixed" stars.

The technical difficulties facing the glass manufacturers were concerned with producing glass that was uniform in optical properties throughout a disc big enough to make into a lens of large aperture. Mixing up sand, potash and lead oxide and heating the mixture in a furnace was not enough. If the temperature was kept low the melt was viscous and the air trapped in the initial dry powder remained to form a multitude of little bubbles. A hotter furnace led to a less gooey melt in which the bubbles could rise. Unfortunately, the heavy lead oxide sank at the same time, so the result was impossibly non-uniform. The solution was found by Pierre Guinand, a Swiss who started his career as a cabinet-maker. From making clock cases Guinand progressed to casting the bells needed for striking clocks, and then to glass-making. After a quarter-century in the business, during which he made glass not significantly better than his competitors, Guinand applied some of his bell-casting skill to his glass working. Bell metal was stirred while molten to get a uniform mixture and in 1798 Guinand began to look for a stirrer for optical glass. The search took seven years. Rods that were strong enough to stir the red-hot viscous raw glass were hard to find. The simple solutions failed. Molten glass is close to being a universal solvent and even a trace of iron or copper from a metal stirrer gives the glass a dark brown or green colour. Inert stirring rods of fireclay are not strong enough and snap off far too easily. The solution is to use a fireclay tube with an iron rod inside it.

Guinand kept his new technique secret, but his reputation as a maker of high-quality flint glass spread rapidly. In 1805 a Bavarian lawyer, Joseph von Utzschneider, who was the financier of a firm of instrument-makers, persuaded Guinand to move to Munich and set up his glass furnaces anew. He was to get a salary, an annual fee for the rights to his invention and a fifth of the profits. Two years later a new ten-year contract was signed, at more than three times the earlier salary, with a provision that Guinand train an assistant nominated by Utzschneider. The assistant chosen was Joseph Fraunhofer, a young man of outstanding genius who was to revolutionize the design and manufacture of high-quality lenses. Joseph was the eleventh

JOSEPH FRAUNHOFER

Joseph Fraunhofer, 1787–1826, with his spectroscope. Fraunhofer was a remarkable technical genius who made major contributions to the theory of lens design and to the engineering problems of large telescopes. While working on the problem of how to eliminate chromatic aberration in lenses – the same problem that had led Newton to invent the reflecting telescope – Fraunhofer made fundamental discoveries about the spectrum of visible light, the Sun and the stars.

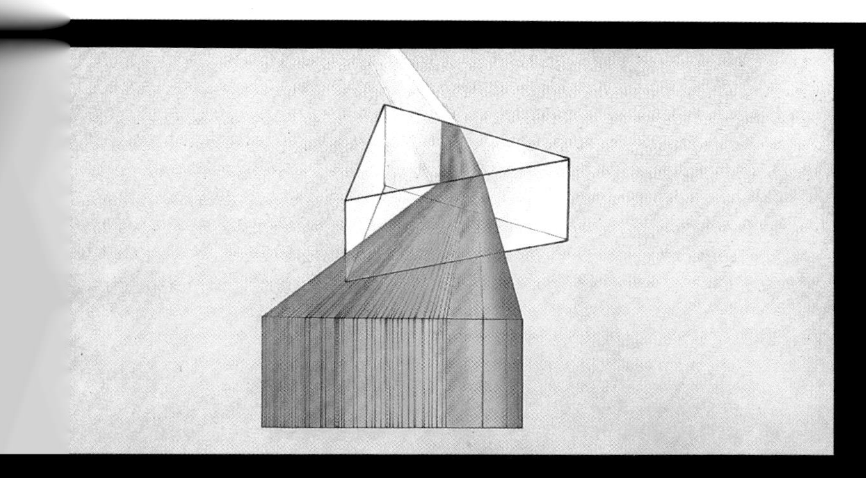

FRAUNHOFER LINES

In Fraunhofer's experiments a beam of sunlight was passed through a narrow slit and then through a prism to be dispersed into a spectrum. This spectrum was then examined using a small telescope and it was found that the band of colours was crossed by hundreds of fine dark lines. Fraunhofer's own drawing of these lines is shown below. Later the lines were shown to relate to the chemicals in stars.

THE HELIOMETER

The objective lens of a heliometer (below) is split into two D-shaped halves to produce a double image. It was designed to measure variations in the apparent size of the Sun. This particular heliometer was made by Fraunhofer according to a design initiated by Bessel, who realized that the instrument could be used to determine the distance between two widely separated stars. The delicate screw device on the heliometer was set against a scale that could be read with a microscope and this made possible the measurement of even the slightest movement between such stars. In 1838, Bessel measured the parallax of 61 Cygni – an angle of only 0·676 arc seconds.

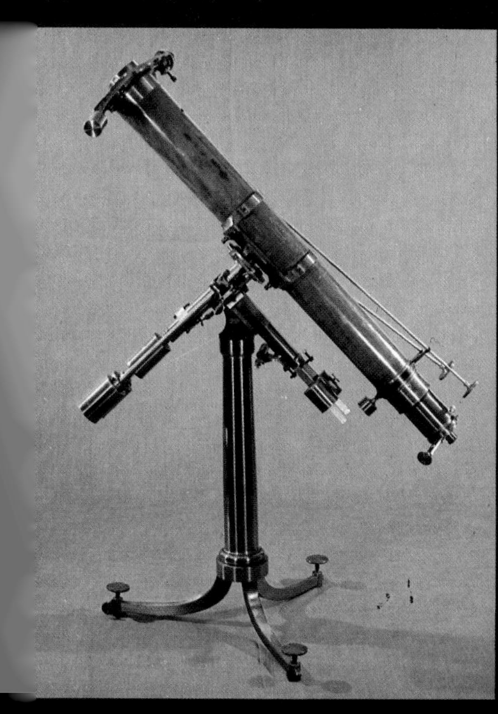

The great Dorpat refractor (right). This telescope was the largest telescope completed by Fraunhofer. The balls near the eyepiece are lead weights which act through their mounting rods to minimize the bending of the telescope tube.

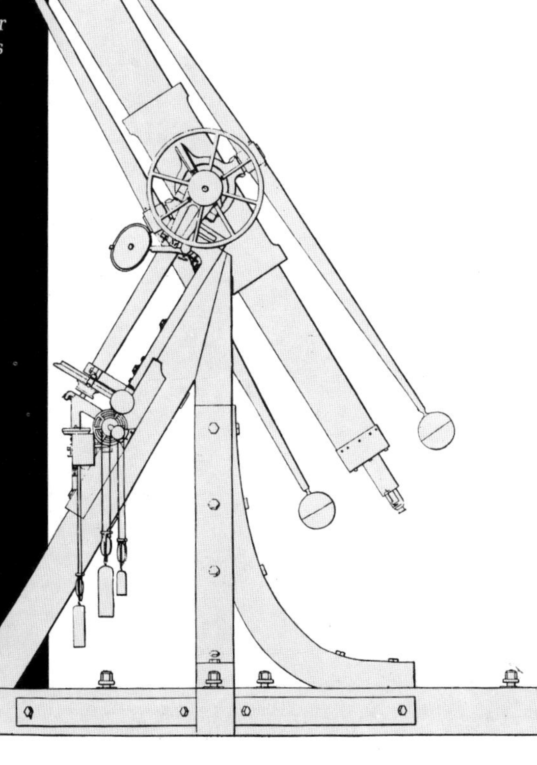

child of Franz Fraunhofer, a glazier, and when his father died the twelve-year-old boy was apprenticed to a decorative glass cutter and mirror-maker, P. Weichselberger. Soon after his seventeenth birthday Joseph Fraunhofer set up in business on his own as a copper plate engraver, but this failed and he returned to the glass industry. In 1806 Utzschneider, who had known the young man for four or five years, offered him a job and a year later directed him to join Guinand. The two men did not get on well together. A lens may be of poor quality if either the glass or the polishing is faulty, and the sixty-year-old Guinand did not take it kindly when his young assistant, who did the polishing, criticized the material with which he was supplied. After two years of argument, Utzschneider intervened to ease the friction. The result of his action was that Fraunhofer became a junior partner and in sole charge of the glass works. Not long after, Guinand returned to Switzerland to set up in competition.

Fundamental to all Fraunhofer's work was his invention of the spectroscope. Newton, Dollond and Clairault had all studied the behaviour of glass prisms by using them to deviate and disperse beams of light, usually sunlight, and had measured the effects in terms of the overall length of the spectrum. Fraunhofer built an instrument based on a theodolite, with which he could measure the deviations with a graduated scale. The first trials of this instrument showed that sunlight was not a uniform band of pretty

F. W. Bessel, 1784–1846, Director of Konigsberg Observatory, was originally a shipping clerk who learned mathematics and astronomy to travel as a navigator.

THE PARSEC

Parallax measurements are the fundamental basis of all measurements of the distance to the stars. The unit of length that results is the parsec, the distance at which the separation between the Earth and the Sun would appear to be one arc second. This is 206,000 times the distance from the Earth to the Sun.

colours, one merging imperceptibly into the next, but was a complex mixture of light and dark, with the general trend of the bright spectrum interrupted by many dark lines. Further experiment and improvements to the spectroscope showed that the angles at which these dark lines fell were always the same, and could be measured precisely. Fraunhofer labelled the strongest lines with letters: A in the far red, then B, C, D through to H at the extreme end of the violet. The lines were always there, as they were a property of sunlight. They were visible in the spectrum not only of the Sun but of the sky, of clouds, and of the sunlight reflected from the Moon and planets.

Fraunhofer's lines were technically important, too, for they meant that the design of an achromatic lens could at last be attacked with confidence. Hitherto the change of refractive index of a particular glass with colour could be measured only in the vaguest manner. The deviation produced by a prism could be measured for, say, the "orange" or the "blue-green", but these were ill-defined. A second observer might have a different view of what colours "orange" or "blue-green" were. Now the properties could be specified accurately, not for "orange" but for Fraunhofer's D line, not "blue-green" but F. Why the lines were there mattered very little. They were there, well defined and available whenever needed.

With better glass in large pieces and a much better knowledge of its properties, Fraunhofer was able to design better lenses. He also improved the polishing and testing methods of the Utzschneider optical workshop. His masterpiece, completed in 1826, was the 24-centimetre ($9\frac{1}{2}$-inch) aperture "Dorpat" refractor made for Wilhelm Struve at Dorpat Observatory (now Tartu in Estonia). Like many other of Fraunhofer's telescopes the big telescope was mounted on an equatorial axis, using a layout in which short offset axles were used – a design that was to be christened the "German" mounting. Another of Fraunhofer's highly successful telescopes was the 16-centimetre ($6\frac{1}{3}$-inch) heliometer designed for Friedrich Bessel at Konigsberg. The heliometer was a telescope specially built to measure angles of the order of the solar diameter. Angles around half a degree were a problem. They were too small to measure by setting the telescope at each extremity and reading the positions off a scale in a divided-circle instrument and too large to measure directly in the telescope's field of view with an eyepiece micrometer. The solution, developed by John Dollond, was to make a telescope with the lens cut into two D-shaped halves, each half mounted in a slideway. Each half-lens gives a complete image, and the two images can be moved, by moving the lens halves, until, for instance, the opposite sides of a pair of solar images just touch. The lens movement needed is a measure of the solar diameter.

Bessel used his heliometer not for solar measurements but to study a close pair of stars, listed in Flamsteed's catalogue as number 61 in the constellation Cygnus and called 61 Cygni. Piazzi had shown that this pair of stars was moving across the sky at an unusually high speed, and Bessel confirmed its motion by his own reduction of Bradley's observations. In place of the old assumption that bright stars were near – a theory upset by Herschel's observation of many ill-matched double stars – Bessel assumed that the stars which seemed to be moving fast were probably those nearest the Earth. His assumption was justified and the quality of Fraunhofer's telescope demonstrated when in 1838 Bessel measured the parallax of 61 Cygni. Measurements spread over a year showed a change in the angle between 61 Cygni and slower moving (presumably more distant) stars of just over half an arc second.

This then was the first positive result at the end of a 300-year-long search for stellar parallax and the key to stellar distances. It was in fact the first in a near triple dead heat. F. G. W. Struve, with the Dorpat refractor and an eyepiece micrometer, determined the parallax of the bright star Vega, and Thomas Henderson, with Jones's copy of the Troughton mural circle, measured that of Alpha Centauri, the brightest star of the southern skies. The three epoch-making measurements were very much the achievement of the achromat. The reflector is much more sensitive than the refractor to errors of alignment, small distortions and other troubles, and even though Herschel's instruments gave better definition than Fraunhofer's and Jones's, they lacked the stability needed for parallax determination. The tiny parallax angle measured by Bessel gave the distance to 61 Cygni – and this was truly astronomical, in spite of the fact that it was one of the nearest stars. A parallax of a whole arc second would be observed for a star 31,000,000,000,000 kilometres (19,000,000,000,000 miles) from the Sun, more than ten thousand times farther away than Herschel's planet Uranus. This enormous distance, called a parsec, is however less than the distance to the nearest star and less than one-third of the distance to 61 Cygni. The Galaxy is indeed an empty place: the hundred nearest neighbours of the Sun stretch out to fill a sphere thirteen parsecs in diameter, and even so, as Herschel realized, this is but a minute fraction of the vast system of which the Sun is so small a component.

CHAPTER SIX

PROFESSIONAL LENSES AND AMATEUR MIRRORS

Merz achromats and the reflectors of Lassell, Nasmyth and Rosse

"Amid the universal applause so justly excited by the brilliant researches of M. Le Verrier, it could hardly be expected that the announcement of similar researches prosecuted independently of them in another country, terminating in similar results, and therefore claiming by implication an equal degree of credit, would be received without some degree of reluctance, or discussed with a total absence of passion, by a people especially sensitive on points of national glory."

Grant's History of Physical Astronomy, *Ch. XII, 1852.*

The measurement of stellar parallax in 1838 marked the beginning of a decade of major astronomical discoveries to which not only professional and amateur astronomers but also theorists made significant contributions. The professional astronomers, led by Bessel and Struve, rapidly equipped their observatories with refracting telescopes based on lenses made by Fraunhofer's successors. Amateur workers, concerned with observation rather than precision measurement, built for themselves a series of large reflecting telescopes. The theorists, the third class of contributor, achieved perhaps the most newsworthy achievement of their art since the days of Newton.

Bessel's measurement of parallax confirmed that the achromatic refracting telescope was the first choice as the working tool of the professional astronomer. The refractor was already used for the routine observation of lunar and planetary positions, but now it could claim a discovery of sufficient magnitude to match the achievements of the clumsy and inefficient reflector. This superiority was further emphasized by the work of Friedrich Struve, who in 1833 moved from Dorpat to set up the new Imperial Russian Observatory at Pulkovo, just outside St Petersburg. Tsar Nicholas's reason for founding an observatory was still the classical need for good longitude measurement. Unlike Greenwich, however, the Pulkovo Observatory was not established as a support service for the Imperial Navy. The commercial importance of Siberia was fast increasing, and Russia was also at that time expanding southeastward into central Asia. The Tsar needed astronomy for map-making rather than navigation. Struve took the 20-centimetre (8-inch) Fraunhofer telescope to Pulkovo, a circumstance which must have caused much regret at Dorpat.

The new observatory was rapidly established on the international scientific map. By 1845 Struve was able to publish a description of the observatory that listed nine brand new telescopes, including six of world class. Many of these were used on the basic programme of longitude determination, whose quality surpassed even that of Greenwich. Struve's own research was on double stars, where he followed up suggestions originating in Herschel's work of the previous century and extended by Felix Savary at the Paris Observatory in the 1820s. Repeated measurement of the separation and orientation of thousands of double star pairs confirmed that in many of them the stars were orbiting one another, and Struve's observations were good enough to show that the orbits were ellipses similar to those found by Kepler two hundred years earlier for the orbits of the planets round the Sun.

These observations were of considerable philosophical as well as astronomical significance. Newton had called his law of gravitation a "universal" law, and later work, especially by Laplace, had demonstrated the law's validity within the solar system. It described the motions of the planets and their satellites from Venus out to the faintest moon of Saturn, though there were some doubts about Uranus and slight suspicion of Mercury's motion. The application of Newton's law to the double stars, pairs separated by much more than the Sun–Saturn distance, and at even vaster distances from the Earth, showed that it did apply everywhere in the known universe: there was indeed a universal law. It is now a fundamental assumption of astronomy and astrophysics that *all* laws are universal, and any worker assuming otherwise is most unlikely to get a sympathetic hearing.

An early portrait photograph (taken by the famous photographer Julia Margaret Cameron) of John Herschel, 1792–1871. He continued in his father's footsteps, mapping the stars of the Southern Hemisphere and cataloguing binary stars and nebulae. His name became publicly known when the New York newspaper The Sun *falsely reported that he had seen living creatures on the Moon through an enormous telescope at the Cape of Good Hope.*

Struve's excellence was recognized by the purchase for Pulkovo of the biggest refracting telescope yet built. Merz & Mahler produced a 38-centimetre (15-inch) aperture achromat and this was mounted by Repsold. While Struve used the new instrument to extend his double star measurements to even fainter pairs, Bessel in Königsberg was concentrating on the motion of the brightest star in the sky, Sirius. In 1844 he announced that Sirius was moving along a path that was a combination of a straight line and an ellipse. Sirius must, therefore, be one half of a double star system – and yet there was no companion to be seen. This confident assertion of the existence of an "invisible star" was very much in keeping with Bessel's development as an astronomer. He had been appointed as director of Königsberg observatory in 1810 and had founded the "observatory" without the apparatus to make observations. For ten years or so he had used his time and outstanding talents to reduce other men's observations, especially those of James Bradley. This had led him to a profound understanding of the errors of observation and the way in which instrumental accuracy could be improved by making observations in deliberate and logical sets followed by an equally logical analysis. When outstanding instruments made by Fraunhofer and others finally arrived at Königsberg, Bessel was able to use this background to make observations of unprecedented precision, including his historic parallax determination in 1838. The new "double" star Sirius took fifty years to complete one orbit, so Bessel had to combine his own data, covering only ten years, with an analysis of a great deal of work by earlier observers.

The reflecting telescope, in contrast to the refractor, was very little used in the early nineteenth century. A Scottish amateur, John Ramage, made a series of reflectors of steadily increasing aperture and length, culminating in a 38-centimetre (15-inch) aperture telescope of 7·5 metres (25 foot) focal length. In 1825 this telescope was taken to Greenwich for testing, and proved something of an anticlimax. The mirror would only produce a reasonable image if its aperture were reduced to 27 centimetres ($10\frac{1}{2}$ inches) – and even then, with half the area covered, it was still far from perfect. The telescope tube was mounted on a turntable and A-frame support (similar to that used by Herschel) and this too had its problems – one of the assessors testing the instrument fell off the high platform from which observations were made.

The lack of progress in the development of the reflecting telescope was underlined when, in 1833, John Herschel took his father's 47-centimetre ($18\frac{1}{2}$-inch) aperture, 6-metre (20-foot) long reflector to South Africa to make observations on star clusters and nebulae visible only in the Southern Hemisphere. The telescope was then forty-nine years old, but was still the best large reflector available. The two telescopes used by John Ramage and John Herschel mark the end of an era: they were the last to be named for their length, not their aperture. Herschel's "20-foot" and Ramage's "25-foot" gave way to Struve's and Bond's "38-centimetre" and "15-inch" telescopes respectively.

Before the invention of the achromat a large-aperture refracting telescope was necessarily a long instrument. As the achromatic refractor became the dominant instrument the length was no longer so important and the emphasis shifted to stress the light-collecting aperture directly.

Henceforth the size of a telescope (and by implication the prestige of its owner) was specified by giving the diameter of the aperture. This process has now reached the point that, while all the users of the "200-inch" telescope on Mount Palomar are aware that it is now the "5-metre", relatively few of them can quote either the focal length of the mirror or the length of the telescope tube in which it is mounted.

The reflecting telescope was rescued from a slide into oblivion by three contemporary amateurs: an Irish politican, a Scottish engineer and an English brewer. The last two, James Nasmyth and William Lassell, had both built modest reflecting telescopes while still apprentices, Lassell's instrument being chiefly remarkable for being the first sizeable (22-centimetre, $8\frac{1}{2}$-inch) reflector to be mounted on an equatorial axis, but it was the Irishman, William Parsons, who started on the construction of a telescope to rival in size those of William Herschel.

William Parsons was born in 1800, the eldest son of the second Earl of Rosse. After studying mathematics at Trinity College, Dublin, and then at Oxford, he graduated in 1822 and went into politics. Although known by the courtesy title Lord Oxmantown, he was not eligible for the House of Lords, and in 1823 he was elected to the House of Commons as Member of Parliament for King's County, the site of the family castle. This was before the reform acts, and so the seat was more or less in his father's gift – an undemocratic approach somewhat ameliorated by the fact that the second Earl was known as "one of the very, very few honest men in Irish politics". Political life does not appear to have roused any enthusiasm or ambition in the young Lord Oxmantown, who instead turned his attention to telescope mirror making. From the first his ambition was to make very large mirrors, and his early researches were concerned with various oblique attacks on how to make a large parabolic mirror out of speculum metal. The best-reflecting alloy of copper and tin is a material so hard and brittle that any large casting will shatter as it cools. William Herschel had been forced to use alloys with more than the optimum quality of copper to achieve a mirror which, even if less reflective than he would like, did at least stay in one piece. Lord

William Parsons, 1800–67, third Earl of Rosse (above), built his telescopes in the grounds of the family castle at Birr in central Ireland. He enthusiastically used his biggest telescope, which had a mirror 180 centimetres (72 inches) in diameter, to examine the structure of nebulae. His fourth son, Sir Charles Parsons, developed the steam turbine (to the considerable embarrassment of the British Admiralty) and later took over Howard Grubb's telescope-making company to form Grubb-Parsons.

(Right) *Nebula M51 – a modern picture compared with the drawing by Rosse himself. This nebula is sometimes called the Whirlpool galaxy and lies in the constellation of Canes Venatici (The Hunting Dogs). The drawing shows the spiral nature of M51, which Rosse was the first to recognize. The discovery attracted a lot of attention as Laplace had already suggested that the solar system had formed from just such a vortex. It is now known that the spiral structure of M51 is not rare.*

Oxmantown worked with the difficult but most shiny alloy. He made a series of mirrors using small pieces of this alloy soldered to a back plate of a special brass he developed himself. In the early 1830s he made first a 38-centimetre (15-inch) diameter mosaic, then one of 60 centimetres (24 inches). These mirrors were ground and polished on a steam-driven machine which also kept the mirror cool, so that it was being worked at a temperature near that at which the telescope would be used. The polishing machine and the telescope that was built to test the 60-centimetre mirror were bigger than needed, both being capable of accepting a 90-centimetre (36-inch) mirror. The 60-centimetre mosaic mirror was a success, and in 1834 Lord Oxmantown abandoned Parliament for astronomy.

William Cranch Bond, 1789–1859, first Director of Harvard College Observatory. Like Flamsteed in England in 1675, Bond had to build up an observatory from a very poorly equipped establishment.

In the mid-1830s Lord Oxmantown made only slow progress, partly because the country was on the brink of civil war, but the pace quickened as his efforts towards a 90-centimetre mirror began to show promise. In 1840 a mosaic of this size was finished and proved a disappointment; the joins between the segments affected the polishing process and the final image was never perfectly sharp. By now Lord Oxmantown had spent a third of his lifetime fighting the intractable speculum alloy, and he finally won through. Turning away from mosaics, he used a complex procedure involving control of the conductivity and porosity of a part-iron, part-sand mould, followed by a slow cooling that took fourteen days to cast a monolithic 90-centimetre mirror blank. This mirror justified all Lord Oxmantown's efforts. For the first time there was an instrument of light grasp and optical quality to surpass Herschel's telescopes of the previous century. Unfortunately, while the new reflector was the best telescope in the world, it simultaneously held the record for being on the worst site. Birr Castle, the home of the Earls of Rosse, is in the same Irish county as the Bog of Allen, and there are only two types of weather: either it is just going to rain or it is raining. In the words of the Astronomer Royal, visiting the site a few years later, conditions varied between "vexatious" and "absolutely repulsive".

In the brief moments of clear sky the telescope showed hitherto unseen detail in star clusters and nebulae, including new observation of a curious class of objects, the planetary nebulae. A planetary nebula gets its name from the fact that in a modest-size telescope it appears as a disc, looking like a planet rather than a star. The largest and best known, M57 in the constellation Lyra, is a disc appearing a little bigger than Jupiter (though less than a ten thousandth of the planet's brightness), and the new telescope showed that there was a faint blue star in its centre, a hundred times fainter than the nebula itself. The observers, however, recorded clouds far more often than stars and only a few new observations were obtained. Lord Oxmantown was undeterred; he set himself to build a bigger machine. This was to be a 180-centimetre (72-inch) aperture giant.

Once the decision to try for double the aperture was taken, the "Leviathan of Parsonstown" was built with quite astonishing speed. In 1841 Lord Oxmantown's father died and he succeeded to the earldom. The new Lord Rosse was now free to use the entire facilities of his almost feudal estate in his telescope building. All the work was done by local labour and was on a necessarily vast scale. In 1842 new furnaces and machinery were ready and five attempts were made to cast a 180-centimetre blank. The first was a success in that it emerged in one piece after a carefully controlled cooling process that took sixteen weeks, but the heavy but brittle disc was accidentally broken before it could be ground. The second casting was rough but usable after much grinding, number three cracked while cooling, number four failed right at the beginning of the casting process and number five was successful. Each attempt required 4 tonnes of alloy, melted in crucibles made in accord with Lord Rosse's detailed instructions and developed specially to withstand the molten speculum metal. The smelting used local fuel and labour and 70 cubic metres (2500 cubic feet) of peat were needed to heat the furnaces for each casting.

When completed in 1845, the new telescope was a triumphant demonstration of Lord Rosse's skill as engineer, foundryman, optical worker and manager of unskilled labour. The figure of the mirror was so accurate that a double star of separation only half an arc second could be resolved, and the 4-tonne mirror was supported on a system of pads and levers so that distortion under its own weight was very much reduced. The instrument was still a long way short of perfection. It was slung between two walls, making it clumsy and difficult to use, and restricting its view to a narrow slice of the sky to the south of its site. No star could be observed for more than an hour, a circumstance that combined with the miserable climate to frustrate much observation. None the less the telescope did achieve a notable success in that it collected enough light to enable the observer to see for the first time a spiral-shaped structure in some of the still mysterious nebulae.

One other legacy of the giant telescope is the "Crab" nebula, a name chosen by Lord Rosse for the first object on Messier's eighteenth-century list, because he felt the nebula looked like, if not a crab, at least a member of the crab family. Since the nebula is not in the constellation of Cancer, the crab, but in Taurus, the bull, the name is misleading, especially to someone trying to find what is one of the most studied astronomical objects in the sky.

In one respect the performance of the telescope misled astronomers. The unprecedented light grasp and image quality enabled the observers to resolve many "nebulae" into clusters of stars. The steady development of the telescope during the eighteenth century had been marked by the resolution of more and more of these diffuse patches of light into aggregations of stars, and Lord Rosse was inclined to the opinion that all diffuse nebulae, including the brightest, Orion, would prove to be star clusters. He was supported in this view by William Bond, who had at last succeeded in equipping an observatory in the United States with a world-class telescope.

The new American telescope was a direct consequence of the Tsar's purchase of the 38-centimetre (15-inch) refractor for Struve at Pulkovo. When the spectacular comet of 1843 appeared in the sky the citizens of Boston asked the director of the newly established Harvard College Observatory for details. The Observatory director, William Bond, pointed out that his observatory had no equipment of world class (he had had to provide most of it himself), and in particular no observatory in the United States could match the magnificent telescope owned by the Russians. Rivalry between the United States and Russia was as fierce then as now and the Bostonians immediately set about raising by public subscription the money to provide a telescope "as large as or larger than" that at Pulkovo. By 1847 Harvard had a matching 38-centimetre telescope from the same makers, Merz, Mahler & Repsold, installed and working in Cambridge, Massachusetts. The Americans saw a rapid return for their investment when William Bond's son, George, discovered Saturn's eighth satellite, even though this added little to the development of astronomy.

The poor output of the Parsonstown telescope was not entirely due to Lord Rosse's preference for making rather than using telescopes, nor to the awful weather. It was, of all telescopes ever built, the most unfortunate in timing. The first observations were made in February 1845, the beginning of the first year in which the Irish potato crop failed. In the next few years half a million Irish died and over a million emigrated. Astronomy was no longer a reasonable pursuit for any landowner with a sense of responsibility for his tenants. Lord Rosse diverted his income to famine relief and returned to politics, this time in the House of Lords.

The second of the amateur telescope-builders of the 1840s was James Nasmyth, the son of a Scottish painter and one of the outstanding figures of nineteenth-century engineering. Although Nasmyth had all the facilities needed and no reluctance to build machines, especially steam hammers, on a giant scale, his astronomical ambitions were more modest. In rapid succession in 1840, 1841 and 1842, he built three telescopes of 25-, 33- and 50-centimetre (10-, 13- and 20-inch) aperture, and used them primarily to study the topography of the Moon. The first of these telescopes caused something of a stir along the banks of the Bridgewater Canal, the main transport artery which ran beside Nasmyth's Manchester factory. The telescope was small enough to keep indoors and the "tube" was of plain timber. One fine summer night Nasmyth decided to do a little observing after retiring to bed, and so, wearing only his nightshirt, he went into the garden carrying the wooden telescope. He was seen by a passing bargee and the next day reports that a ghost had been seen in Mr Nasmyth's garden, carrying a coffin, were spreading like wildfire.

James Nasmyth, 1808–90, was a Scottish engineer and a self-made man. The success of his engineering inventions allowed him to retire at the age of forty-eight and devote his time to astronomy.

If the 25-centimetre telescope enjoyed a brief notoriety, it was the 50-centimetre that would achieve enduring fame. Nasmyth had an eye for unorthodox but convenient design. He was the first to turn a piston engine "upside down" with the cylinder at the top, a design which was so advantageous that nowadays it is unusual to find any other configuration. The same inventiveness applied to the telescope came up with a three-mirror layout. The converging beam of light from the primary mirror was reflected off two secondaries to a final focus on the axis of the elevation trunnion. As a result, Nasmyth could sit comfortably at the telescope without the need to climb scaffolding or lie at an awkward angle to see through it. The drawback was that the light lost by reflection from three speculum mirrors was very great; but this was of little concern to

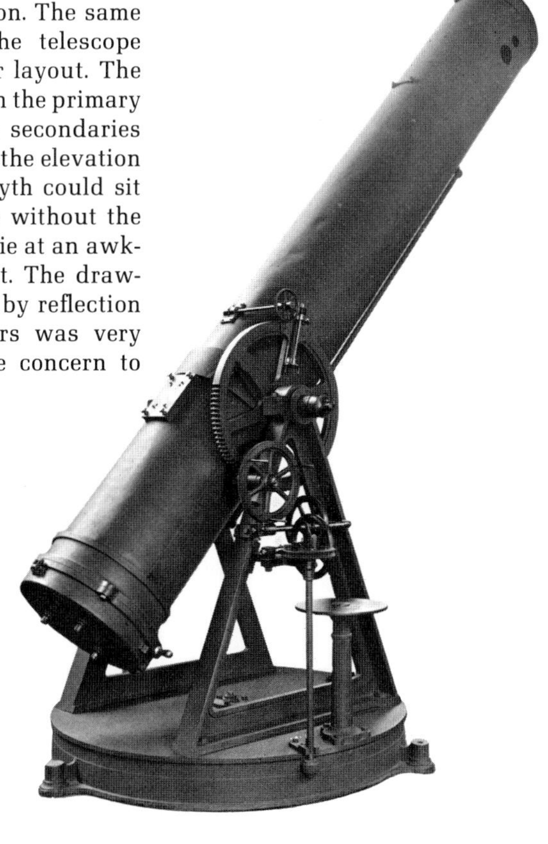

James Nasmyth's 50-centimetre (20-inch) reflector (right), which he mounted on a large turntable with the eyepiece placed so that he could sit and observe in comfort. By turning small handwheels the instrument could be directed to any part of the sky. The optical design uses a primary and secondary mirror similar to those of a Cassegrain telescope and a third, plane mirror tilted at 45 degrees to reflect the beam to the eyepiece.

Nasmyth, who used the telescope chiefly to observe one bright object, the Moon, rather than faint nebulae or star clusters. The loss of light and the difficulty of accurate guidance of a telescope of this style meant that Nasmyth's ideas were not taken up by his contemporaries (who were well aware that the standard map of the Moon, by Beer and Madler of Berlin, was drawn using only an 8-centimetre [3-inch] refractor). Today, however, with highly reflecting mirrors and computer-controlled guiding, there is scarcely a large telescope under construction or being designed that does not include a Nasmyth focus. Nasmyth's achievements were recognized at the time in that both his steam hammer and his drawings of the Moon were awarded medals at the Great Exhibition of 1851.

Nasmyth's 50-centimetre (20-inch) telescope could be driven in altitude and azimuth to point to any part of the sky. Lord Rosse's 180-centimetre (72-inch) monster was mounted on a universal joint that allowed for drives in two elevation angles, a system that approximated to an equatorial drive over the limited section of sky accessible. William Lassell, the third amateur builder and a friend of both Rosse and Nasmyth, built reflecting telescopes which were on true equatorial mountings. The first of these, of 22-centimetre (9-inch) aperture, has already been mentioned. In 1844 Lassell decided that the time had come to build a new instrument and, after a visit to Lord Rosse, he assembled the casting and polishing machinery for a 60-centimetre (24-inch) mirror. The telescope was finished in 1846. Its first and most important observation, however, was made as a result of quieter endeavours than those of the foundry and the machine shop. While Rosse, Nasmyth and Lassell built their telescopes, two theorists, Urbain Le Verrier in Paris and John Adams in Cambridge, worked on a problem set by Herschel's planet Uranus.

Soon after the discovery of Uranus in 1781 sufficient data were collected to calculate an approximate orbit. It was then possible to calculate where the planet had been before its discovery and so to check whether it had been seen by earlier observers. In all, nineteen old observations were found, including two by Flamsteed and six by Lemonnier, both of whom had assumed that the planet was a star. These pre-discovery data were then used to check the computed orbit: and a problem came to light. In 1821 when Laplace's assistant, Alexis Bouvard, tried to determine a definitive orbit, he found it impossible to reconcile the pre-discovery observations with the newer measurements. Bouvard assumed that all the old data were inaccurate and rejected them. This not unreasonable approach proved insufficient, as within a few years the planet had strayed from Bouvard's supposedly precise orbit. This deviation was not due to insufficient care in the calculations: one consequence of the genius of Lagrange and Laplace was that France in particular bristled with mathematicians able to do the sophisticated celestial mechanics required. Laplace's epoch-making text the *Traité de Mécanique Céleste* treated at length the problem of planetary motion. The difficulty of the work lies in the fact that while it is possible to calculate a precise path for a solitary planet orbiting a star, it is impossible to deal exactly with two planets in orbit round a star, because each planet's gravitation affects the other's orbit. Only a series of better and better estimates is possible.

In order to be certain that Uranus was off course, it was first necessary to calculate the varying gravitational effects of the other planets, especially those of the massive Jupiter and Saturn. Much of the groundwork for this was done by Bessel. He had measured the positions of the moons of Jupiter and Saturn with his customary precision, and from their orbits the masses of the two planets were accurately determined. As confidence grew that Uranus's orbit could not be explained within the limits of current knowledge, the search began for an explanation. Two possibilities were canvassed. Either Newton's law of gravitation was very slightly modified at large distances, or Newton was correct, and there was an outer unknown planet disturbing the path of Uranus by its gravitational pull. The Astronomer Royal, George Airy, favoured the first explanation. He analysed the Greenwich observations to show that not only was the angular position of Uranus out of step with theory but so also was the distance of Uranus from the Sun. Bessel was the senior among those who suggested that the error in the position of Uranus might be the result of the gravitational field of an eighth and more distant planet. Finally, in mid-1841, two people decided to do some calculation as soon as they had the time. One was Bessel himself and the other John Adams, an undergraduate at Cambridge. Bessel's intention did not reach fruition: although Flemming, one of Bessel's assistants, checked all the available measurements of Uranus's position, both Bessel and Flemming died before the work could be taken any further.

In 1843, as soon as he had graduated, John Adams started work to try to calculate where a planet of unknown position and mass would need to lie in order to upset the Uranus orbit by the observed amount. The calculation of the effect of one

(Far left) *Rosse's 180-centimetre (72-inch) reflector at Birr Castle, popularly known as "The Leviathan". It took twelve years to build at a cost of £12,000 using entirely local labour. The telescope was mounted on a pair of hinges fixed behind the main mirror. The hinges were at right angles, one to allow up-and-down motion, the other side-to-side. Technically this is called an altitude-altitude mount and Rosse's telescope is almost the only instrument to have used the system.*

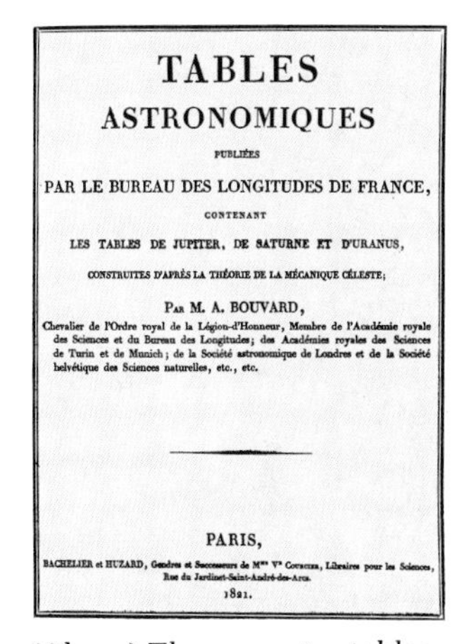
TABLES
ASTRONOMIQUES
PUBLIÉES
PAR LE BUREAU DES LONGITUDES DE FRANCE,
CONTENANT
LES TABLES DE JUPITER, DE SATURNE ET D'URANUS,
CONSTRUITES D'APRÈS LA THÉORIE DE LA MÉCANIQUE CÉLESTE;
PAR M. A. BOUVARD,
Chevalier de l'Ordre royal de la Légion-d'Honneur, Membre de l'Académie royale des Sciences et du Bureau des Longitudes; des Académies royales des Sciences de Turin et de Munich; de la Société astronomique de Londres et de la Société helvétique des Sciences naturelles, etc., etc.

PARIS,
BACHELIER et HUZARD, Gendres et Successeurs de Mme Ve Courcier, Libraires pour les Sciences, Rue du Jardinet-Saint-André-des-Arcs.
1821.

(Above) *The computer tables for the major planets published by Alexis Bouvard in 1821. In the following years it was noticed that the orbit of Uranus deviated from the tables predictions which suggested, prior to the discovery of Neptune, that there might be an unknown planet disturbing the motion of Uranus.*

known planet on another is difficult; to work backwards from the effect to the cause is worse. Indeed in 1840 it was not even clear that it was possible. Adams's first step was to do some preliminary calculation, and when these proved encouraging, in February 1844 he asked James Challis, the professor of Astronomy at Cambridge, to get for him an up-to-date set of figures on Uranus's position. Challis obtained these from Airy. Then Adams withdrew to his college, where he had a fellowship, for a further eighteen months of analysis and calculation. In September 1845 he reached a solution. The discrepancy in Uranus's position could be explained by the influence of an unknown outer planet, and the result of all Adams's calculation was a prediction of the orbit of the planet and of its position on 1 October, 1845.

Urbain Le Verrier, 1811–77, the Frenchman who predicted the position of the planet Neptune as lying beyond Uranus using Newton's laws of motion. His calculation led to the discovery of the planet on 23 September, 1846.

Being young and reticent, Adams did not send his prediction to the Royal Society for publication, but sent it to Challis in a private letter. Challis gave Adams a letter of introduction to Airy, after which the two senior British astronomers successfully frittered away the gift with which they had been presented. Adams made a trip to Greenwich to see Airy, but the Astronomer Royal was out when he first called, and managed, perhaps accidentally, to snub the young theorist when he made a second visit. Adams left a copy of his results and returned to Cambridge, after which Airy wrote to enquire whether the theory would explain the errors in distance as well as those in angular position. Adams, upset by the indifference with which his unprecedented work had been received, decided not to reply.

A few months before Adams completed his analysis, Dominique Arago, the Director of the Paris Observatory, suggested to one of his colleagues, Urbain Le Verrier, that he might explore the hypothesis of an outer planet as a disturber of Uranus's motion. Le Verrier moved quickly, effectively and, in contrast to Adams, publicly. In November 1845 he published a paper confirming that neither Jupiter nor Saturn could cause the observed discrepancies. In June the following year he published a calculated position for the planet, a position only one degree different from that obtained by Adams. George Airy, now having two matching predictions, made moves – cautiously. First he wrote to Le Verrier, as he had to Adams, enquiring whether the calculation explained both errors in angular position and in distance from the Sun. Le Verrier replied saying that the new theory accounted for both the discrepancies and he also asked Airy whether the Royal Observatory could make a search for the planet.

Ironically, the best telescope in Britain for the task was that at Cambridge. Airy had been professor at Cambridge before becoming Astronomer Royal and had designed and built the 30-centimetre ($11\frac{1}{2}$-inch) refractor presented to the University by the Duke of Northumberland. Airy, therefore, wrote to Challis in mid-July 1846 expressing the view that the search for the new planet should be given the highest priority, and offering Challis an assistant from Greenwich to help with the work. Challis decided to carry out the search himself, choosing a method that was reliable but slow. On four nights, 29 and 30 July and 4 and 12 August, he noted the positions of all the stars in the search area. After a brief check that the first thirty-nine stars in the set had the same positions on 12 August as they had on 30 July, Challis put the rest of the data to one side and went off to do something else.

The next developments were a third paper by Le Verrier and two days later another letter to Airy from Adams. The paper by Le Verrier went beyond solely predicting the position of the hypothetical new planet. After making a reasonable estimate of the density of the planet the author was able to say how big it would appear from Earth. The diameter was predicted to be three arc seconds, so the planet should show in the telescope as a definite disc and not as simply a point of light, as does a star. Le Verrier was naturally eager to see the search widened and, as the Paris Observatory was ill-equipped and rarely observed anything, he wrote to Johann Galle at Berlin asking if he would see what he could do. Galle got permission from the Director, Johann Encke, to make observations starting that night, 23 September. He had the advantage over Challis that the observatory had recently charted the positions of all the stars in the search area, and within an hour Galle and his assistant, d'Arrest, found one that had moved. This was the new planet. Its motion was soon confirmed, Galle wrote back to Le Verrier and the Paris and Berlin Observatories announced their triumph.

Meanwhile, back in Cambridge, Challis read Le Verrier's prediction that the planet would show a disc. On 29 September he searched once again and saw one star that did show a small disc. If he had stayed up all night Challis could have seen the planet's motion, but he decided to confirm his observation the next night. On 30 September the critical area of the sky was covered by the Moon – and the following day Challis heard of Galle's discovery.

This was by no means the end of the saga. The next event was a letter by Sir John Herschel in the copy of the journal *The Athenaeum* published on 3 October. In the letter Herschel set out an account of Adams's work, pointing out that Adams's prediction had been made first, and that Challis had already started a search before Le Verrier wrote to Galle. Herschel's letter started the very devil of a row. Arago, an extreme left-wing politician as well as Director of the Paris Observatory, was furious at this attempt by the English establishment to appropriate some of the credit due to one of his staff. He promoted the French claim, seeking to rename the planet, already christened Neptune, "planet Le Verrier". This name he justified by comparison with Halley's comet (among others) and even tried to turn back the clock, in pursuit of consistency, by calling Uranus "planet Herschel". Adams and Le Verrier remained aloof from the battle, while Challis apologized to anyone who would listen.

It is possible to understand Airy's lack of enthusiasm for the search for Neptune. He did not know Adams, but he did know that mathematics at Cambridge was in a sorry state – so bad that a group including John Herschel and others had tried to get some reforms accepted by the University Senate. Challis's letter of introduction of Adams to Airy was little more than lukewarm – "I should consider the deductions from his premises to be made in a trustworthy manner." Finally, Airy was busy and thought that if the man on the spot, Challis, was doing nothing, then Adams was probably wrong, or optimistic, or both. On the other hand, it is almost impossible to comprehend Challis's inaction. Living within the close society of the university, he must have known that Adams was an undergraduate of outstanding brilliance. Equally, if he had wanted to question or check any part of Adams's calculation it is scarcely a couple of kilometres from the observatory to Adams's college, St John's. Given all this and one of the best telescopes in England, Challis sat on Adams's results for nearly a year doing nothing. When Le Verrier confirmed the computation and Airy wrote urging that a search be started immediately, Challis took nearly three weeks to begin observation. Then, having obtained data that hindsight would show to include the first deliberate sighting of Neptune, he left the data unreduced. Last, and hardest to accept, he saw an object showing a disc on the night of 29 September and went to bed! The fact that the Moon would be in the way the next night is utterly insignificant compared with this bathos. There can have been few experimental scientists born with souls so dead that they could act in such a way. Any man with a little enthusiasm for his subject would have sat up all night worrying over the image in his telescope. It was only poetic justice that Adams, many years later, was appointed Director of the Observatory – Challis's position in 1846.

Ten days after the news of Neptune's discovery reached England, William Lassell studied it with his new 60-centimetre (24-inch) reflector and observed that the new planet had a satellite. The next year, when the planet was again visible, he confirmed the observation and the new moon was christened Triton. Lassell followed this by an independent discovery of Saturn's eighth moon, already seen at Harvard Observatory, and the rediscovery of Uranus's two moons, Umbriel and Ariel, lost since William Herschel's observations sixty years before. With the encouragement that came from these achievements, Lassell, unlike Lord Rosse, decided to move his telescope to a far better site and took himself and his 60-centimetre reflector to Malta.

During the Neptune affair Airy had been busy with the design and construction of a new refracting telescope specifically for the measurement of the Moon's position, still the prime task of the Greenwich Observatory as it had been for nearly 200 years. Airy had had a most difficult time revitalizing the observatory after the retirement of his predecessor, John Pond, in 1835. The observatory staff had drifted into anarchic disarray. The First Assistant was dismissed for a range of misdemeanours including drunkenness and falsifying basic astronomical data. Airy also tried to dismiss another assistant, Richardson, but the observatory paymasters, the Admiralty, would not agree. Ten years later, Richardson was on trial for wilful murder of his incest child.

The new regime at Greenwich was busy and its duties were defined to the last decimal point. In the first eight years the assistants (including Mr Richardson) made nearly 70,000 observations. Unlike most of Airy's predecessors, these observations were immediately fed into the computing section and reduced to useful astronomical data. Airy went further and had the old observations of his predecessors Bliss, Maskelyne and Pond similarly reduced. These "reductions" are necessary, numerous and dull. The observed position of a star must be corrected for refraction, aberration, nutation and precession. Observations of the Sun or Moon must also be corrected for the latitude of the observatory and, as the measurement is made by setting the cross wire on the edge of the solar or lunar disc, for the resulting offset of the centre, an angle that varies from day to day and is itself affected by refraction. Standard forms were designed by Airy for each stage of the calculations. His methodical approach to paperwork was such that a friend once remarked, only partly in jest, "If Airy wiped his pen on a piece of blotting paper he would duly endorse the blotting paper with the date and particulars of its use, and file it away amongst his papers."

All this organization and data reduction led to a clearer understanding of the observatory's instruments. The two mural circles which Pond had used for most of his best work were not sufficiently steady to give the precise measurements now needed of star transits (the time at which a star is precisely on the observatory meridian). The existing special-purpose transit telescope, built for the Royal Observatory by Troughton in 1816, had become untrustworthy. It was also too small to show the asteroids, now included with the planets in the Greenwich programme. Airy persuaded the Admiralty to fund a new transit telescope and he set himself to design this, his third major instrument (after the Cambridge Northumberland telescope and the new Greenwich refractor). The basic telescope was an achromatic refractor of 20-centimetre (8-inch) aperture and 3·5-metre ($11\frac{1}{2}$-foot) focal length. Although Airy was familiar with the successful transit instrument built by Troughton for Groombridge in 1806, he worked to a very different design philosophy. Groombridge's transit circle is an obvious descendant of Ramsden's Palermo circle – a nest of many conical and cylindrical tubes forming a light but rigid frame in which the telescope was not so much supported as buried. Airy believed that this design was far from the most stable possible: stability came instead from having as few components as practicable. To this end, the new transit telescope was designed to support itself: it was a simple tube fixed to a heavy axle. The resulting telescope weighed 0·85 tonnes and was carefully counterpoised so that the load did not rest on, and so wear out, the two pivots. The design was a great success, except for a minor detail. Airy had used an existing granite pier to support one pivot and built another, of a different stone, for the other. These two piers expanded differently in response to changes in temperature, and this turned out to be a persistent nuisance.

The telescope was mounted 5·8 metres (19 feet) to the east of Bradley's earlier transit telescope, and the "Greenwich meridian" moved with it. In 1851 celestial measurement was the duty of the Admiralty; terrestrial measurement, via the Ordnance Survey, that of the army. The two forces were just about to embark on that most ill-administered military endeavour of British history, the Crimean War, and it is not surprising that the army did not immediately adopt the shift of the meridian. The correction finally filtered through in 1947.

A caricature of George Airy, Astronomer Royal from 1835 to 1881, taken from a contemporary magazine. Airy was a mathematician by training and used his mathematical skills to analyse both the motion of the planets and the design of new instruments. The diffraction pattern of a telescope lens is still called an Airy disc and if a beam is supported to give the minimum deflection the supports are said to be at "Airy's points"

Airy spent fourteen years from 1872 to 1886 on calculations of the motion of the Moon. Four years later, in 1890, he found a mistake in one of the first steps of his argument. He was then aged eighty-nine and commented that he had not the heart to start the calculations again. He died in 1892.

CHAPTER SEVEN

CHANGING THE RULES

Spectroscopy and photography make all previous equipment obsolete

"Then it was [in 1862] that an astronomical observatory began, for the first time, to take on the appearance of a laboratory. Primary batteries, giving forth noxious gases were arranged outside one of the windows; a large induction coil stood, mounted on a stand on wheels, so as to follow the position of the eye end of the telescope, together with a battery of several Leyden jars; shelves with Bunsen burners, vacuum tubes, and bottles of chemicals . . . lined its walls. . . . In February 1863 the strictly astronomical character of the observatory was further encroached upon by the erection, in one corner, of a small photographic tent, furnished with baths and other appliances for the wet collodion process."

Sir William Huggins in Publications of Sir William Huggins' Observatory, *Vol. 1, 1899.*

The astronomical discoveries of the 1840s were very much in the tradition of the previous two centuries. The positions of the stars and planets were recorded with precision, while larger objects, such as star clusters and nebulae, were described and, if of sufficient interest, drawn. This approach to observation was to be overturned by two developments which took place away from the telescope. The first of these was the photographic plate, which came to replace the human eye as the chief detector of starlight. The second was the realization of the great advantages that could be achieved by using a prism deliberately to spread the carefully collected starlight into a spectrum. The two techniques, photography and spectroscopy, were at first used independently but they were eventually to be combined to provide a new astronomical tool of such power that a whole new subject grew from their union – the science of astrophysics, in which astronomers could interpret the physical nature of astronomical objects. The pattern of the development of spectroscopy and photography had many features in common. Both took some fifty years to span the interval between the first systematic measurements and the arrival of a technique of general application; both evolved through a phase of data gathering when confusion was common; and both techniques were only slowly accepted by the majority of astronomers.

During the eighteenth century a few scientists had noted the effect of light on various chemicals and mixtures. By the end of the century, particularly as a result of the work of Karl Scheele, a Swedish chemist, and Jean Senebier, the Chief Librarian of Geneva, it was clear that many silver compounds were blackened by sunlight, and that the black colour was due to the formation of minute particles of metallic silver. (In very finely divided form, metals appear black rather than shiny.) This was even exploited in one of the most horrifying pieces of advice ever given in an "Isn't Chemistry Fun" book – Weigleb's *Naturliches Zauberlexikon* (Dictionary of Natural Magic) published in 1784. The book advised that, to blacken the face, "The face is moistened with aqua fortis [nitric acid] in which fine silver is dissolved . . . then the Sun is allowed to shine on the face, thus one becomes for some time a black man." Let alone the use of very strong acid, silver salts are poisonous.

In a less drastic experiment, Johann Ritter employed a silver compound to demonstrate that sunlight contains invisible radiation which falls into the region beyond the violet when sunlight is spread out into a spectrum of colours. The year before, 1800, William Herschel had discovered invisible rays lying beyond the red end of the spectrum, radiation that could be detected with a thermometer and which we now call infra-red. The "ultraviolet" part of the Sun's spectrum carries too little energy to affect a thermometer in the same way, but it is dramatically effective in producing chemical change. Ritter was exposing the white compound silver chloride to light, and he found that the ultraviolet radiation blackened it in less than a thirtieth of the time taken by the yellow rays in the middle of the visible spectrum. The blackening of silver chloride by sunlight meant that it was possible to make a silhouette of any opaque object, simply by letting its shadow fall on paper impregnated with the compound. These silhouettes were not even close to being photographs, however; in the first place, it was not possible to make a stable, permanent record and second it took a great deal of light – some hours of direct sunshine – to produce a modest blackening.

A daguerreotype of Louis Daguerre, 1789–1851, taken in 1851. He was a painter of "dioramas" – translucent paintings that could be given added drama by changes in lighting. The aim of his early experiments was to reduce the time needed to produce the realistic pictures his art necessarily required.

The first successful attempt to produce a permanent image – that is, an image that was "fixed" – did not start out to use either silver or silver compounds deliberately. The inventor concerned, Joseph Niépce, exploited instead another of Senebier's results, the discovery that some natural compounds are made insoluble by light. The intention was not to make a photograph as an end in itself, but to find a way to speed up the preparation of lithographic printing plates. After a great deal of experiment, lasting over ten years, Niépce succeeded in recording the view from his window as a permanent image. The image was focused on to a thin layer of bitumen deposited on a pewter plate. Wherever the camera image was sufficiently bright, the bitumen became insoluble. When Niépce carefully washed the exposed plate to remove the still soluble parts of the picture, it left a permanent image in which the dull metal of the plate showed as the shadows, with the bitumen remaining on the brighter regions. The exposure time was eight hours in bright sunshine, and in order to reduce this Niépce made two improvements. He started to use silvered copper plates in place of pewter, and he sought out a better camera. The silvered plate had several advantages. Because of its copper base, the plate was harder and thus more suitable as a base for a printing process, while the shiny silver surface reflected rather than absorbed any light which passed through the bitumen coating. This increased the amount of light available to affect the bitumen. In addition it was possible to enhance the contrast of the final plate. Niépce treated it with iodine, to blacken the exposed silver surface, and then removed the remaining bitumen, beneath which was a pristine shiny surface. The result was a picture with a much greater contrast than the bitumen and pewter process could achieve. The search for a better camera brought Niépce into contact with Louis Daguerre.

Daguerre was an artist-showman, who had started experiments to try to record a photographic image by using paper coated with silver chloride. The experiments were not particularly successful, but they gave Daguerre the background to negotiate a partnership with Niépce, who by 1829 was running out of funds. Daguerre took the next step towards a successful photographic process in 1831, when he discovered that Niépce's iodine-treated silver plate was itself sensitive to light; using this, it was possible to record, but not fix, an image without using bitumen. Joseph Niépce died in 1833 and was succeeded in the partnership by his son, Isidore. Two years later Daguerre made a second important discovery: he found, entirely by chance, that a plate that had been exposed to an image too weak to produce a visible blackening none the less carried an invisible "latent" image which could be "developed" by exposure to mercury vapour. This solved one of the crucial difficulties of early photography, for it enabled the exposure time to be cut to a mere twenty minutes. After a further two years, in 1837, Daguerre managed to "fix" the developed image by washing the plate with a strong solution of common salt.

Daguerre, now very much the senior partner of Daguerre and Niépce, spent the next year trying to capitalize his new invention. When that failed he turned to the government for aid. He found the ally he needed in Dominique François Arago, the Director of the Paris Observatory, who was both scientist and politician. In his presentation of the invention to the Chambre des Deputés, Arago stressed the scientific potential of the new process, though he did suppress the scientific side of his nature in favour of the political on more than one occasion. In Arago's report to the government Niépce's exposure time rose from eight to twelve hours, and Daguerre's fell from twenty minutes in sunshine to "scarcely ten in dull winter weather". Unfortunately, he went further. A few months after he had won government support for Daguerre, Arago persuaded Hippolyte Bayard, the inventor of a rival process, not to publish his invention. Later on, in a row over priority, Arago "forgot" that Bayard had shown him photographs in May 1839.

Daguerre and Arago made the first attempt at astronomical photography in 1839, but with only limited success. Using the equipment of the Paris Observatory, they did succeed in blackening a plate by moonlight, but the image was too blurred to show any detail. The first success was achieved not in Paris but in New York, where the daguerreotype process was taken up with enthusiasm. John Draper, the Professor of Chemistry at the University of the City of New York, was among the early enthusiasts and took his first daguerreotype within a couple of days of the arrival, on 20 September 1839, of the first steamship carrying an account of the process. After a brief excursion into portrait photography in partnership with his fellow professor Samuel Morse, who had met Daguerre while visiting Paris to patent his electric telegraph, Draper turned to scientific work. He succeeded in photographing the Moon in 1840, using a twenty-minute exposure to record an image 25 millimetres (1 inch) in diameter. In 1841, he recorded the solar spectrum in sufficient detail to show the dark lines that Fraunhofer had discovered.

The daguerreotype was further developed in the early 1840s, when it was found that silvered plates sensitized not with iodine alone but with iodine-bromine or iodine-chlorine mixtures were far more sensitive to light, enabling exposure times to be reduced by a factor of twenty or thirty. This was a step of considerable commercial importance as it made portrait photography practicable, though it was not of immediate relevance for the next application of the Daguerre process to astronomy. This was the photography of the Sun, where the problem was to build a sufficiently fast shutter to avoid severe overexposure. The first successful daguerreotype of the Sun, using an exposure of one-sixtieth of a second, was taken in Paris by Hippolyte Fizeau and Léon Foucault in April 1845 at Arago's suggestion. The result confirmed that the edge of the Sun's disc is only about half as bright as the centre, a fact that had been called into question by a series of measurements made by Arago himself. As with the original invention, the gain in speed of the improved daguerreotypes was first applied to astronomy in America, this time by William and George Bond at Harvard, using the new 38-centimetre (15-inch) refractor. The Bonds were assisted by two professional photographers, John Whipple and J. W. Black, and in 1849, 1850 and 1851 they succeeded in taking much improved daguerreotypes of the Moon with exposures of only a minute. Of greater significance, the group took plates of stars for the first time, photographing first Vega and then Castor, the brightest star in the constellation of Gemini. The recorded images were not very sharp: Castor is a double star, whose separation in 1850 was six arc seconds, but the photograph shows the pair of stars as an elongated single

John Draper, 1811–82, emigrated to the United States from Liverpool, England, in 1833. Both John Draper and his son Henry were professors of chemistry and both worked on astronomical photography.

Paris Observatory in the mid-nineteenth century. The chief instruments it housed were 30- and 38-centimetre aperture refractors.

Groupe V. — FEUILLE N° 48.
MÉDAILLE D'OR: MARSEILLE 1883

L'HISTOIRE DE LA PHOTOGRAPHIE

SERIE ENCYCLOPÉDIQUE GLUCQ
des Leçons de Choses Illustrées
Ouvrage adopté par la VILLE de PARIS comme Récompense dans ses Écoles.

Vers 1570, un physicien de Naples, nommé Porta, inventa la **CHAMBRE OBSCURE**, c'est-à-dire une chambre fermée n'ayant qu'une seule petite ouverture munie d'une lentille à travers laquelle passe un rayon de lumière qui reproduit sur une glace, mais en le renversant, le paysage environnant.

Vers 1566, l'alchimiste Fabricius, ayant fait tomber du sel dans une dissolution d'argent, vit s'y former un gros flocon blanc de **CHLORURE D'ARGENT**, qui, à la lumière, devint bien vite gris, puis violet, puis enfin noir. Cette sensibilité du chlorure d'argent à la lumière forme la base de la Photographie.

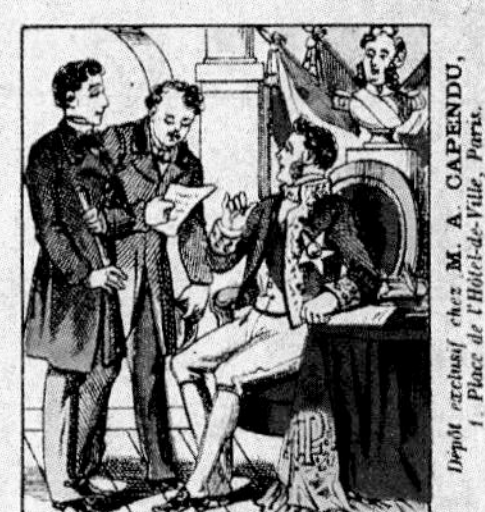

Du reste, tout est sensible à l'action du soleil : les rideaux des fenêtres qui passent de couleur en faisant le désespoir des maîtresses de maison, les robes, les étoffes, les tentures, les papiers ; tout change de ton à la lumière. Le chlorure d'argent est simplement plus sensible que le reste.

Un magistrat, M. Bayard, voulant que les excellentes pêches de son jardin portassent sa marque authentique, y collait son initiale découpée en papier blanc. Le soleil rougissait les pêches tout autour, et le papier une fois enlevé, l'initiale se détachait en blanc, écrite ainsi par le soleil lui-même.

En 1815, Nicéphore Niepce chercha, le premier, à obtenir des images gravées chimiquement par le soleil. Il étendait du bitume de Judée, qui est sensible à la lumière, sur des plaques d'étain, appliquait une estampe sur la plaque et exposait le tout au soleil. L'estampe se décalquait sur la plaque.

. A ce moment, vivait à Paris, un peintre nommé Daguerre, inventeur du *Diorama*, et qui, lui aussi, s'occupait de la reproduction des images données par la chambre obscure. Nicéphore Niepce et Daguerre s'associèrent ensemble en 1829 et mirent en commun leurs travaux et leurs découvertes.

Daguerre, ayant un jour laissé une cuiller sur une plaque recouverte de sel d'argent, fut fort surpris, en revenant, de voir la plaque toute noircie par la lumière mais restée blanche aux endroits protégés par la cuiller dont l'image était fidèlement reproduite. Telle est l'origine du *Daguerréotype*.

Niepce et Daguerre résolurent de céder à l'État la propriété de cette merveilleuse découverte pour en faire profiter tout le monde, et allèrent trouver le ministre M. Duchâtel. Cette invention qui, depuis, a enrichi tant de gens leur fut payée moyennant une rente de 6,000 fr. pour Daguerre et une rente de 4,000 fr. pour Niepce.

En 1848, Niepce de Saint-Victor, cousin de Nicéphore Niepce, inventa le cliché négatif sur verre. Officier de la garde municipale de Paris, il s'amusait à faire le portrait de ses soldats, se servant, comme laboratoire, de la salle de police où il n'y avait jamais personne. Le cliché de verre remplaça la plaque métallique de Daguerre.

En 1851, le français Legray inventa le **COLLODION**, (dissolution de coton poudre iodure dans un mélange d'alcool et d'éther.) On verse cette liqueur sur le cliché de verre que l'on baigne ensuite dans le bain d'argent. Le cliché s'y recouvre alors d'une pellicule d'iodure d'argent sensible à la lumière.

Le cliché de verre, une fois recouvert de sa pellicule d'iodure d'argent sensible à la lumière, n'a plus qu'à être placé dans la chambre obscure pour reproduire l'image placée devant l'objectif. C'est ainsi que le génie militaire relève aujourd'hui les détails de fortifications et les profils des terrains.

Grâce à la photographie, on obtient l'image exacte des variations par minute ou par seconde des taches du Soleil, ou des diverses phases du passage de Vénus sur le Soleil. La photographie astronomique nous fait toucher du doigt les volcans de la Lune et les continents de la Planète Mars.

La photographie rend les plus grands services à la société. On fait, de gré ou de force, le portrait de tous les gens suspects ou condamnés à la prison. Plus tard, si un crime vient à être commis, le portrait du coupable sert à le faire arrêter et aide puissamment l'action de la justice.

La photographie permet d'étudier, au moyen de projections et d'agrandissements, toutes les merveilles microscopiques de la création. En projetant sur le mur le **PORTRAIT PHOTOGRAPHIÉ** d'une puce, on peut voir cette désagréable petite bête grossie à la taille d'un mouton, par exemple.

Par contre, la photographie a permis, pendant le siège de Paris en 1870, grâce à des appareils de réduction, de condenser 500 dépêches sur une petite carte de visite en gélatine que l'on roulait sous l'aile d'un pigeon voyageur, qui s'envolait, emportant nos nouvelles à nos amis.

La photographie a droit à notre reconnaissance éternelle, puisque c'est elle qui place à notre foyer le portrait des parents aimés que nous avons perdus, des amis qui sont absents ou des enfants qui sont loin de nous et dont nous baisons les traits chéris pour nous soulager le cœur.

Dépôt exclusif chez M. A. CAPENDU, *1, Place de l'Hôtel-de-Ville, Paris.*

Auteur-Éditeur de la série encyclopédique des Leçons de Choses Illustrées

GLUCQ. — 115, Boulevard Sébastopol, Paris. —

Typ.-Lith. de Ch. PELLERIN à Épinal. (Déposé)

(Above) *A pictorial history of photography from an encyclopedia of 1833. It outlines the principle of the camera and the history of photography. The story starts with the development by della Porta of the camera obscura, a darkened box with an aperture for projecting an image of a distant object on to a screen. This "camera" was used as an aid to accurate drawing. By the 1800s it was established that materials change colour when exposed to light and that silver salts are particularly sensitive in this respect. A number of the pictures deal specifically with the work of Niépce and Daguerre and the establishment of photography as a research tool, particularly in astronomy.*

(Above) *A view of the Great Exhibition held in London in 1851. The telescope on view is an instrument belonging to Lord Rosse. Astronomy in the mid-nineteenth century was dominated by refracting telescopes of aperture 20 to 25 centimetres (8 to 10 inches), similar to the one shown. However, the exhibit that created a sensation was Bond's photograph of the Moon that was taken with a 7·5-centimetre (3-inch) aperture refractor.*

(Right) *An early photograph taken in February 1839 by John Herschel. The view through a window is of William Herschel's 12-metre (40-foot) telescope. The photograph is accompanied by John Herschel's own notes on photography.*

(Above) *The earliest extant "photograph". It was taken by Nicophore Niépce in 1826, using a plate made by coating a sheet of pewter with bitumen. The view is from a window looking out on to a sunlit yard (in Gras, near Chalon-sur-Saone, France). The image is hard to unravel, not only as a result of age but also because during the eight-hour exposure the Sun moved round to shine from both sides of the view. The contrast in the photograph is poor too, as the shadows of the bare metal of the plate are merely highlighted against the hardened bitumen. In order to make his pictures more distinct, Niépce exposed them to vapours of iodine and fixed the images by using silver compounds. This picture, being the first permanent record of a view from nature, was submitted to the Royal Society in London in 1827.*

image. John Whipple's skill was acknowledged by a wider circle, when in 1851 his pictures of the Moon were awarded a prize medal at the Great Exhibition in London – a higher award than that given to Nasmyth for his drawings of the same object.

The last pioneering use of the daguerreotype was at the solar eclipse of July 1851. The path of the eclipse lay across Sweden and East Prussia, and the staff of Königsberg Observatory attended in force. With a telescope of only 60-millimetre ($2\frac{1}{2}$-inch) aperture Berkowski obtained two successful records of the corona during the brief interval of total eclipse when the Sun's disc was covered by the Moon. One of the two daguerreotypes was analysed to obtain the coronal intensity distribution, not in 1851, but in 1951 after resting in the archives of the Swiss Federal Observatory for most of the century.

The announcement of Daguerre's invention and its immediate commercial success stimulated a great deal of effort, chiefly in England and France. Goddard, an Englishman, and Claudet, a Frenchman living in England, improved the daguerreotype itself. Bayard, Lassaigne, Verignon, Blanquart-Evrard and Martin in France, and Fyfe, Talbot and Reade in England worked on processes which used silver compounds on a paper support instead of the silvered copper of the daguerreotype. John Herschel also tried his hand at photography, using glass plates in a process that was complex and insensitive. The majority of the paper processes were unable to match the high definition of the daguerreotype, and the exception, Louis Blanquart-Evrard's albumen process, was extremely slow. John Herschel's interest led him to invent much of the vocabulary of the craft: positive, negative and snap-shot were all his terms, as was the word photography, though that had been suggested a few days earlier by the German astronomer, Mädler. Nevil Maskelyne, the grandson of the Astronomer Royal, attempted to use Talbot's process (the calotype) to photograph the Moon in 1846 using a telescope belonging to an acquaintance in Bath, a few kilometres from Talbot's home at Lacock Abbey. He hoped to record an image 12 centimetres (5 inches) across, but he was not successful.

The process that rendered the daguerreotype obsolete was invented by Frederick Archer in 1850, and used glass plates coated with a thin film of collodion (guncotton). The recipe for the production of a plate started by covering a clean glass plate with a mixture of collodion and potassium iodide dissolved in ether. The ether was allowed to evaporate and, while the collodion was still tacky, the plate was immersed in a solution of silver nitrate. This reacted with the potassium iodide to precipitate the insoluble silver iodide. The resulting silver iodide-impregnated collodion was about as sensitive as Daguerre's silvered copper, provided that it was used while still wet. Once exposed and developed it would dry to give a permanent negative. Blanquart-Evrard's albumen process paper proved ideal as a material with which to make positive prints from collodion negatives. Soon after Archer's announcement of the wet collodion plate it was further improved, again by using mixtures of silver iodide and bromide, to give a plate that was about ten times the sensitivity of the daguerreotype.

The new process was first used in astronomy by Warren de la Rue, a friend of Nasmyth and Lassell, and like them a successful businessman. De la Rue had built himself a 33-centimetre (13-inch) aperture reflector and this he modified to take Archer's wet plates. The first results were photographs of the Moon taken with exposures of ten to thirty seconds. The success of these photographs was remarkable, as de la Rue had no drive fitted to the telescope, and he had to guide the instrument by hand in order to hold the image steady on the plate. The lack of a good drive made longer exposures impracticable and de la Rue postponed further work until he had built and equipped a new observatory. The lack of a good telescope drive also delayed the use of the wet collodion process at Harvard, where George Bond was the chief enthusiast. A new clock drive was finally installed and in April 1857 Bond and Whipple turned again to photographing the stars. The improvement was considerable: plates taken of Mizar and Alcor (the pair of stars in the Great Bear which the Arabs had used as a test of good vision) showed the brighter star in three to five seconds and the fainter in eighty seconds. The limit was similar to the faintest stars visible to the naked eye – stars only a fortieth of the brightness of the Pole Star, which Bond and Whipple had been unable to capture on a daguerreotype.

The first large-aperture telescope to be made expressly for astronomical photography was built by Lewis Rutherford in New York, working with Henry Fitz, an optical worker who was already established as a telescope-maker. The need for a special telescope is a consequence of the different colour sensitivity of the eye and the photographic plate. The eye is most sensitive to the green, yellow and orange parts of the spectrum, while the wet collodion plate is most affected by the blue, violet and ultraviolet. Achromatic lenses which aim to give a good visual image are designed to bring the colours at the middle of the spectrum – green to orange – to a common focus; corrections for the violet, to which the eye is insensitive, and the ultraviolet, which is invisible, are not included in the design process. Rutherford found that the ordinary achromatic lens of his telescope had a visual focus more than 2 centimetres (about 1 inch) away from the setting for the most rapid photography and that no setting gave a really sharp picture. He therefore set out to make a lens specially corrected for the colours to which the plate was most sensitive.

This proved a difficult and time-consuming process. The lens had to be polished to produce a sharp image in invisible rather than visible light, so every

A picture of the Moon taken in 1865 by Lewis Rutherford, 1816–92, one of the pioneers of celestial photography.

test required the lens to be fitted in the telescope, new wet plates prepared and a series of photographs taken of bright stars. The 28-centimetre (11-inch) aperture "photographic achromat" took a long time to polish and was completed in December 1864. The effort was successful and, for the first time, photographs were taken of stars too faint to be seen by the naked eye – stars ten times fainter than on the Harvard plates, which used a considerably larger-aperture telescope. Rutherford's plates of the Pleiades showed some 50 to 75 stars. They were the crucial material needed to test whether it was possible to measure star positions accurately from their images on a photographic plate. The incomparable Bessell had determined the positions of the stars within the Pleiades thirty years before, and the measurements from Rutherford's plates proved to be of similar accuracy. The plate had several other advantages over direct measurement at the eyepiece: once taken, the plate could be measured at any time – night or day, clear or cloudy; and, as all the stars on the plate were recorded at the same time (unlike a sequence of measurements on individual stars) the corrections due to effects such as refraction and telescope flexure were the same for all the stars and so need only be calculated once. Rutherford also took photographs of the Moon, the first attempts yielding exposures that showed sufficient detail to justify enlargement to 50 centimetres (20 inches) in diameter. The exposure, for the full Moon, was only a quarter of a second.

Like Lord Rosse, Lewis Rutherford was a man with a greater interest in building instruments than in using them, and the success of his photographic telescope persuaded him not to do astronomy but to start work on a photographic microscope. Benjamin Gould used Rutherford's lenses, first in Boston and then in Argentina, mainly to photograph star clusters. By that time, however, Warren de la Rue in England had built his new observatory and started a regular programme of astronomical photography. De la Rue continued to use his reflecting telescope,

A picture of the Moon by Warren de la Rue, 1815–89, also taken in 1865, this time in London and using a reflecting telescope. De la Rue also pioneered photographic studies of the Sun, including eclipses.

which avoided the difficulties of ultraviolet correction because a curved mirror reflects light of all wavelengths to exactly the same focus. Its efficiency was much lower than a comparable-sized refractor, however, because of the mirror's poor reflectivity. The telescope was used to take an eight-year-long series of plates of the Moon, a sequence which was of particular relevance to a controversy started by Julius Schmidt, the director of Athens Observatory, who announced in 1866 that one of the lunar craters had disappeared.

De la Rue was also concerned with another aspect of astronomical photography, which was again a long programme. John Herschel had advocated the commencement of a daily sequence of photographs of the Sun, a programme which required a special telescope – a "photoheliograph" – with once again a lens specially corrected for photographic use. With an aperture of only 9 centimetres ($3\frac{1}{2}$ inches), the lens was much easier to make than Rutherford's lens, and it was ground by a London optician, A. Ross. The telescope was not applied immediately to the solar patrol. De la Rue took it to Spain in July 1860 to photograph a total eclipse of the Sun. The object of the expedition was to collaborate with Father Angelo Secchi of the Collegio Romano and take photographs of the solar prominences from two stations 400 kilometres (250 miles) apart. Secchi used an orthodox 15-centimetre (6-inch) aperture visual refractor made by R. A. Cauchoix of Paris and both observers obtained several photographs during the total phase of the eclipse. The plates showed absolutely conclusively that the prominences were attached to the Sun and were not, as had

Solar prominences, photographed in Spain at the 1860 eclipse of the Sun. Prominences are enormous eruptions of luminous gas spouting up from the solar surface. The size of the eruptions shown can be estimated from the fact that the Sun is 1·4 million kilometres (875,000 miles) in diameter.

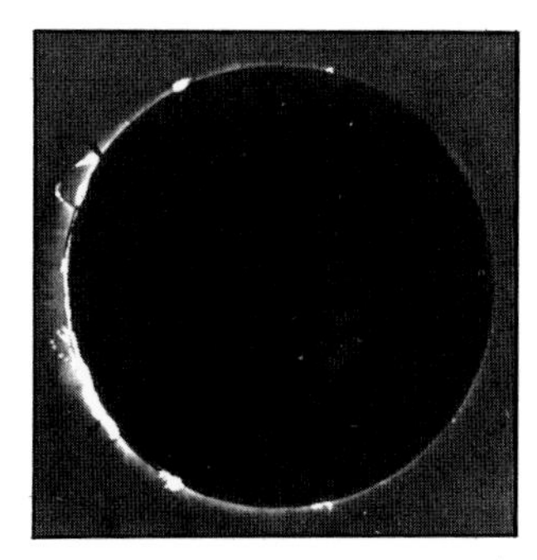

The photoheliograph at Kew. It was built by Warren de la Rue and used for photography of the Sun. It was used for the eclipse of 1860.

been suggested, either effects within the Earth's atmosphere or the result of some unknown lunar phenomenon.

When the photoheliograph returned it was set up at Kew, an observatory funded by the British Association and sited about 8 kilometres (5 miles) from de la Rue's home and private observatory at Cranford. The Kew observatory was chiefly concerned with measurement of the Earth's magnetism and it needed the telescope because of the discovery that small variations in the Earth's magnetic field are related to the number and position of spots on the Sun. The periodic variation of the number of sunspots, with an eleven-year cycle, had also been discovered by Heinrich Schwabe and confirmed by Rudolf Wolf in 1852. The photoheliograph recorded one such cycle before it was superseded by a better telescope, and the work was transferred to Greenwich.

The photographic plate evolved from the study by chemists of the absorption of light. In contrast, spectroscopy developed from the study by physicists of the emission of light. The foundation of the subject was the work by Fraunhofer on the spectrum of the Sun. In the decades which followed, the spectrum of every laboratory source that emitted light was investigated. Flames, sparks, Geissler tubes (the electrical discharges which were the precursor of the neon sign), red hot lumps of iron and coal – anything that gave out light was examined by the spectroscopist. The most obvious feature of all the experiments was that there were two types of spectrum. The simpler of the two was that emitted by hot solids. In this class were both the light emitted by a red-hot wire heated by an electric current and that from the white-hot piece of lime in the flame of a limelight (used for theatre lighting). For these light sources the spectrum formed when a prism dispersed the light was a continuous band of colour. For comparatively low temperature sources the spectrum is brightest at the red end; as the temperature is raised to "white-hot", the band extends further and further towards the blue and violet. Adding extra prisms to spread the spectrum out more disclosed no fine detail (bright or dark lines) comparable with that seen in the solar spectrum. This became known as a continuous spectrum. The other class of spectrum, that seen on looking at light sources like flames and sparks, was very different. The spectrum was a complex mixture of bright and dark. Some colours were completely absent, others were intensely bright. In this case the addition of more prisms revealed a mass of exceptionally fine detail.

A star spectroscope mounted on the Lick Observatory 91-centimetre (36-inch) aperture refractor. Starlight passes through a prism (mounted on the turntable at the bottom left of the picture) and the spectrum is observed through the eyepiece (at the extreme left). The Lick telescope has been used to take more than 30,000 spectrograms of about 300 stars.

Auxiliary experiments had shown light to have all the characteristics of a wave motion. The waves moved very fast, 300,000 kilometres per second (186,000 miles per second), and the wavelength was minuscule, a few hundred millionths of a millimetre (a nanometre). In particular the colour was related to the wavelength. At the extreme red end of the spectrum the wavelength was about 750 nanometres, in the green 500 and in the violet 400. It was a good deal more precise to work in terms of wavelength rather than colour, as two points in the spectrum that differ in wavelength by less than one nanometre could be distinguished in the spectroscope but were not of discernibly different colour. Some light sources appeared to emit light at very few wavelengths: the spectrum of a flame which was coloured by adding common salt was dominated by light emitted in the orange yellow, and close examination showed this light to consist of only two wavelengths (589·0 and 589·6 nanometres). Other sources emitted light at different wavelengths and maps of the spectra were drawn with each bright wavelength indicated by a line. Some of these spectrum "lines" appeared to be identical in wavelength with some of the Fraunhofer lines in the Sun's spectrum, while others quite clearly were not. The main source of confusion (a parallel to the photochemists' problem that sunlight darkened some chemicals but bleached others) was that the continuous spectra seemed independent of the chemical constitution of the source, while the "line" spectra were all different. A hot platinum wire emitted a spectrum the same as that given out by a hot lump of carbon, but the spectrum of a spark between two platinum wires was quite different from that of a spark between two carbon rods.

The first step forward was made in 1833 by David Brewster, editor of the *Edinburgh Philosophical Journal*, who showed that a tube filled with nitric oxide gas would absorb several of the range of wavelengths emitted by a hot solid. This evidence, that gases could produce dark absorption lines similar in appearance to the Fraunhofer lines, led him to study the Sun's spectrum at midday and near sunset, when sunlight traverses more of the Earth's atmosphere. He found that some of Fraunhofer's lines, in particular the strong lines labelled A and B, were much stronger (darker) at sunset, while others, including C and D, remained of the same strength regardless of the Sun's attitude. The Fraunhofer lines A and B must therefore be due to absorption by the Earth's atmosphere and the others due to absorption occurring at the Sun itself.

For the next quarter of a century the physicists stumbled around in search of a coherent explanation of their experiments. The daguerreotypes of the solar spectrum, by Draper and independently by Edmond Becquerel, showed that the Fraunhofer lines were present in both the ultraviolet and the infra-red as well as the visible. Léon Foucault, in Paris, did some crucial experiments using a new light source, the electric arc, but these went largely unnoticed. George Stokes, mathematics professor at Cambridge, put together a substantially correct theory but failed to publish it. Anders Ångstrom, "Keeper of the Observatory" at Uppsala, also came near to the true explanation, but it was the professor at Heidelberg, Gustav Kirchoff, who published the definitive statements which were immediately and widely accepted.

Kirchoff worked with a colleague, Robert Bunsen, the professor of chemistry, and they carried out a series of experiments on flames containing sodium salts, using the type of gas burner that still bears Bunsen's name. They started with the knowledge that the flames emitted two bright spectrum lines in the yellow-orange whose wavelength corresponded precisely with that of the pair of dark Fraunhofer lines in the spectrum of the Sun called the D lines. However, when they placed the flame in a beam of sunlight and looked at the light coming through it, they found that the emission from the flame did not necessarily "fill in" the two dark lines. If the flame was relatively cool, the dark lines became even darker: the flame was absorbing part of the light remaining even in the darkest parts of the solar lines. Only if the flame was very hot was it possible to produce a combination of Sun and flame in which

SPECTROSCOPY

A mid-nineteenth-century spectroscope and the associated apparatus needed to observe emission spectra. The spectroscope is being used to study the light emitted by substances placed in the flame (7) on the right. The fraction of the light that passes through the slit and lens mounted at opposite ends of the tube (6) is then dispersed by the prism (2). The spectrum is focused by a further lens and passes down the tube (3) to be examined using the eyepiece on the extreme left. The second slit and lens system, mounted in the tube (4), is illuminated by a second burner (5) and produces a beam of light, part of which is reflected off the near face of the prism. This provides a reference for measuring the position of the spectrum lines.

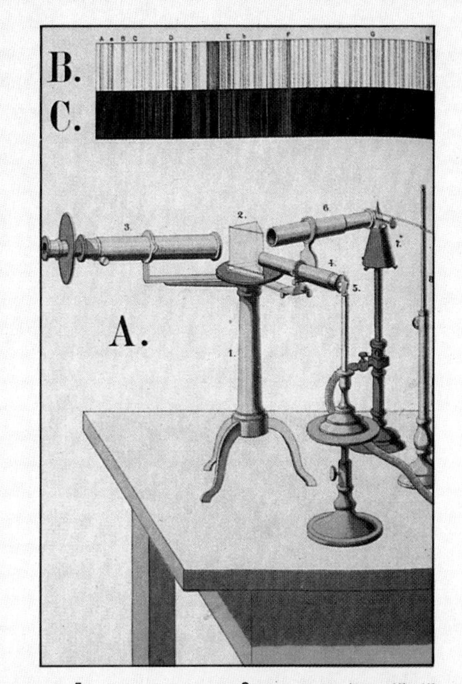

(Above) A Victorian illustration of the spectrum of the Sun, seen in ordinary conditions and at total eclipse. The lower "line emission" spectrum is produced by the solar corona. The lines marked are due to hydrogen. The main feature of the upper spectrum is dark lines due to absorption by atoms in the cool outer layers of the Sun, that are superimposed on a continuous background. The absorption lines represent different elements.

Although light can be dispersed into a spectrum with the aid of a prism, it is not possible to measure wavelengths with such a device. The alternative, pioneered by Joseph Fraunhofer, is a diffraction grating (below), a set of thousands of fine parallel lines ruled on an optical surface. This allows both dispersion and wavelength measurement and was used in the basic astronomical researches of the Swedish astronomer Anders Ångstrom, who gave his name to the Angstrom unit (A), a ten billionth of a metre. In this photograph a laser beam has been dispersed using a diffraction grating.

the D lines were brighter than the neighbouring continuous solar spectrum. The conclusions, published in 1859, were that any source could both emit and absorb radiation. Whether absorption or emission took place depended on the temperature. A moderately hot gas in front of a very hot continuous-spectrum source would give a series of dark absorption lines; the same gas in front of a cooler background would give a bright-line spectrum in emission superimposed on that background. The hot gases of a flame or an arc emitted a line spectrum, and the wavelengths of the lines depended on the atoms and molecules present in the gas. Such a gas would absorb light from a suitable background source, but could only absorb at those wavelengths at which emission lines were observed: at other wavelengths the gas would be transparent. The conclusions were held to be of general validity, and applied not just to line spectra but also to the sources of continuous spectra. All of these were opaque and therefore absorbed all wavelengths, so the corollary was that they must emit all wavelengths when heated sufficiently. The chemical constitution of the solid was unimportant.

Gustav Kirchoff, 1824–87, made fundamental contributions to many branches of physics, including the analysis of electrical circuits, as well as his work with Robert Wilhelm Bunsen on radiation of light.

The acceptance of Kirchoff's laws immediately led to a model of the Sun with a hot opaque core emitting a continuous spectrum surrounded by a cooler gaseous envelope whose absorption caused the solar Fraunhofer lines. The coincidence between the D lines and the lines emitted by hot gas containing sodium atoms showed that the Sun's atmosphere contained sodium. Bunsen returned to his chemistry laboratory to use the knowledge that each element showed a characteristic spectrum as a critical aid in his discovery of the hitherto unknown elements caesium and rubidium. Kirchoff and his assistant K. Hoffman carried out a detailed comparison of the Fraunhofer spectrum with that of a range of different elements. They concluded that the Sun's atmosphere contains at least a dozen elements known on Earth, including sodium, calcium, magnesium, chromium and iron.

Kirchoff's work was extended by Ångstrom and his colleague Thalen. Ångstrom had one experimental advantage over Kirchoff in that he used an electric arc as his laboratory source, an emitter that is a much nearer match to the Sun than is a high-tension spark. Ångstrom accounted for about 800 of the Fraunhofer lines, adding hydrogen and manganese to Kirchoff's list of elements present in the Sun, while Thalen identified titanium lines in the solar spectrum. However, the Uppsala work is chiefly commemorated in the "Ångstrom Unit", a tenth of a millionth of a millimetre (0·1 nanometres). Ångstrom, unlike most of his contemporaries, used a device called a diffraction grating instead of a prism to disperse sunlight into a spectrum. The grating is a series of very fine lines ruled close together on a glass or metal plate, and it disperses the different wavelengths through angles which are simply related to the wavelengths themselves. As a result, Ångstrom was able to measure accurate wavelengths for all the lines. Ångstrom's map and list of the solar lines was used as a standard by everyone else and, as he was much in favour of a measurement system based on powers of ten, Ångstrom used the metre as his basic unit and chose a sub-multiple such that ten billion (mathematically, ten raised to the power ten) was a convenient working unit. He called the unit a "tenth metre", but it was soon christened "Mr Ångstrom's Unit", then an "Ångstrom unit (A.U.) and finally just an "Angstrom" "Å". Many spectroscopists still use Angstroms, although the nanometre – 10 Angstroms – is now more commonly used.

The problem of assigning the Fraunhofer lines to the right element is not an easy one, because, when the solar spectrum is examined in detail, there are tens of thousands of lines. Only very careful measurement will decide whether a rare element with few lines in its spectrum is truly present or whether the few faint Fraunhofer lines concerned only coincide in wavelength by chance, and belong to one or more other elements. This difficulty of assignment was further underlined by spectroscopic studies of the eclipsed Sun. At the 1868 eclipse, when four expeditions all studied the spectrum of the solar prominences, these proved to emit a spectrum with only a few bright lines, most of which belonged to hydrogen. The prominences were therefore gaseous in character and dominantly constituted of hot hydrogen. An orange-yellow line was assumed to be the well-known sodium D lines, the close pair being unresolved by the spectroscopes in use. The lines emitted by the prominences were so bright that one observer, Pierre Janssen, decided that they would be visible even if there were no eclipse. Surprisingly, the same idea had occurred to two men who were not at the 1868 eclipse, Norman Lockyer and William Huggins, and Lockyer and Janssen published their findings on the same day. The new technique also enabled Lockyer to prove the orange line was not due to sodium, the difference in wavelength being small but significant – 1·6 nanometres in 590. There was no element known on Earth that emitted a simple spectrum showing this line and Lockyer suggested that there must be a new element responsible. Edward Frankland, the Professor of Chemistry at Imperial College, christened the new discovery helium.

The application of the spectroscope to stars rather than the Sun was pioneered by William Huggins and Angelo Secchi, in a pair of complementary programmes. Secchi studied the spectra of a large number of stars (about 4000), using a telescope with a prism mounted at the top end of the telescope, just outside the objective lens. Every star seen through the eyepiece thus appeared not as a point but as a small spectrum. Huggins, in contrast, mounted an orthodox spectroscope at the eye end of his telescope and studied the spectra of a few of the brightest stars in great detail. Secchi found that he could classify the stars he examined into four

A medal showing J. Norman Lockyer, 1836–1920, and Pierre Janssen, 1824–1907. Both were directors of "astrophysical" laboratories rather than classical observatories.

classes. First were the blue-white stars like Vega in Cygnus, for which the spectrum was relatively simple – a uniform continuum with few dark lines appearing in absorption. Of these few lines four due to hydrogen were by far the strongest. The second class, yellowish white in colour, like Capella and Arcturus, showed spectra similar to the Sun, with a large number of dark lines. The redder stars split into two groups: the majority, similar to Betelgeuse, showed a "banded" spectrum, with a series of regular flutings due to some unknown absorber; the remainder also showed a fluted spectrum, but one that was known to be associated with carbon. The overall picture was one of astounding uniformity – virtually all the stars that were the same colour as the Sun showed similar spectra. The idea that the sequence from blue-white to dull red was just that to be expected as a white hot star slowly cooled took such a firm hold that even now, when it is known to be wrong, the hot, blue-white stars are called "early" type and the cool, dull, red ones "late" type stars.

Huggins's detailed studies, assisted once again by a professor of chemistry, William Miller, rapidly established that elements such as hydrogen, calcium, sodium and iron were to be found in the stars – the universe was made up of the well-known terrestrial elements. The change of view in a generation was complete. In the 1830s Auguste Comte, the founder of positivism, had used the idea of determining the chemical constitution of the stars as the definitive example of the totally and eternally impossible. Huggins's first paper was published in 1863. In the same year Huggins tried, unsuccessfully, to photograph a stellar spectrum. Two other major discoveries followed soon after. The telescope, with attached spectroscope, was directed at the nebula in Draco. To Huggins's surprise, the nebula did not have a star-like spectrum, but showed a single bright line. After checking that his apparatus had not fallen out of adjustment, Huggins realized that the line was real, and that the nebula must therefore be a cloud of glowing gas and not a cluster of stars. The long debate as to whether all nebulae would be resolved into stars if only the telescope were sufficiently powerful was over. It was rapidly shown that the Orion nebula was gaseous, and that Bond and Rosse had been too optimistic about the performance of their two telescopes (at the time the largest refractor and the largest reflector in the world) when they thought they had resolved it into stars. The other important discovery was that some of the stellar spectra had all the lines very slightly displaced in wavelength compared with terrestrial light sources. This had to be due to the motion of the star towards or away from the Earth. In the former case, all the waves of light are compressed, and the wavelengths arriving at Earth are all slightly shorter than when they left the star; the entire spectrum seems shifted towards the blue end. If a star is moving away, the wavelengths are stretched, and the spectrum is shifted to the red. Motion of stars across the sky (proper motion) had long been known. Now the spectroscope could detect motion in the line of sight, and Huggins was able to measure velocities of tens of kilometres a second for some stars.

The lessons of photography and spectroscopy, where astronomy of the highest class had been carried out by observers with very modest telescopes – de la Rue's 9-centimetre ($3\frac{1}{2}$-inch) photoheliograph, Secchi's 10-centimetre (4-inch) prism on a 15-centimetre (6-inch) Cauchoix refractor, Huggins's 20-centimetre (8-inch) refractor – were not learned by the astronomical establishment. To them Urania, the muse of astronomy, was cold and distant, concerned with the smooth and silent motion of the stars, not a grubby figure in an apron, standing at the laboratory sink and doing the washing up.

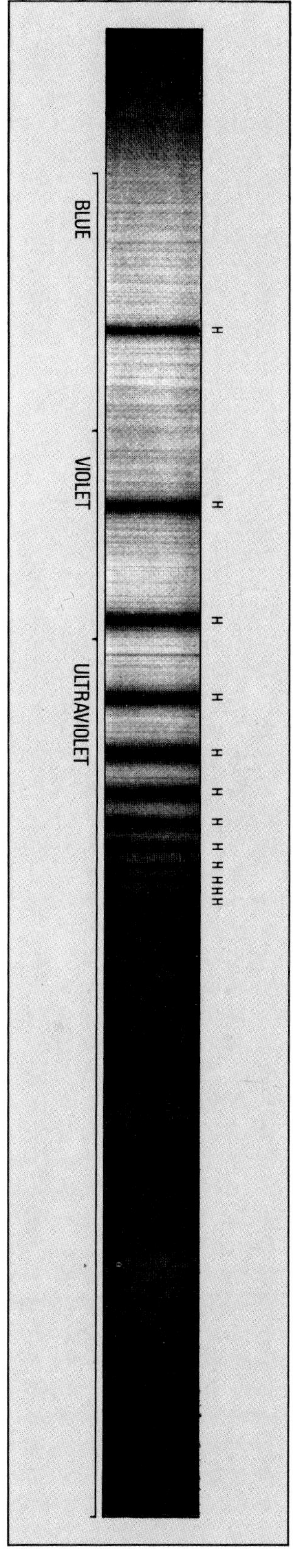

Huggins's spectrum of Vega (Alpha Lyrae). The prominent series of lines in the spectrum, marked with the letter "H", are the "Balmer series" and are all due to hydrogen, the most abundant element in the universe. The long-wavelength end of the visible spectrum, green, yellow, orange and red, did not affect early plates and so was not recorded.

CHAPTER EIGHT

THE BRIEF REIGN OF THE BIG REFRACTOR

The interval from the Newall 63-centimetre to the Yerkes 101-centimetre telescope: 1869–97

"The table of 530 star catalogues which is appended, I have endeavoured to make exhaustive of the labours of astronomers in this direction. I cannot suppose it to be perfect; but the omissions will, I trust, be few and unimportant."

E. B. Knobel in
Memoirs of the Royal Astronomical Society,
volume 43, 1877.

For at least thirty years after Airy's transit circle was installed at Greenwich, the main efforts of astronomy continued to be directed to the precise measurement of the position of the Moon, the planets and the stars. Neither Loomis's *Practical Astronomy*, published in 1881, nor Campbell's *The Elements of Practical Astronomy* of 1891 even mention spectroscopy or photography. The chief development was that measurement of star brightness was added to the measurement of position.

In classical times the stars visible to the naked eye had been arranged in order of "magnitude": first magnitude for the brightest stars down to sixth magnitude for those that were only just visible. Ptolemy's catalogue worked to a slightly finer scale, giving the brightness to a third of a magnitude. The next step, taken by Bayer 1500 years later, was to label the stars in each constellation in order of brightness: Alpha for the brightest, then Beta and so on down the Greek alphabet. Bayer did not make any measurements himself but relied on the ancient tables, a circumstance which led Delambre to comment that seldom had anyone acquired immortality so cheaply.

The first systematic attempts to measure rather than estimate the brightness of the stars were made in the 1830s by John Herschel at the Cape of Good Hope and by Steinheil and Seidel at Munich. Herschel devised an ingenious and complicated device which used a small lens, a prism and a quantity of string to produce a tiny image of the Moon, whose brightness could be varied to match the star under observation with the main telescope. Steinheil's instrument was based on a telescope with the lens split into two D-shaped halves mounted so that they could be focused separately. The image of a bright reference star was deliberately set out of focus and the size of the smudged image adjusted until the circular patch appeared to be as bright as the smaller patch formed by the slightly out-of-focus image of the fainter star whose magnitude was being measured. A pair of prisms placed in front of the half-lenses acted as mirrors (not to give spectra), and allowed the two halves of the telescope to look at widely separated stars. Steinheil's photometer was used by Seidel and was sufficiently sensitive to set on a sound basis the older naked-eye observations, which showed that the atmosphere is not perfectly transparent. Seidel found that a star thirty degrees above the horizon was dimmed by about a quarter of a magnitude compared to a star of similar brightness at the zenith. At ten degrees above the horizon, which meant looking obliquely through six times as much air as at the zenith, the loss of light was more than a magnitude.

In parallel with these measurements Friedrich Argelander, Bessel's pupil and the Director of the Bonn Observatory, did the job which Delambre thought Bayer should have done in 1603: he reassessed the brightness of all the 2700 "naked eye" stars visible from Bonn. In this work, Argelander did not use a photometer, but instead he estimated the brightness of the stars to the nearest half-magnitude, and fitted his results to the old magnitude scale entirely intuitively. Soon after the publication of Argelander's list in 1843 the photometer users established a more reproducible system, starting from two separate groups of observations. The first set, based on the work of Seidel and Herschel, showed that an average star of the first magnitude was roughly a hundred times as bright as one of the sixth magnitude. The second observation, which was not explicitly stated for some years, was that equal increments in brightness, as seen by the eye, correspond to equal multiples of the energy reaching that eye. This meant, for instance, that to arrange candles to give a scale that appears to progress in even steps one must use a sequence of one, two, four, eight, sixteen . . . candles not one, two, three, and so on.

When this response is applied to the observation that first to sixth magnitude, a difference of five magnitudes, corresponds to a factor of a hundred in energy, it follows that one magnitude must be a step which, when multiplied by itself five times, gives a hundred. This number is the fifth root of a hundred, 2·51188 (roughly $2\frac{1}{2}$), and was put forward as the defining step for one magnitude difference in brightness by N. R. Pogson, an assistant at the Radcliffe Observatory, Oxford. The number defines a step size for a ladder going down to ever fainter objects: a seventh magnitude star transmits 2·512 times less energy to the observer than a sixth magnitude star; eighth magnitude is roughly $2\frac{1}{2}$ times weaker still. An observer with a 12·5-centimetre (5-inch) aperture refracting telescope can just see a star of magnitude twelve and a half; this is a hundred thousand times fainter than a star at the zero point of the system. The top end of the scale is occupied by the brightest stars, which proved to be not a very homogenous group. The new measurements showed that Spica, the brightest star in the constellation Virgo, was almost precisely a first magnitude star, so Vega, $2\frac{1}{2}$ times brighter in terms of energy received at the Earth, had to be given a magnitude zero, one less than Spica. For the four stars brighter than Vega the magnitude goes negative, as it does for several of the planets, the Moon and the Sun. Steps in brightness of less than $2\frac{1}{2}$ are dealt with by using fractions of a magnitude: Castor and Pollux, the two brightest stars in Gemini, are of magnitude 1·6 and 1·2 respectively, intermediate between Spica (magnitude 1) and the Pole Star (magnitude 2). Argelanders's intuitive scale proved to be equivalent to one magnitude for a change in brightness by a factor of 2·3 rather than 2·5, and unfortunately both scales continued in use for many years, so that well into the twentieth century it was necessary to record the sort of magnitude used next to every measurement.

While the accuracy of the early photometers was modest, rarely better than a tenth of a magnitude (a ten per cent change in brightness), the precision of position measurement was formidable. This was clearly illustrated by two analyses; Urbain Le Verrier investigated the motion of Mercury, while his old rival John Adams treated the motion of the Moon. Mercury, the innermost planet, moves in a highly elliptical orbit, varying between 46 and 70 million kilometres (29 and 44 million miles) from the Sun. The long axis of this ellipse rotates very slowly round the Sun, the position of the point of closest approach, the perihelion, moving through an angle of 56 arc seconds a year. Ninety per cent of this is due to the precession of the equinoxes and is the price paid for measuring celestial longitudes from a moving reference point. The remainder, after this has been deducted, is 5·7 arc seconds a year and of this Le Verrier was able to explain 5·3 as due to the gravitational action of the other planets, especially Venus. He was left with a small discrepancy of 0·4 arc seconds a year that he was unable to account for. Adams tackled the problem that the Moon's motion about the Earth was not steady but slowly accelerating. He was able to show that an earlier analysis by Laplace was wrong and that there was again a discrepancy between theory and observation: the predicted acceleration giving a change of position of 3·3 arc seconds in a century, while observation showed a change of 6·6 arc seconds a century. The important point in both cases was that these extremely small discrepancies, 40 arc seconds a century for Mercury and 3·3 for the Moon (for which a much longer series of observations was available), were both accepted as real and not just uncertainty in the measurements. Le Verrier, following the route he had taken to find Neptune, attempted to solve his problem by postulating an additional inner plant, "Vulcan". Adams was immediately plunged into another row, this time with a group of theorists who felt that the great Laplace could not be wrong. Ironically, both discrepancies turned out, much later, to be due to effects entirely outside the bounds of Le Verrier's or Adams's celestial mechanics – Einstein's

F. W. A. Argelander, 1799–1875, compiled his catalogue of a third of a million stars with the aid of an elderly Fraunhofer telescope of modest size – a "comet seeker" of only 60 centimetres (24 inches) focal length.

relativity for Mercury and the pull of the Earth's ocean tides for the Moon.

In spite of these infinitesimal discrepancies, the majority of astronomical measurements made in the third quarter of the nineteenth century were concerned more with quantity rather than quality. This arose partly as a response to the many millions of stars visible with the new generation of refracting telescopes. The leader in the quantity business was Argelander at Bonn. Assisted by Schönfeld and Krüger, Argelander used an elderly Fraunhofer telescope of very modest (9-centimetre, 3½-inch) aperture to record the position and estimated magnitudes of all the brighter stars visible from Bonn. The result was the *Bonner Durchmusterung*, a record of the position and brightness of over 324,000 stars, given both as a list and on a series of charts. This looks at first glance like one of the world's wilder exercises in obsessive data gathering, requiring a Teutonic thoroughness in keeping with Argelander's East Prussian upbringing, if not his Finnish ancestry. In practice the Bonn catalogue is a fundamental research tool, still in daily use. The reason for this is that an astronomer working on a particular star must be able to tell his colleagues which star it is. Reading the co-ordinates from the telescope scales is not adequate and it is very much easier to use a chart. Telescopes like Rosse's or Nasmyth's reflectors did not have precise scales fitted, and the problems of correcting for alignment errors, flexure, refraction and so on for lesser instruments are time-consuming and tedious.

For the brightest stars the star names or Bayer's Greek letter labels suffice, though names are falling out of fashion: Betelgeuse, Rigel, Bellatrix and Mintaka are more often called Alpha, Beta, Gamma and Delta Orionis. (Convention requires that the genitive case of the Latin constellation name be used – star Alpha of Orion is Alpha Orionis, usually abbreviated to α Ori.) For fainter stars still visible to the naked eye Flamsteed's numbering of all the stars in a constellation, working steadily from right to left, is still in use after 300 years. Thus, Alcor, the fourth magnitude star next to Zeta Ursae Majoris, is also known by Flamsteed's label, 80 Ursae Majoris. The next third of a million stars are known by their number in the Bonn list: for instance BD + 4° 3561 is a very dull red star of magnitude 10·3 in the constellation of Ophiucus, a star notable only for being the nearest star observable from the Northern Hemisphere (also known as Barnard's star). All three systems – Bayer, Flamsteed and Argelander – are useful. The two older systems endure because they immediately indicate that the star under discussion is either very bright or fairly bright. Thus, both Bradley's "aberration" star Gamma Draconis and Bessell's "parallax" star 61 Cygni are visible to the naked eye and can be seen in the stated constellations. The star names are much less helpful: Gamma Draconis has two – Rastaban and Etamin – while Zeta Ursae Majoris is one of three stars Mizar (the others are Beta Andromedae and Epsilon Bootis).

The increasing number of customers for refracting telescopes were served by a growing number of manufacturers. In Germany, Merz, the surviving partner of Merz and Mahler, was the dominant firm for very large telescopes, but Fraunhofer's successor, Ertel, Repsold & Sons and later Steinheil, were all busy. Just as in Germany Repsold mounted many Merz lenses into telescopes, so in France a partnership grew between the Henry brothers as lens-makers and Gautier as telescope engineer. The situation in England was transformed by the abolition of the tax on glass and also by the French Revolution of 1848, which led to the foundation of the Second Republic and also to a skilled French glass-maker, George Bontemps, leaving France to join the English glass-making firm of Chance Brothers. The availability of good blanks of both flint and crown glass stimulated Thomas Cooke, already established as a maker of telescopes of modest size, to attempt larger instruments. All these workers, however, were

Durchmusterung

des nördlichen Himmels

zwischen 45 und 80 Grad der Declination

auf der Interims-Sternwarte

der Königlichen Rheinischen Friedrich-Wilhelms-Universität

zu Bonn

in den Jahren 1841 bis 1844

ausgeführt

und mit Hülfstafeln zur Reduction der scheinbaren Oerter der Sterne auf die mittlern zu Anfange des Jahres 1842 versehen

von

Dr. Friedrich Wilhelm August Argelander,

o. ö. Prof. d. Astronomie und Director der Sternwarte, Ritter des Stanislaus-Ord. II. Cl. und des Wladimir-Ord. IV. Cl., Mitglied der Königlichen und der Astronomischen Gesellschaften zu London, sowie der Finnischen Gesellschaft d. Wiss. zu Helsingfors, Corresp. der Academien d. Wiss. zu Berlin, Palermo und Petersburg.

Bonn,

bei Adolph Marcus.

1846.

The frontispiece of one section of Argelander's catalogue, the Bonner Durchmusterung, *in which he recorded all the brighter stars of the Northern Hemisphere. The star positions were both listed in tables and drawn on charts. The need for more precise measurement than the initial rapid survey led Argelander to set up, in 1867, an international programme for the revision of his catalogue. He made a special effort to observe and catalogue variable stars and appealed for the help of amateur astronomers who were studying them.*

exceeded both in precision of manufacture and sheer size by the work of an American portrait painter, Alvan Clark.

Clark had given up painting for lensmaking after looking through the 38-centimetre (15-inch) Harvard refractor. He had made several small lenses and mirrors as a hobby, and the fact that he could detect the small residual errors of manufacture in one of the best lenses Europe could offer convinced him that he could do as well. He was entirely correct, but no one would buy lenses from a maker completely without a reputation. Americans either bought telescopes from Henry Fitz in New York or imported them, and Europe was well supplied by established makers. The turning point came when Clark made contact with an English amateur astronomer, the eagle-eyed W. R. Dawes, who proved to be an enthusiastic advocate of the excellence of Clark's lenses. When Clark visited Europe in 1859 news of his reputation and of the orders he received finally filtered back to America. Soon after his return he was asked to build a 47-centimetre (18½-inch) refractor for the University of Mississippi. The glass blanks for the lens were made by Chance Brothers and reached Boston in 1861 to find Clark's new factory ready, complete with a fireproof safe in which the valuable glass discs could be stored at night.

The new lens was completed in January 1861 and immediately showed its quality. Preliminary testing at the factory, which involved looking at bright stars, showed Sirius to be double, in agreement with Bessel's prediction. The companion star was not spectacularly close to its bright companion, the separation being eight arc seconds. It was, however, ten magnitudes fainter than Sirius itself, showing that Clark had produced a lens which gave so sharp and well-defined an image that a nearby star only a ten thousandth as bright could be identified. Unfortunately January 1861 was a few months before the beginning of the American Civil War and by the time the telescope was complete there was no way in which an instrument ordered in one of the Southern States could be delivered by a manufacturer in the North. Instead, the Chicago Astronomical Society bought the telescope and installed it at the University of Chicago.

The 47-centimetre telescope, although bigger than the Harvard and Pulkovo instruments, never held the title as the largest in the world. The success of Chance Brothers in casting big discs of glass produced three large English refractors: 53, 61 and 63 centimetres (21, 24 and 25 inches) in aperture, all built for wealthy amateurs. Two of the three instruments had no influence on the development of the telescope. The smallest, completed in 1862 for J. Buckingham, was very little used by its busy owner, while the 61-centimetre aperture telescope, built for the

The American Civil War prevented the delivery of Clark's 47-centimetre (18½-inch) refractor to the University of Mississippi, virtually the only major observatory in the southern states. The depressed conditions in the south after the war and the technological superiority of the north meant that fifty years later the northern observatories still outnumbered the southern by ten to one.

Reverend Craig and the earliest of the three to be completed, was an utter disaster. The mounting of the telescope was clumsy and awkward and the lens, made by Thomas Slater, so bad that he seems never to have made another. The third and largest of the three pairs of lens blanks was purchased by R. Newall, and given to Thomas Cooke to be ground, polished and mounted in a telescope. The strain of the prestigious project told on Cooke, who died in 1868 just before the telescope was completed. The new instrument represented a dramatic increase in size, 63 centimetres in aperture and 8·8 metres (22 feet) in focal length, and this was underlined when, several years later, Newall offered it to the observatory at the Cape of Good Hope. At that time the biggest telescope at the Cape was only 1 centimetre larger in aperture than the "finder" telescope mounted on Newall's giant.

The 63-centimetre Newall refractor was followed in 1869 by another of larger aperture, this time for the United States Naval Observatory in Washington. Alvan Clark had already demonstrated his skill to the Observatory staff by repolishing the lenses of some of the existing telescopes and was commissioned to build the biggest telescope he could supply for the sum of $50,000. The result was a 66-centimetre (26-inch) aperture refractor, which again demonstrated the superb image quality of Clark's lenses. It was used by Asaph Hall in 1877 in a successful search for a possible moon for Mars – he found two moons and both were extremely faint objects. The inner, Phobos, was $13\frac{1}{2}$ magnitudes fainter than Mars itself; the outer, Deimos, was a magnitude fainter still. The magnitude differences correspond to brightness ratios of a quarter of a million and over six hundred thousand, and Clark's 66-centimetre lens had to be very nearly perfect for Hall to be sure that the moons were real and not due to some minor error in the figuring.

Not all the discoveries were made with large telescopes. Wilhelm Tempel, using a very modest instrument, found that there was a hazy nebula associated with the Pleiades star cluster. Many astronomers, including Tempel, Pogson and Schiaparelli, searched for minor planets, and in 1868 the hundredth was discovered by Watson and christened Hecate. The most controversial observations were those of Mars made in 1877 by Giovanni Schiaparelli, the Director of the Milan Observatory, using a 22-centimetre ($8\frac{1}{2}$-inch) Merz telescope. Secchi, Lockyer, Dawes and Proctor had already drawn careful maps of Mars using 15- to 20-centimetre (6- to 8-inch) aperture telescopes, but they all had to contend with the turbulence of the Earth's atmosphere, which destroys the sharpness of the image. A casual glance at Mars through a good telescope shows a few ill-defined smudges on a pinkish-orange disc. At the pole there may be a whitish polar cap, while at lower latitudes the smudges are usually grey. By watching patiently the observer may have the luck to hit on a moment when the atmosphere is unusually still, and he will get a

Alvan Clark, 1804–87, seated on the left in his workshop, set up as a lens-maker after examining the 38-centimetre (15-inch) Merz telescope at Harvard that was a copy of an instrument at Pulkovo. Thirty-four years later his reputation was such that he was to make a 76-centimetre (30-inch) lens for Pulkovo.

brief glimpse of a sharper image. Proctor's map of Mars, drawn in 1867, showed the light and dark areas and named just over fifty different features, most of them irregular shapes several hundred kilometres across. Schiaparelli found the dark areas to be joined by narrow dark lines – the notorious canals of Mars. These canals were only 100 kilometres (60 miles) or less in width, and ran straight for thousands of kilometres. The straightness and the man-made implication of the word "canal", a mistranslation of the Italian word canali, a channel, led to vast and furious speculation about life on Mars. Very few observers could see Schiaparelli's canals, but his observations were defended with vigour. When Asaph Hall, with a telescope three times the size of the instrument in Milan, said he could not detect any of Schiaparelli's fine detail, it was assumed that Alvan Clark's big telescope had developed some defect, the main suggestion being that the giant lens had sagged under its own weight.

Several years later, after Schiaparelli had equipped Milan with a 50-centimetre (20-inch) aperture telescope, again from Merz, he published observations reporting that the canals were double and some of them were only 30 kilometres (20 miles) wide – a tenth of an arc second across as seen from the Earth. The origin of these illusory canals can best be understood by spending a sunny hour at a swimming pool. The ripples on the water make a turbulent, transparent optical surface, and can focus the sunlight, often into bright lines, on the bottom of the pool. The Earth's atmosphere has a similar effect, focusing the low contrast markings on Mars into lines. After that, the insidious subjective effects of seeing creep in. Once a "canal" has been seen, any recurrence of the same illusion is confirmation, while

Drawings of Mars (above) made by Giovanni Schiaparelli in 1877 when Mars was closer to the Earth than is usual. Schiaparelli was a pupil of F. G. W. Struve before moving to Milan, and his researches included meteors and double stars, as well as the discovery of the asteroid Hesperia.

The United States Naval Observatory (below), established in 1844, was the premier observatory in the country for much of the nineteenth century. Simon Newcomb, the director from 1861 to 1897, produced an analysis of planetary motion that was accepted for almost a century.

Edward Pickering, 1846–1919 (below), *was director of Harvard College Observatory for more than forty years. During this time he initiated a comprehensive survey of stellar spectra, grouping stars into their special classes. This gave rise to early notions on stellar evolution, a problem solved in the 1930s.*

(Right) *The telescope built by Gautier with optics by the Henry brothers, which was used as the prototype for the "Carte du Ciel" survey. This great star-mapping exercise was initiated in 1889. The telescope tube encloses two refractors, the upper a visual instrument for guiding, the lower a photographic refractor.*

(Below) *A measuring instrument being used to determine star positions on the plates taken in the Carte du Ciel programme. The programme was established in order to record the entire sky; around the world a set of almost identical telescopes were used simultaneously.*

any other pattern is "bad seeing". Later in his career, Schiaparelli studied Venus and came to the opposite conclusion from his studies of Mars. He announced that all the apparent markings that others claimed to have seen on Venus were illusory, and were due entirely to the seeing conditions.

The latter decades of the nineteenth century saw astronomers grappling with the balance between quantity and precision. One compromise led to a consortium of thirteen observatories agreeing to remeasure the positions of all 324,000 stars in Argelander's catalogue, the larger number of observers allowing the positions to be determined more accurately. Schönfeld extended Argelander's original catalogue to the southern regions of the sky not visible from Bonn, and Gould, working in Argentina, completed the survey. In France, Arago adopted a less international policy and directed the checking of all the star positions determined by Lalande a century earlier.

The distances of the stars were also subjected to intense study, and many series of parallax measurements were made. The precision of the better measurements slowly improved to between a tenth and a twentieth of an arc second, but the majority of these measurements showed only that the stars concerned were too distant to have a measurable parallax, an observation which was just as time-consuming as one that gave a usable result. Dunsink Observatory in Ireland was particularly unfortunate: in one campaign involving 450 determinations of stellar parallaxes, only one gave a measurable distance to the star observed. The biggest surprise was that Canopus (Alpha Carinae), Deneb (Alpha Cygni) and Rigel (Beta Orionis), three of the brightest stars, proved to be at least 20 parsecs away; they must therefore be many thousand times brighter than the Sun. The pursuit of accurate measurement was further stimulated by the two transits of Venus in 1874 and 1882, which produced a great deal of international travel and another flood of discordant results. In twenty years, thirty-four determinations were made of the Earth-Sun distance using a host of different methods and the extreme results differed by 3 million kilometres (2 million miles). A summary of the set by Simon Newcomb of the US Naval Observatory led to an estimate of 149·9 million kilometres (93·1 million miles) with an error of 400,000 kilometres (250,000 miles).

Four astronomers developed good photometers: Zöllner at Leipzig, Edward Pickering at Harvard, Pritchard at Oxford and Gould in Argentina. Zöllner concentrated on the planets and their satellites, while Pritchard, Pickering and Gould all carried out major surveys of the brightness of many thousands of stars. Pickering built three photometers in succession, the first using a pair of 5-centimetre (2-inch) lenses, the second 10-centimetre (4-inch) and the third 30-centimetre (12-inch). Just as better measurement of star positions had supported earlier suggestions that some of the "fixed" stars moved, so the photometers confirmed that many stars altered in brightness. Pickering was able to distinguish five separate types of "variable" stars. The first of these, and the rarest group, were the "new stars", or novae, which burst unexpectedly to great brilliance from previous obscurity. Tycho's star of 1572 and Kepler's of 1604 were examples, and the first quantitative measurements were obtained in 1876 for a star which exploded in Cygnus, rising suddenly to third magnitude and then fading back to insignificance at the rate of a magnitude every four days. The second class of variable stars showed large but slightly irregular slow periodic variations: the typical example is Mira (Omicron Ceti), which oscillates between third and ninth magnitude in a cycle that takes nearly a year. Third came "irregular" stars, whose brightness fluctuates rather less and without any discernible rhythm. Betelgeuse (Alpha Orionis) is of this type. The last two types are much more regular in behaviour. The first is the "Cepheids", named after the typical example Delta Cephei, a star which oscillates rather less than a magnitude in a regularly repeating cycle taking 5 days 8 hours 53 minutes. Last of all are the eclipsing binary stars such as Algol (Beta Persei). Here a pair of stars is rotating one about the other in an orbit such that, seen from the Earth, one star eclipses the other at regular intervals.

In parallel with the compilation of the catalogues of brightness and position, all based on visual measurement, the 1880s marked the long-overdue emergence of the photographic plate as an essential astronomical tool. The critical step was the development by R. L. Maddox and Charles Bennett of the gelatine-based photographic emulsion, leading by 1879 to a dry, durable plate which was easy to use and even a little more sensitive than the old wet plate. In the next few years a handful of workers in Europe and America all produced photographic observations which showed that the ability of the plate to accumulate starlight over a long exposure made it a better detector of light than the eye. Draper in the United States in 1880, Janssen in France in 1881 and Common in England in 1883 all took plates of the Orion nebula, recording a wealth of detail entirely beyond the capacity of any observer using eye and pencil. In addition, the photograph was unarguably accurate, so that debate about whether or not the structure of the nebula was stable no longer depended on the skill of the astronomer as artist. The crucial sets of observations, however, were made by Isaac Roberts in Maghull, just north of Liverpool, and by Paul and Prosper Henry in Paris, both using telescopes specially designed for photographic work. The two photographic telescopes were each mounted as one of a pair of telescopes on the same axis, the second telescope being a visual refractor used to guide the photographic half of the pair. Both French and English telescopes were used to photograph the Pleiades, and plates taken in 1885 by the Henrys with a 3-hour exposure showed over 1400 stars in the cluster. The next year Roberts, with his bigger telescope, was able to record the large and complex nebulosity interlaced between the stars, showing a host of detail which had never been seen by eye observation. A later plate by the Henrys, again of the Pleiades, showed over 2000 stars in the cluster: for comparison, six are visible to the naked eye and in 1610 Galileo's telescope had shown him thirty-six, while in 1866 Rutherford had only been able to photograph seventy-five members of the group. The difference between eye and plate was driven home by a photograph of a small section of the Milky Way. The Henrys found roughly 5000 stars on their plate where older methods, working by hand and eye, had recorded 170.

The photographic plate was also applied to spectroscopy, first by Draper at

Harvard and then by Huggins, and again new data resulted. Huggins's plate of the spectrum of Vega showed the ultraviolet region as well as the visible. It confirmed and extended a formula invented by Johann Balmer, a Swiss schoolmaster, which related the wavelengths of the four visible spectrum lines due to hydrogen. No terrestrial light source showed more than four hydrogen lines, so Huggins's ability to record fourteen was particularly striking. His interpretation of line shifts in stellar spectra as due to the velocity of the stars towards or away from us in the line of sight was extended, particularly by Hermann Vogel, the Director of the new Observatory at Potsdam. The bright line spectra of nebulae and of the Sun's corona both proved to contain lines not matched in any known terrestrial spectrum, and, following the precedent set by helium, two new elements were postulated – nebulium and coronium. Neither element proved real, though it took eighty years to show that the spectrum of "coronium" was due to iron atoms at a very high temperature and very low pressure.

The study of the Sun also benefited from the application of photography. Henry Rowland, the first professor of physics at Johns Hopkins University, recorded the solar spectrum with a precision that made his work the standard set of wavelengths for many years. In true nineteenth-century style he catalogued thousands of Fraunhofer lines not once but twice. The near infra-red spectrum of the Sun was photographed by Captain W. W. Abney after several years' work to develop a plate that would record longer wavelengths than the normal gelatine emulsion. Janssen, in 1885, obtained an outstanding photograph of the Sun which showed the detailed structure of the surface (granulation), ending the need for verbal descriptions such as "rice grains" or "willow leaves". In 1953, the picture was still described as the "best photograph of solar granulation in existence".

The most courageous idea to emerge from photographic and spectroscopic studies was put forward by Norman Lockyer, Abney's colleague at South Kensington and Janssen's old rival. He proposed that at the high temperatures found in stars and in laboratory high-voltage electrical discharges the atoms themselves broke into pieces. A century of successful chemistry had been built up based on the assumption that an atom was unique and indivisible, and therefore this theory was guaranteed a hostile reception. The idea that a calcium atom, for example, would split up when heated was not only revolutionary, it also proved to be correct (though nowadays the process is called ionization, and it is recognized that the atom breaks into a positive ion and an electron). Like many pioneers, Lockyer had a rather abrasive personality, and it is alleged that his only communication with his fellow worker, William Huggins, was via Edward Pickering at Harvard, although their two London observatories were not far apart.

The study of individual stellar spectra had been greatly extended by Vogel, whose researches led to a second classification scheme similar to that proposed by Secchi, while the range of stellar types was extended by the discovery of a new class of very hot stars by Wolf and Rayet in Paris. The alternative approach of applying photography to record large numbers of spectra simultaneously, was pioneered by Edward Pickering at Harvard. Like Secchi before him, Pickering mounted a prism at the objective of a telescope, but instead of a visual examination he photographed the spectra. He recorded in all about a million stars, with several hundred on each plate. The preliminary analysis led to a classification into some twenty groups, lettered A, B, C, D. . . . Later and more thorough work showed that some classes were accidents of exposure, such as class C, in which all the spectrum lines were double, and other classes proved unnecessary – for instance, E and G were indistinguishable – and finally the order was changed. The main change affected class B stars, where it turned out that the hydrogen lines were weak because the stars were very hot, and the hydrogen atoms had ionized. At first it had been assumed that the lines were weak because the stars were not hot enough. The final sequence ran O B A F G K M (generally remembered by the mnemonic "Oh, be a fine girl – kiss me") and proved to be the classification that endures to the present day. The results of the Harvard survey, in which over 200,000 stars were classified, was published as *The Henry Draper Catalogue* since the programme was funded by Draper's widow as a memorial to her husband's pioneering work. Harvard catalogue numbers, prefaced HD for Henry Draper, are used as much if not more than the BD numbers of the *Bonner Durchmusterung*.

In spite of the successes of photography the majority of the big telescopes continued to be designed and built for visual use. Fourteen refracting telescopes of more than 50-centimetre (20-inch) aperture were built in the last twenty years of the nineteenth century and of these only two, one in Paris and the other at Greenwich, were specifically "photographic" telescopes. Two others, the 91-centimetre (36-inch) telescope at the Lick Observatory and a 71-centimetre (28-inch) instrument at Greenwich, managed to compromise. At Lick, Alvan Clark supplied a third lens of 84-centimetre (33-inch) aperture, which could be added to correct the 91-centimetre doublet for photographic use. For the Greenwich telescope Howard Grubb contrived an ingenious design in which the components of the visually corrected doublet lens could be separated and reassembled with the front component turned round to give a combination corrected for photography. The persistence of the visual refractor was partly conservatism and partly to support planetary work and studies of double stars, where photography was no match for eye observation. The eye can pick out the moments of good seeing and can, for instance, discern the separation of a pair of stars separated by less than the seeing blur, or pick out a faint moon close to a bright planet. The photographic plate can accumulate light throughout a long exposure, but must necessarily be limited to a sharpness of definition set by the average seeing during the exposure.

The 1880s saw the construction of four telescopes larger than Clark's 66-centimetre (26-inch) at the United States Naval Observatory: Howard Grubb made a telescope and its 69-centimetre (27-inch) lens for Vienna, Clark made a 76-centimetre (30-inch) lens mounted by Repsold for Pulkovo, the Henrys matched Clark with another 76-centimetre lens mounted by Gautier for the wealthy banker R. Bischoffsheim at Nice, and then Alvan Clark outstripped his competitors with the 91-centimetre refractor for the Lick Observatory. This last project was clearly getting near the limit: the manufacturers of the glass blanks for the lens went bankrupt and the construction of the observatory and the telescope cost so much ($575,000) that there was very little money left to carry out astronomy. The telescope established several precedents, the most important being that it was the first big telescope to be sited at high altitude well away from city lights and pollution. It also marked the arrival of a new firm of telescope engineers, Warner & Swasey of Cleveland.

The two chief research programmes undertaken by the Lick Observatory exploited contrasting aspects of the big telescope's performance. In the first it was used as a high-resolution visual instrument for a survey of bright stars to determine how many of these were single and how many double. The second programme was a study of the motion of the Sun itself, deduced not from the proper motions of nearby stars but from spectroscopic velocity measurements of them. (Stars in the direction of the Sun's motion should appear to be coming towards us.) This programme exploited the unequalled amount of starlight collected by the telescope to

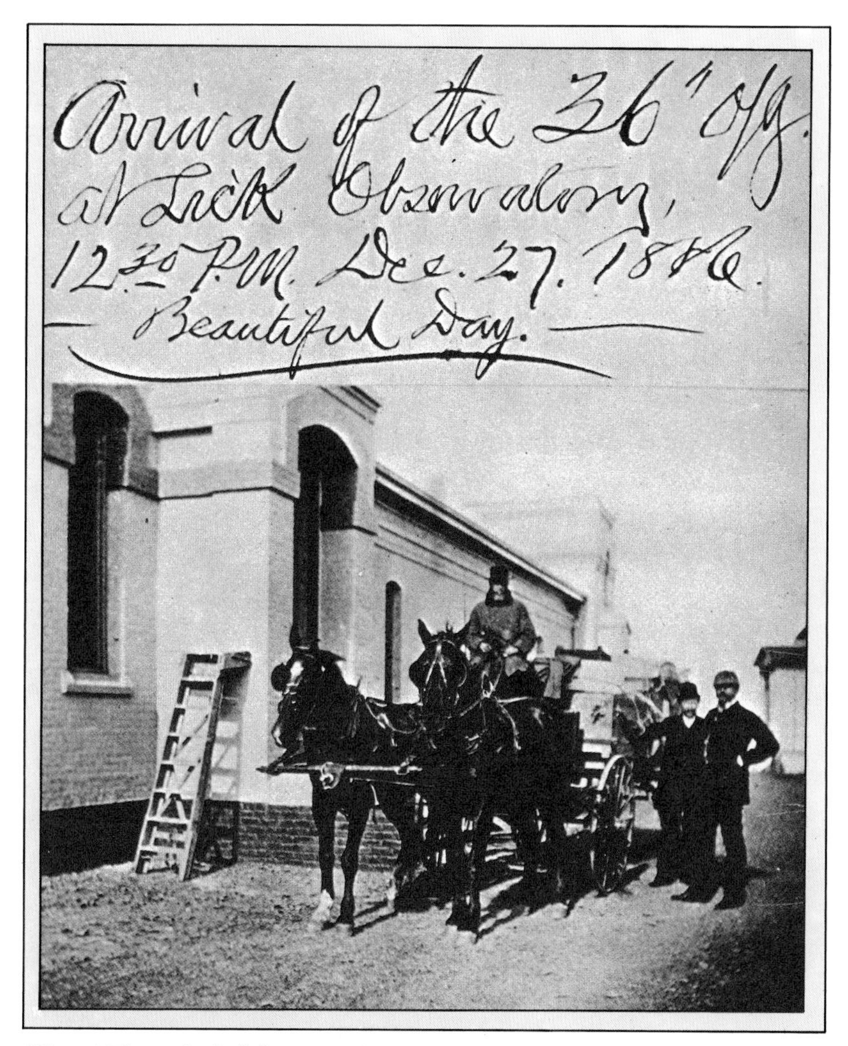

(Above) *The arrival of the 91-centimetre (36-inch) lens at Lick Observatory. No one knows why James Lick bequeathed the funds for the observatory. It is thought that Lick had never seen a telescope and was entirely ignorant of astronomy. The telescope was used, in its early years, for double star studies and the measurement of stellar velocities. In 1888 the lens was mounted in its telescope* (below) *at the Lick Observatory on Mount Hamilton.*

The pier and mounting of the 1-metre (40-inch) Yerkes refractor (above) *completed in 1897, which is still the largest refracting telescope in the world. The entire observatory floor rises and falls to keep the eyepiece within reach and the opening ceremony had to be postponed when the 37·5-tonne floor slipped from its mounting and fell some 15 metres (50 feet).*

photograph spectra in greater detail than had previously been possible.

The success of a large telescope in southern California led to a plan by the University to acquire a rival instrument. Lens blanks just over a metre (40 inches) in diameter were ordered by Alvan Clark & Sons, and Warner & Swasey were asked to design a new and even larger mounting. At this point the University found that it could not raise the funds needed to go through with the project and Clark was left with two glass discs for which the Paris glass-maker Mantois wanted $20,000. A crisis was averted by the emergence of a new phenomenon, a 24-year-old assistant professor at Chicago named George Ellery Hale. Hale was a highly competent and original young astronomer who had an absolute genius for talking wealthy men who knew little about astronomy into paying for giant telescopes. In this case it was Charles Yerkes, a Chicago railroad millionaire, who was cajoled, persuaded, entreated and otherwise bounced into parting with a third of a million dollars to enable the big telescope to be built. Yerkes Observatory is not on an isolated mountain as is the Lick Observatory on Mount Hamilton, partly because Hale had visited both Mount Etna and Pikes Peak and had been caught in crashing thunderstorms on both sites. Instead, the observatory was established at Lake Geneva, 130 kilometres (80 miles) northwest of Chicago. The 101-centimetre (40-inch) telescope proved particularly useful for the measurement of stellar parallaxes. The combination of great focal length, nearly 19 metres (62 feet), and the simple construction of the refractor is ideal for this type of astronomy. The results were of unprecedented accuracy, and it became possible to measure star distances out to about 75 parsecs, sixteen million times the Earth–Sun distance.

The discovery of the companion to Sirius using Alvan Clark's first big lens roughly marked the point at which the big refractors took over from the reflectors of Rosse, Lassell and Nasmyth. In the last forty years of the nineteenth century refracting telescopes supplied the observations needed to put our knowledge of the solar system and the nearby stars on a firm foundation. The distances to the nearby stars, their motion and chemical constitution were all determined using refracting telescopes. The motions of the planets and their satellites were fully determined and well matched by theory, the only exception being the discrepancy in the motion of the perihelion of Mercury. The study of the planets themselves had reached a limit set by seeing, and little more was achieved until observations were made from above the atmosphere by spacecraft almost a century later.

The Yerkes Observatory. Charles Yerkes originally agreed to pay for a telescope lens and mounting. Eventually, George Hale talked him into parting with $349,000 – enough (in 1890) to pay for the whole observatory.

CHAPTER NINE

GLASS MIRRORS ARE BETTER THAN GLASS LENSES

Reflectors are not all bad;
and the refractor reaches its limit:
1898–1938

"The reflector is so seriously influenced at times by air currents and changes of temperature as to be an instrument of moods and Dr Common has accordingly compared it, somewhat ungallantly, to the female sex."

The Great Star Map,
by H. H. Turner, 1912.

The reflecting telescope was not completely ignored during the late Victorian era, though it rarely competed on equal terms with the big refractors, as the professional astronomer had a strong preference for the demonstrated success of the refractor. The dominance of the lensed telescope was increased not only by its own merits but also by the well-publicized failure of two attempts to build large reflectors, first for Melbourne in 1870 and then for Paris in 1877.

The critical steps towards a reflecting telescope that could surpass the Lick and Yerkes refractors were taken only a few years after the completion of Lord Rosse's little-used giant. Two men, Steinheil in Munich and Foucault in Paris, picked up a technique which had been demonstrated at the Great Exhibition in London in 1851 and reinvented by a German chemist, Justus von Leibig, a few years later. This was the art of producing by chemical manipulation a highly reflecting film of silver which would coat any piece of glass used in the process and turn it into a mirror. The new type of silver layer, unlike the seventeenth-century mercury amalgam "silvering", was very thin and, with care, very uniform. The result was that a glass disc polished to make a concave mirror could be silvered on the front surface to make the glass highly reflective without losing the precision with which it had been polished.

The freedom that came from this simple process was considerable. First of all, optical workers were well used to grinding and polishing glass, so there was no need to learn the esoteric art of polishing speculum metal. Second, although glass was fragile and difficult to cast, it was much easier to handle than the wilfully perverse speculum alloys. Third, and most important, both silver and speculum metal tarnish in use, but while the silver layer can be dissolved away and replaced with a fresh coating, a speculum alloy surface has to be repolished. Any error in this repolishing either ruins the mirror or makes it necessary to generate a new and accurate surface by starting the whole polishing process again. Steinheil built the first silver-on-glass telescope in 1856, a modest instrument of 10-centimetre (4-inch) aperture. The next year Foucault polished and silvered a 33-centimetre (13-inch) mirror which was mounted by F. W. Eichens for use at the Paris Observatory. The telescope is noteworthy for two separate developments: the new silver-on-glass mirror and a new equatorial mounting with the telescope on neither a "German" cross axis nor an "English" yoke, but held in a short, stiff fork.

In spite of the success of this pioneering effort and the breadth of his other re-

THE FOUCAULT TEST

This is a very simple method for examining the image quality of an optical surface (or system) and indicating those parts of the surface that are defective. The test can be applied to any optical system in which a point object is (or should be) re-imaged as a point. It is illustrated below for a simple concave spherical mirror.

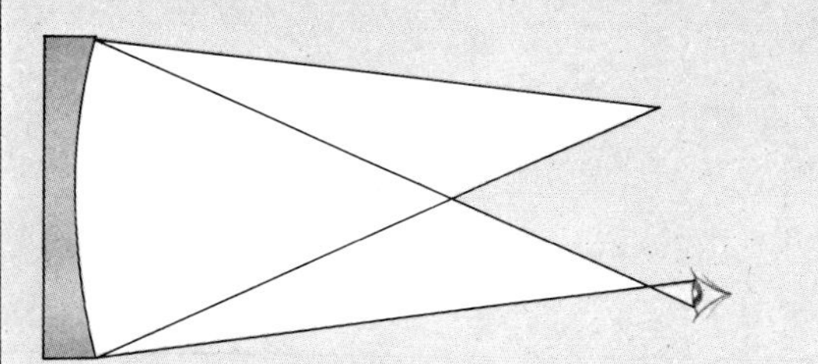

1 The starting point (above) *is to use the mirror to give an image of a small light source, usually an illuminated pinhole. The observer then places his eye near the focused image and looks not at the image but at the mirror. Light from the pinhole reaches the eye from every part of the mirror, which therefore appears uniformly bright to the observer* (above, right).

2 The next step (below) *is to place a knife edge near the final image and slowly push it across the field of view. The blade will intercept the light from the mirror and, if it is offset from the focused image, will cut off the light from one side of the mirror before the other* (below, left). *The mirror will then appear part bright, part dark, and the dark edge will move steadily across the field* (below, right).

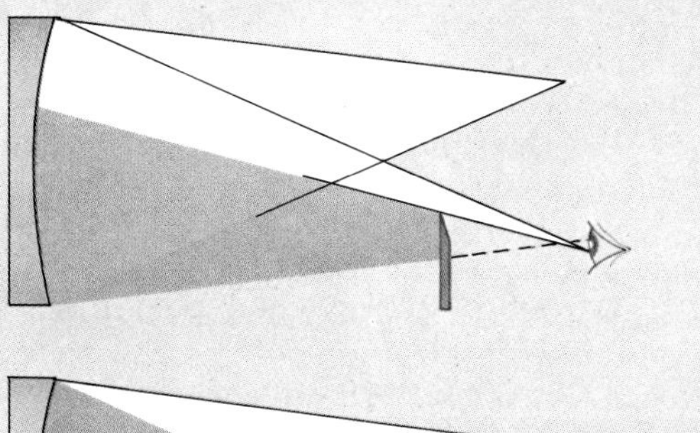

3 If the knife edge is set precisely at the focus (above), *and the mirror is perfect, the blade will cut off the light from the whole aperture simultaneously and the field of view will switch rapidly and uniformly from bright to dark. With a little care an intermediate condition, in which the mirror is a uniform grey, can be seen* (above, right).

4 If the mirror is not a perfect sphere but has some hills or hollows, these deflect the light so that it does not come precisely to the focus. The knife edge will then intercept these rays before or after reaching the focus produced by the main part of the mirror (below, left). *In this case the defective part of the mirror will either darken before the main cut-off or remain bright until the knife edge is further advanced. Thus one side of an unwanted hump will appear noticeably dark and the other light* (below, right).

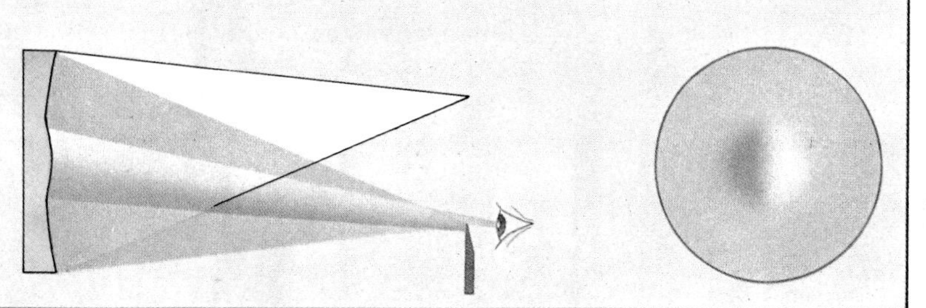

searches, Foucault is best remembered for another contribution to the development of astronomical optics. From the point of view of the telescope-maker, the invention of the "Foucault test" was, and is still, of primary importance. This test allowed any optician to see, for the first time, which part of a defective lens or mirror was responsible for a poor image and in addition whether the defect was due to a local high spot or a hollow in the polished surface. The apparatus needed was of trivial cost – a light source, a pin hole in an opaque screen and a sharp knife edge for which a present-day amateur mirror-maker uses a razor blade. With this simple equipment it is possible to measure unevenness in the mirror surface due to polishing defects whose height or depth is much less than a wavelength of visible light (0·5 micrometres).

Soon after Foucault and Eichens had completed their 33-centimetre telescope, William Lassell decided to build a bigger instrument than the 60-centimetre reflector with which he had discovered Neptune's moon, Triton. Since Lassell had behind him the experience acquired while building his earlier metal mirror telescope he chose to continue with speculum mirrors and cast a pair of 1·2-metre (48-inch) diameter metal discs. The completed telescope was set up in Malta and used for studies of faint planetary satellites and of nebulae too far south to have been seen by William Herschel. It was the largest equatorially mounted reflector yet built and

Lassell's 1·2-metre (48-inch) telescope, erected in Malta in 1861. Observations were made at the Newtonian focus. Two additional assistants were required, one to work the three winches that carried the observer close to the eyepiece and the other to act as the telescope drive.

(Above) *The first attempt at the casting of the mirror for the Melbourne telescope. The third attempt proved to be successful but was the last large mirror to be made with the speculum alloy.*

(Below) *The 1·2-metre (48-inch) Melbourne telescope of which G. Ritchey wrote: "I consider the failure of the Melbourne reflector to have been one of the greatest calamities in the history of instrumental astronomy."*

gave images of high quality. None the less it was clearly rather a disappointment. When Lassell returned to England after three years in Malta he never bothered to re-erect the telescope.

The three-year-long series of observations with Lassell's big reflector came at a crucial stage in the history of the ill-starred Melbourne telescope. The original plan to provide a large telescope in the Southern Hemisphere was formulated in 1849, soon after the reports of the observations made by Lord Rosse and his colleagues were published. The Royal Society and the British Association recommended "the establishment of a telescope of very great optical power" and set up a committee including Lord Rosse, Nasmyth, Lassell, Airy, Adams, John Herschel and the entire council of the Royal Society. Three sites were considered – the Cape of Good Hope, Tasmania and Australia. The advantages lay with the Cape, where there was already an observatory, and Tasmania, where transported convict labour could be acquired to build an observatory at no additional cost to the Exchequer. Similarly, three types of telescope came under review – a 75-centimetre (30-inch) refractor from Merz, and a 1·2-metre (48-inch) or a 1·8-metre (72-inch) reflector from Thomas Grubb. The refractor was expected to cost £20,000, the larger reflector £15,000 and the smaller £5000. The committee picked the cheapest solution, but the British Treasury was by then paying for the Crimean War and even the 1·2-metre reflector failed to win the necessary capital.

Nothing at all happened for the rest of the decade, but in 1862 the citizens of the Australian colony of Victoria announced that they were prepared to pay for a telescope themselves. The Australians asked the advice of the Royal Society and the committee was reconvened. They dusted off the old 1·2-metre Grubb design and decided that it was just what they wanted. The new silver-on-glass mirrors were considered and rejected as too little tried to be used with confidence. William Lassell offered his 60-centimetre telescope as a gift, but it was rejected as too small; later he offered his 1·2-metre and that too was turned down, on the grounds that it was too cumbersome. The Australians went ahead with Thomas Grubb's telescope and in 1868 it was assembled at Grubb's Dublin factory and ready for inspection by a sub-committee comprising de la Rue, Lassell and Robinson (Lord Rosse's chief observer). The result of the inspection was a two-to-one view that the new instrument was a masterpiece of engineering and "perfectly fit for the purpose for which it was designed". William Lassell, who had given up using his own 1·2-metre reflector

(Left) *The 1·2-metre (48-inch) aperture silver-on-glass reflector of the Paris Observatory. It was sheltered by a removable building when not in use and it used Newtonian optics in contrast to the Cassegrain layout of the Melbourne instrument. Otherwise the telescopes differed only in the degree of their failure. The Melbourne telescope was "a calamity" while that at Paris was only "comparatively useless".*

George Ellery Hale, 1868–1938 (below), *was born in Chicago and, when a schoolboy, took up astronomy as a hobby. Later, while an undergraduate, he was allowed by Edward Pickering to use the facilities at Harvard College Observatory. Hale's activities included setting up the International Astronomical Union and the National Science Foundation.*

in 1864, was satisfied but less enthusiastic about the new telescope.

When the instrument was finally set up at South Yarra as the proud and prestigious "Great Melbourne Telescope" it proved a long-drawn-out and much publicized failure. There were three chief reasons for this, two of which were strongly linked: the optical design and the lack of a dome. Thomas Grubb had designed the telescope as a Cassegrain, with the upper end of the telescope tube as light as possible. The advantage of this was that the eyepiece, at the back of the primary mirror, moved only a modest distance as the telescope tracked across the sky. The disadvantage was that a lightweight "top end" meant a small secondary mirror and so a large secondary magnification. It followed that the telescope was very sensitive to flexure and vibration, as any motion of the tube resulted in a greatly magnified motion of the image. This vibration-sensitive instrument was then set up not in a dome but in a building with

a removable roof, so that in use it was completely exposed to the slightest puff of wind. The third reason for failure was that while the design of the complete system (telescope plus building) was unsatisfactory, the mirrors were an utter disaster. The speculum mirrors were initially manufactured in Dublin and shipped to Australia with a protective shellac coating on the polished surface. On arrival, it was found impossible to remove the shellac without wrecking the surface. The mirrors had to be repolished. Shipping the mirrors back to Dublin was time-consuming and impracticable, so the Director of the Observatory, R. J. Ellery, set about learning how to polish very large speculum mirrors.

It was not surprising that Ellery never achieved the necessarily near-perfect technique needed. Rosse, Lassell and Grubb were all men with a gift for construction rather than observation, and had all dedicated a large fraction of their lives to optical work, starting with small mirrors and working up to larger ones over tens of years. Thomas Grubb had been polishing speculum mirrors for thirty-two years when he completed the Melbourne telescope, and poor Ellery had to try and match his skill in a few months. He also had to defend his observatory against the critics of the project, who had seized on the fact that the telescope had been "a masterpiece of engineering" in Dublin but did not work in Australia, and had mounted a long and acrimonious public attack on the observatory and its staff.

While Ellery struggled with the last large speculum mirrors ever supplied for a telescope, the first large silver-on-glass mirror was being polished for the Paris Observatory. This mirror was for a 1·2-metre (48-inch) aperture telescope that had only one feature in common with the Melbourne telescope – it proved equally unsuccessful. Here the faults of the instrument seem to have been more evenly distributed, since in later years both the mirror was repolished and the mounting was scrapped and a new one built.

In spite of these failures, which, to say the least, led funding committees to show a great lack of confidence in reflecting telescopes, a few workers persisted with the design. A mirror has the enormous advantage over a lens in that it neither suffers from chromatic aberration nor does it absorb the ultraviolet light so vital in photography. The disadvantages, as learned from the Melbourne and Paris telescopes, are that the reflector is very sensitive to the effect of misalignment of the mirrors, to flexure and vibration of the telescope tube and to temperature effects. In contrast, the refracting telescope is much more tolerant of such troubles. The first team that was prepared to persevere with a large-aperture reflector was in England with A. A. Common as both telescope engineer and astronomer and George Calver as mirror-maker.

The Calver–Common telescope, based on a 91-centimetre (36-inch) silver-on-glass mirror, was mounted in an equatorial fork and used as a photographic telescope. The chief innovations in Common's telescope were the achievement of a smooth drive by relieving the bearings of almost the entire weight of the telescope, and the invention of an adjustable plateholder. The bearing load was diminished by submerging a hollow steel float in mercury, while the plateholder had an eyepiece and crosswire attached so that a star just off the edge of the plate could be watched and, if it drifted away from its starting position, it could be brought back by moving the plateholder. The end result showed that it was possible to build a telescope that was sufficiently smoothly and accurately driven to allow very good photographs to be taken: one of the Orion nebula – a 1½-hour exposure – won the Royal Astronomical Society gold medal.

The next step in the rehabilitation of the reflector was marked by the delivery in 1885 of a 50-centimetre (20-inch) aperture instrument for Isaac Roberts. Roberts was a tough customer who specified a very high performance and then required Howard Grubb (Thomas Grubb's son and successor) to meet the specification. The young Grubb was no doubt anxious to restore the family firm's reputation as credible builders of reflecting telescopes after the Melbourne affair, but it took two years of additional work after delivery to satisfy Roberts that the new instrument was acceptable. Roberts's persistence paid off in that he was able immediately to obtain a photograph of the great nebula in Andromeda, a photograph which showed for the first time that it too had a spiral structure. The success of the telescope may be measured by the fact that Grubb managed soon after to sell a telescope (albeit a refractor) to Melbourne Observatory, a sale that must have had to overcome much customer resistance.

The critical steps that persuaded others to take up the construction of large mirrors resulted from a combination of English and American efforts. Although Common attempted the construction of a 1·5-metre (60-inch) telescope, it proved beyond his facilities and was still incomplete when he died in 1903. In 1881 Common had published a description of his 91-centimetre (36-inch) reflector, introducing the paper with a list of ten conditions vital to success. The last of these was "a suitable locality for the erection of the telescope". The telescope was set up in suburban

London and Common remarks about his site requirement, "with this condition we are not now concerned". None the less a telescope built by Common became the first large reflector on a good site when another British amateur, Edward Crossley, bought a similar instrument and presented it, in 1895, to Lick Observatory in California. Howard Grubb repolished the mirror and, on arrival at Lick, J. E. Keeler, one of the six staff astronomers, stiffened the mounting. The result of a good photographic telescope on a good site was epoch-making. Keeler found on his plates, in addition to stars, an immense number of very small spiral nebulae. Astronomers now had to admit the possibility that these spirals might be as big as the great nebula in Andromeda and only appear small because they were at enormous distances from the solar system. If this was accepted, even if only as a working hypothesis, then the universe was vastly bigger than had previously seemed possible. Further work on these faint systems would, however, have to wait for the development of larger telescopes.

By the turn of the century George Hale was in a position to bring together several arguments in favour of large reflectors. His experience at Yerkes showed the refractor to have reached a limit: the big lenses were so thick that they absorbed a significant part of the light they collected and they were so heavy that support only at the edges was barely adequate. In contrast the big mirrors could be evenly supported right across the rear of the blank, and the silver coating reflected well in the photographically important violet and ultraviolet, where flint glass was particularly absorbent. In addition even the traditional "achromat" was not entirely achromatic. The doublet lens of the big telescopes reduced, but did not entirely eliminate, the variation of focal length with colour. The Lick 91-centimetre refractor, for instance, still showed a variation of focus of several centimetres in the overall focal length of 17 metres (56 feet). Hale was also fortunate in having on the staff at Yerkes one of the few American optical workers with experience of polishing and silvering fairly large mirrors – George Ritchey, who was to assume much the same reputation for polishing twentieth-century mirrors as Alvan Clark had held for nineteenth-century lenses.

The end result of all this evidence, together with a 1·5-metre blank that Hale had already persuaded his father to buy, was a grant from the Carnegie Institution to allow him to build a 1·5-metre reflecting telescope. This was to be sited on Mount Wilson in southern California, where a new observatory had just been estab-

An aerial view of Mount Wilson Observatory in southern California. On the left are three solar telescopes, first the horizontal Snow telescope and then the two vertical instruments. In the centre and on the right are the domes of the reflectors.

lished with Hale as its director. Hale's confidence in the reflector was such that he persuaded his friend J. D. Hooker to fund an even bigger telescope before Ritchey had, in fact, finished the 1·5-metre. This was at first to be a 2·1-metre (84-inch), but Hale persuaded Hooker to provide more money to allow the construction of a 2·5-metre (100-inch) – an enormous leap beyond Common's rebuilt 91-centimetre (36-inch), the largest reflector actually working effectively on astronomical research.

The 1·5-metre telescope, designed by Ritchey, drew heavily on the Common reflector at Lick. The telescope was fork mounted and the bearing load was relieved with a mercury float. The main difference, reflecting Keeler's experience, was a much stiffer truss system to support the secondary mirror.

The principal reason for building the Mount Wilson telescopes was to research into the evolution of the stars. This rather Darwinian approach to astronomy grew from the recognition that the Harvard classification – O B A F G K M – represented a temperature sequence. Norman Lockyer had proposed that a star began life from an aggregation of dust and gas which heated up as it contracted, and then when contraction was nearly complete, began to cool down again. A Danish astronomer, Ejnar Hertzsprung, had shown from a study of the brightness of stars whose distances were known by the parallax method, that there were two types of cool red stars. The brightness difference must result from a difference in size, so he called the brighter red stars "giants" and the fainter "dwarfs". Soon after the completion of the 1·5-metre telescope this work was extended by Hertzsprung and by Henry Russell of Princeton. The first step in their research was to work out the intrinsic brightness of those stars whose parallax was known. They expressed this in terms of "absolute magnitude", which is a star's magnitude as it would appear if the star were positioned a distance of ten parsecs from Earth. It allows instant comparison of one star's intrinsic brightness with another. When absolute magnitude was plotted against the spectroscopic classification the plot was in effect one of brightness against temperature. The surprise was that the vast majority of points on the graph (94 out of the 100 nearest stars) formed a diagonal line. This was called the "main sequence". The minority of stars which fell off this line were either Hertzsprung's red giants or a new type, called a white dwarf. Most of the sheet of graph paper was, however, empty of stars. The conclusion drawn from this graph is that the

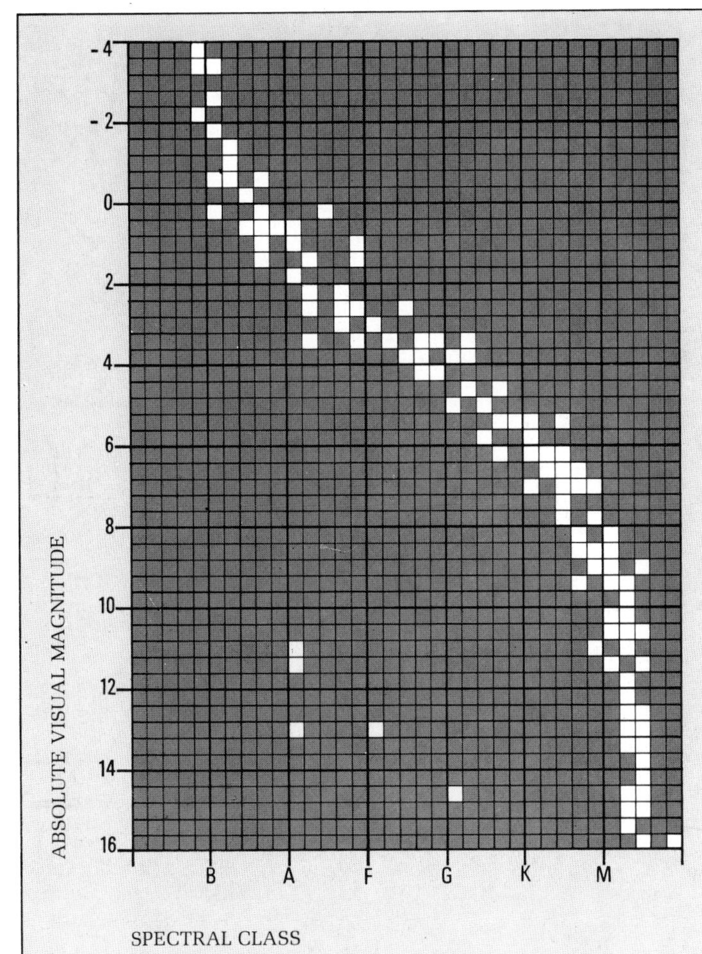

THE HERTZSPRUNG-RUSSELL DIAGRAM

This displays the relationship between the energy radiated by a star and the star's temperature. The diagram (left) shows the hundred nearest stars – most of which fall on a diagonal line, "the main sequence". Three facts are required for each star plotted: the magnitude, parallax and spectroscopic class. The first two determine the "absolute" brightness of the star and the latter is directly related to its temperature. The set of typical stellar spectra (below) range from very hot (class B at 25,000°C) to the much cooler class M at 3000°C.

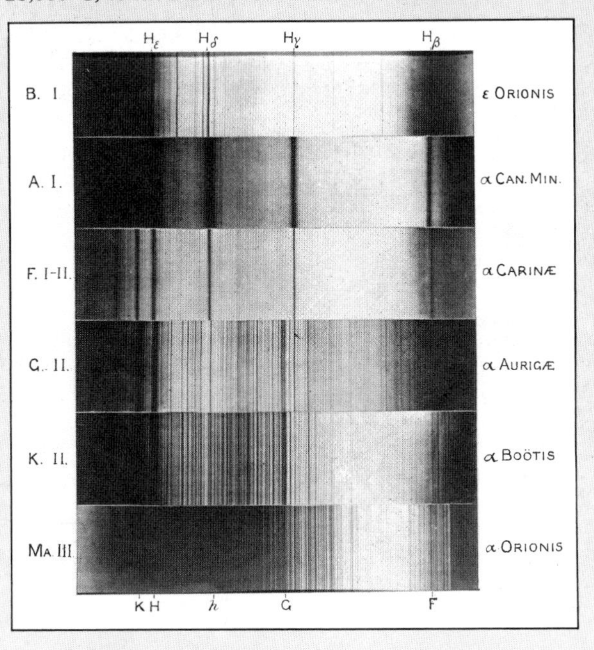

The 2·5-metre (100-inch) Hooker telescope, which was completed in 1918. The complete instrument weighs a colossal 100 tonnes and the majority of this load is carried by the two mercury floats enclosed within the big drums at each end of the polar axis. The telescope is driven by means of a 5-metre (200-inch) diameter gear wheel at the lower end of the axis; the motive power is derived from the slow fall of a 2-tonne drive weight. The yoke mounting of the instrument was chosen by Pease, Hale's colleague, for its great rigidity and strength. The disadvantage of the system is that the telescope cannot swing over to point along the polar axis, and so cannot observe the region of the sky near the Pole Star.

majority of all stars at the same temperature are the same brightness, and more surprising must all be the same size. There is not a range of very small – small – medium – large – very large stars all equally hot. A star with the same spectral class as the Sun is not only the same temperature but it is also the same size. The equivalent discovery for a human population would make nearly all Red Indians 150 centimetres (4 ft 11 in.) tall, all Japanese 170 centimetres (5 ft 7 in.) tall and nearly all white men 190 centimetres (6 ft 3 in.) tall with a few Red Indian giants and some European dwarfs.

The discovery of the main sequence brought an enormous sense of order to the study of the stars, though it did not solve the problem of where stars obtained the energy needed to keep on radiating for millions of years. Russell was forced to postulate two types of energy-producing matter – "dwarf-stuff" to keep stars on the main sequence shining and "giant stuff" for the enormous cool stars not on the main sequence.

A second ordering of the starry universe was achieved by J. C. Kapteyn at Groningen, who repeated Herschel's study of the distribution of the stars, using photographic techniques to get accurate star counts in selected sample areas of the sky. Kapteyn had the advantage that since Herschel's time two more components of our galaxy had been recognized: interstellar dust had been discovered by E. E. Barnard in 1909, and interstellar gas by J. Hartmann in 1904. Kapteyn was therefore able to distinguish between dark regions of the sky where there were no stars and dark regions such as the "Coalsack" in the Milky Way, where the view of distant stars was blocked by clouds of impenetrable dust. The resulting picture of the galaxy, "Kapteyn's universe", was a round, cushion-shaped aggregation of stars, densely packed at the centre and thinning away as empty space outside the galaxy was reached. This island universe was not very different in shape from Herschel's "pocket watch" and about three times the size. It was 17,000 parsecs across and about 3400 parsecs thick, with the Sun close to half-way between top and bottom; and only 650 parsecs from the centre. This universe held fifty billion stars and had an estimated mass eighty billion times that of the Sun. Kapteyn took no account of the less numerous constituents of space, the globular clusters of stars or the spiral nebulae, neither of which existed in significant numbers compared to the stars.

The 1·5-metre telescope at Mount Wilson contributed data to both Hertzsprung and Russell's graph (now called the Hertzsprung-Russell or HR diagram) and provided observations for Kapteyn, who was a regular visitor. The telescope proved to justify Hale's hope to see further and more clearly than ever before, and in other work the telescope was pushed to its limits. Eleven-hour exposures showed stars of twentieth magnitude, and some spectra were photographed using exposures that occupied several consecutive nights in order to allow comparison of solar and stellar spectra in detail. The quality of the new results and their quantity forced a shift in the centre of gravity of astronomy. The clear mountain skies of California were vastly better for astronomy than the clouds above the industrial cities of Europe. While Hale was building a new observatory on Mount Wilson, Lockyer in England was irascibly overseeing the closure of the observatory at South Kensington. Coal fires and gas street lights so polluted the London sky that his second observatory in the country at the resort of Westgate-on-Sea was no longer worth keeping open.

The 2·5-metre (100-inch) Hooker telescope had a great deal in common with the 1·5 metre: the proportion of mirror diameter to focal length was kept at 1:5, the bearing load was relieved using a steel-in-mercury float, the tube was a similar latticework and the plate glass mirror blank was cast by the same French company. The glass disc was less than perfect, but before it could be replaced by a better one the Great War began, so Ritchey and Hale decided to go ahead with the polishing using the blank they had. The main innovation was to return to Jesse Ramsden's "English" yoke to carry the telescope on its polar axis. This gave the much stiffer mount needed for a telescope weighing 100 tonnes, but meant that the instrument could never see the stars near the celestial north pole. The telescope was completed in November 1917, after a series of crises that gave Hale more than one nervous breakdown. The first night that the telescope was fully functioning it was turned to point at Jupiter and Hale took the first look through the eyepiece. The result was horrifying – a great smudge of light with half a dozen different images of the planet in a confused jumble. It was fervently hoped that this might be just a temperature effect. The dome had been open during the day and the 4-tonne mirror was still cooling from the afternoon warmth to the cooler night-time temperature. The only thing to do was wait, and Hale and his colleague, Walter Adams, paced the observatory floor until 2.30 am, when they tried again. Jupiter was by then too far to the west, so they used Vega as a test object. The image was a brilliant single round speck: the telescope a success.

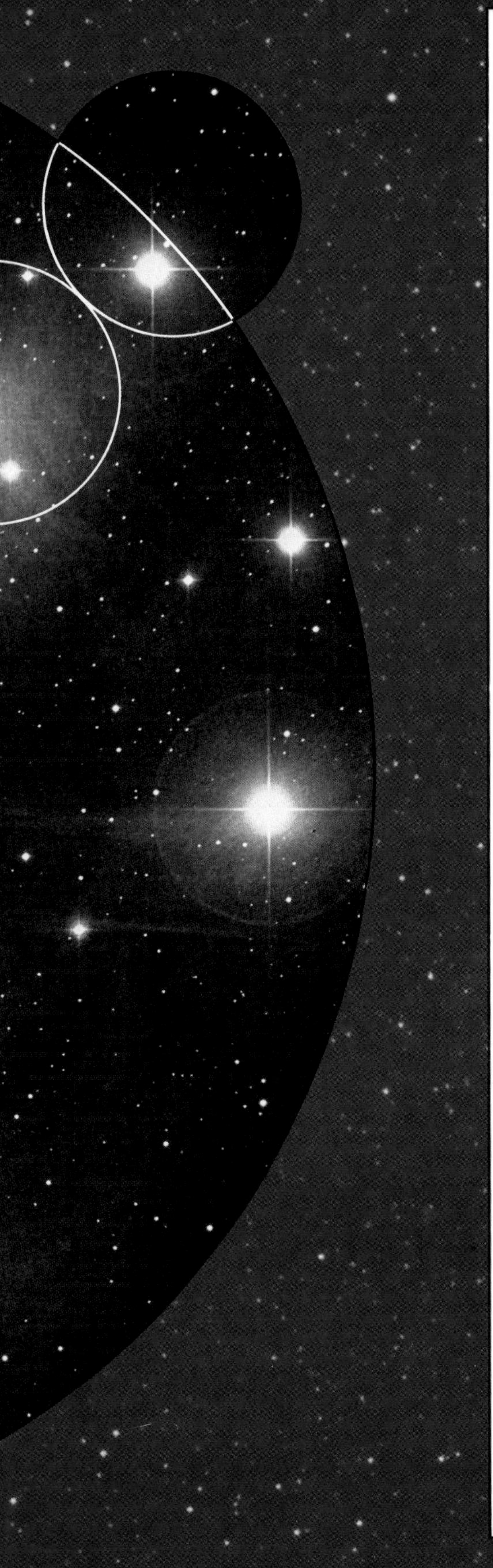

FIELD OF VIEW

Improvements in telescope design have been aimed at both obtaining a larger field of view so that a greater portion of the sky can be viewed at one glance and achieving a greater resolution – the ability to separate two adjacent objects. Here, with the Pleiades as the object – a cluster of hot, bluish-white stars in the constellation of Taurus of which only the six brightest members can be seen with the naked eye – a comparison of three classic telescope's images has been illustrated. Galileo's most powerful refractor had a field of view of 17 arc minutes (smallest circle) and was able to resolve no more than five stars at a time. (As a guide, the full Moon, top left, has an angular diameter of 30 arc minutes.) The Ritchey-Chrétien design provides a field of view of up to 1°40′ (large circle) and can resolve about thirty members of the cluster plus several thousand background stars. The 1·2-metre (48-inch) Schmidt telescope has a field of view of 6° and produces a photographic plate containing images of some 150 members of the cluster plus more than 20,000 background stars. Note: the images have been magnified – the blue-and-black background image spread across these two pages, which has been taken from a Schmidt Sky-Survey plate of the Pleiades measuring 35 centimetres (14 inches) across, represents a field of view of only 3°, half the telescope's normal view (see small diagram below, right, D).

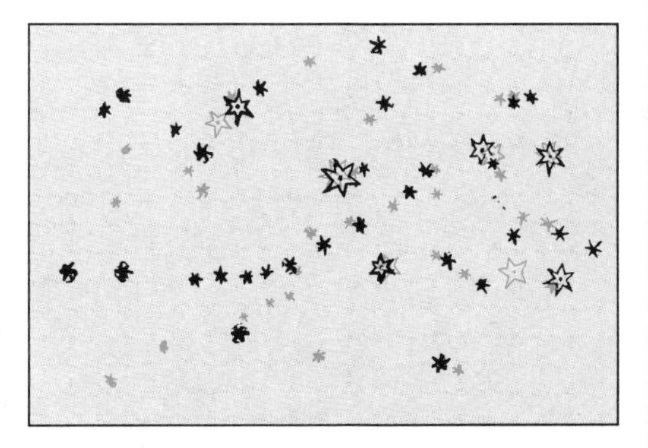

Galileo's chart of the Pleiades (above), *published in* The Starry Messenger *in 1610, shows thirty-six stars in the cluster. The true positions of the stars have been superimposed in blue. The evident errors of Galileo's chart are largely due to the fact that he could never see more than a tiny fraction of the whole pattern at one glance (see left). The fields of view of the three telescopes, A, C and D, and the size of the Moon, B, are shown below.*

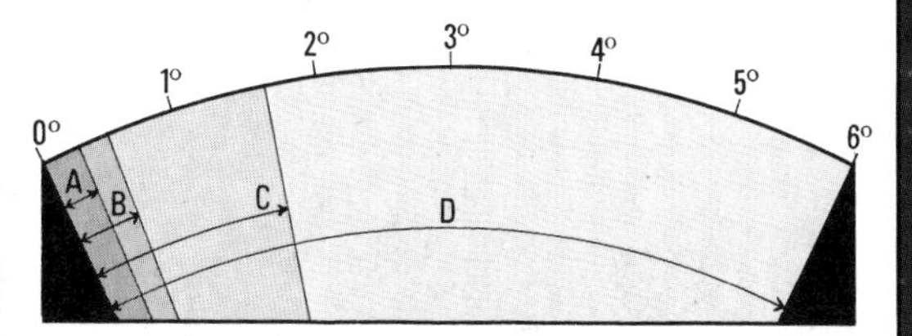

Henrietta Leavitt (below) *was one of a team working, under the supervision of W. and E. Pickering, on the stars in the Magellanic clouds. Her discovery that the period of some variable stars is directly related to their brightness is of fundamental importance in establishing the size of the universe.*

Hale matched the unequalled quality of his two big telescopes with two outstanding astronomers and with the two major "non-stellar" problems; the globular clusters and the spiral nebulae. Harlow Shapley worked on the clusters, usually with the 1·5-metre, and Edwin Hubble on the nebulae, using the bigger Hooker telescope. The clusters and spiral nebulae were both classes of object with curious distributions in the sky. While the stars were concentrated in a band, the Milky Way, the spirals did the opposite and were much more common in the directions farthest from the Milky Way. The globular clusters were even more irregular in their distribution: almost half of the hundred or so known lay in one-tenth of the sky around the constellation Saggitarius, and none at all in the opposite direction. The key which unlocked both puzzles proved to be the same. It was one of Edward Pickering's five classes of variable stars, the Cepheids.

A study at Harvard by Henrietta Leavitt had shown that the maximum brightness of a Cepheid variable was directly related to the period of its fluctuation – the slower it oscillated, the brighter it was. This meant that the Cepheids were like standard candles: if the period could be measured the absolute magnitude could be deduced and, by comparing this with its apparent magnitude in the sky, the distance to the star could be found. The problem was that one Cepheid of known distance was needed to fix the whole scale and no such star is near enough to give a measurable parallax. Ejnar Hertzsprung had tried to determine the distance of the nearby Cepheids by reversing a previous method for determining the motion of the Sun: now that the Sun's motion was known from nearby ordinary stars, Hertzsprung could deduce Cepheid distances from their proper motions and spectroscopic velocities. Shapley extended this analysis and thereby found the distance of the variable stars in a dozen or so clusters. Then other patterns began to emerge – the brightest stars in each cluster were much the same absolute magnitude from one cluster to another, and the clusters were all much the same size, so the distance measurements could be extended to cover almost all the clusters. Shapley found that the clusters occupied a flattened spherical space 80,000 parsecs in diameter. He made the giant step of assuming that the centre of the galaxy was in the same place as the centre of the cluster distribution. This controversial picture made the galaxy vastly bigger than before and moved the Sun from near the centre to the outer suburbs.

The possibility that the galaxy was as vast as Shapley had claimed meant that the spiral nebulae were almost certainly

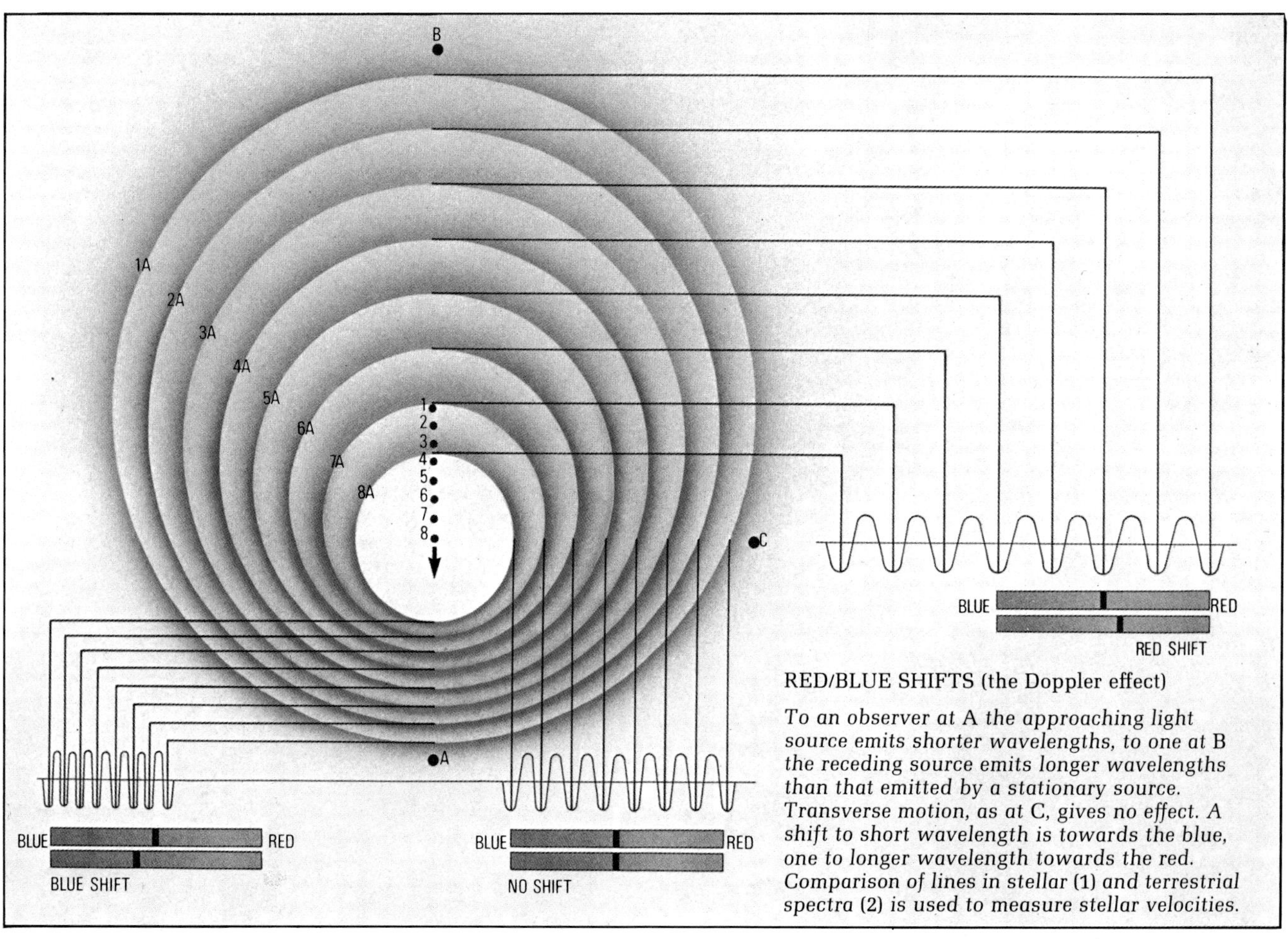

RED/BLUE SHIFTS (the Doppler effect)

To an observer at A the approaching light source emits shorter wavelengths, to one at B the receding source emits longer wavelengths than that emitted by a stationary source. Transverse motion, as at C, gives no effect. A shift to short wavelength is towards the blue, one to longer wavelength towards the red. Comparison of lines in stellar (1) and terrestrial spectra (2) is used to measure stellar velocities.

within its boundaries. The evidence that this was not the case was principally gathered (at the same observatory) by Edwin Hubble, supported by Milton Humason. Hubble first classified the nebulae by shape – spirals of differing openness, nebulae with no signs of spiral arms at all and so on. Then, while Humason measured their velocities spectroscopically, Hubble observed the brightness of the variable stars in the nearest large nebulae and the sizes and brightness of the smaller and more distant nebulae in order to work out their distances. The results, published in 1929, were dramatic. By observing Cepheid variable stars in the Andromeda spiral, Hubble showed that it lies well beyond our galaxy; the fainter, and presumably more distant, spirals must lie yet farther away. Humason's spectra revealed a new law, now known as Hubble's Law: the farther away a galaxy lies, the greater its speed of recession from us (shown by a "red shift" in the spectrum). The universe now acquired a new unit – a million parsecs or Megaparsec – and, as the velocity of expansion could be worked backwards to find a start time, an age as well. The "nearby" cluster of galaxies in Virgo was measured to be 2·2 Megaparsecs distant and one in Ursa Major 72 Megaparsecs from us. Hubble estimated that the faintest galaxies on his plates would prove to be about 150 Megaparsecs from the Earth. The most curious aspect of the work was that our own galaxy appeared to be considerably bigger than any other.

The success of the Mount Wilson telescopes led to the construction of a whole series of big reflectors in the interval between the two world wars. No one matched the 2·5-metre, but a series of instruments in the 1·8 to 2·1-metre range were built by Warner & Swasey, Grubb (now Grubb-Parsons) and by a German company, Zeiss, who had only recently emerged as large telescope-makers. The instruments followed, in general, the design pioneered in 1918 by a 1·8-metre for the Dominion Astrophysical Observatory near Victoria, British Columbia, where the telescope was offset from a single polar axis and balanced by a massive counterweight. The silver-on-plate-glass mirrors, braced latticework tubes and Cassegrain optics were common to all. The main structural change was the disappearance of mercury floats, superseded by ball bearings, and the optical system became significantly more complex. The new large-aperture mirrors meant firstly that telescopes were large enough to carry plate-holders or small spectrographs at the focus of the primary mirror (the prime focus) and, secondly, that they gathered so much light from bright stars that they could exploit auxiliary instruments which were too heavy to be carried on the telescope tube itself. The solution was to use interchangeable components in place of the old Cassegrain secondary mirror. The secondary could be exchanged for another of different curvature to give a beam of starlight that, after two or three more reflections, came to a focus at the foot of the polar axis. This arrangement, called a coudé focus after the French word for elbow, made it possible to use spectrographs several metres in length. These gave extremely detailed spectra that were used to study such phenomena as stellar magnetic fields and stellar rotation. At the other extreme the secondary mirror could be removed altogether to give a system with only one reflection (at the primary mirror) and therefore of maximum efficiency, allowing the telescopes to search for ever fainter and more distant stellar objects.

THE COUDÉ FOCUS

Large spectrographs are too unwieldy to mount on a telescope. One must thus use additional mirrors to bend the light beam from the primary mirror to reach a fixed focus F. Here mirror 3 reflects the light along the declination axis.

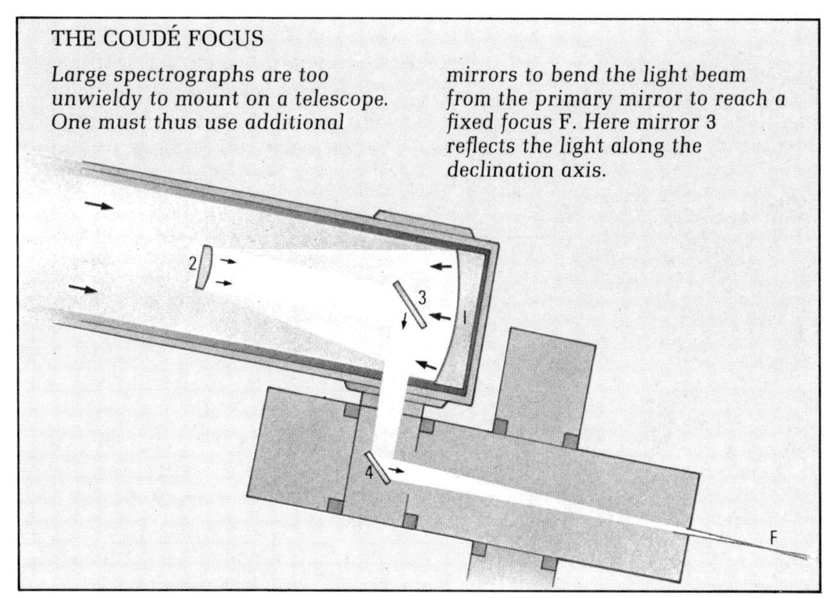

The interpretation of the data gathered by the new generation of big telescopes was much assisted by the developments in atomic physics. Scientists' understanding of the spectra of atoms and simple molecules was enormously advanced by the emergence of the "quantum" theory in the 1920s, a theory that describes both the internal structure of atoms and the bonds that hold atoms together as molecules. A whole range of laboratories contributed, including both "pure" spectroscopic groups and a new type of group, dedicated to the new discipline of "laboratory astrophysics". The former included the National Bureau of Standards and F. Paschen's laboratory at the University of Bonn; among the latter were the Mount Wilson laboratories and that established by A. Fowler at South Kensington in place of the old observatory. One result of this collaboration was the identification of the dominant features of the spectra of the

The Orion nebula (above) is a cloud of glowing gas whose light is derived from very hot stars within the cloud. The edges of the cloud extend to cover an area of the sky greater than the full Moon.

(Below) The disc for the 5·2-metre (200-inch) Mt. Palomar telescope. It was poured at the Corning Glass Works in New York and transported in a specially designed aluminizing chamber to California.

cool stars as belonging to molecules of titanium and zirconium oxides. Another was the recognition of the element responsible for the lines ascribed to "nebulium" in the spectra of the gaseous nebulae. Ira Bowen showed in 1927 that these lines were emitted by ionized oxygen and nitrogen atoms. The lines occur only at very high temperatures and very low pressures, conditions that cannot be attained in laboratories on Earth. The wavelengths of nebulium can be deduced though from other laboratory measurements of oxygen and nitrogen spectra. The pressure in the nebulae could then be determined from the spectrum, and was found to be so low that the best vacua ever attained in the laboratory are still hundreds of times more dense than the nebular gas. The nebulae are visible simply because the infinitesimal density of the gas is offset by its vast extent: the fourth-magnitude Orion nebula for example, is visible to the naked eye: it is 4·5 parsecs across and contains enough gas to make 700 Suns.

Towards the end of the 1930s, two new discoveries emerged to assist the engineer. The first was a method of depositing aluminium, rather than silver, on to glass mirrors. This produced a coating that was much more durable than silver – a very significant gain because the big and fragile mirrors had to be lifted out of their mountings for recoating much less frequently than before. The other event was the arrival of a new glass – Pyrex – that was very much less sensitive to temperature than the traditional mirror material, plate glass. As mirrors now weighed many tonnes and took a long time to cool, the temperature sensitivity of plate glass set an effective limit to telescope size. The first use of the new glass in a large telescope was in the 1·9-metre (76-inch) telescope for the Canadian David Dunlap Observatory, soon followed by the 2·1-metre (82-inch) at the McDonald Observatory of the University of Texas.

These developments spurred Hale to set in motion his fourth successful attempt to build the world's biggest telescope. This time it was a doubling in size over the Hooker telescope, to a spectacular unexplored region – a 5-metre (200-inch) aperture. The telescope was to be Hale's swan-song. He supervised the design and the search for a new site, better than Mount Wilson, which by 1930 was much affected by the city lights of Los Angeles. The mirror was cast in 1934 and delivered to Pasadena in 1935, the year in which the main construction contracts were placed. Sadly, Hale died in 1938 and never saw the completion of the giant telescope erected on Mount Palomar, an instrument now known as the Hale telescope.

CHAPTER TEN

LIMITATIONS ON BIG REFLECTORS

No good for the sun, and a very small field

"If we wish to know what a star really is we must approach it closely, and this is possible only in the case of the Sun. Indeed, because the Sun was regarded as important, offering so many opportunities to increase our knowledge of its nature, the observatory was conceived primarily for solar research."

Ten Years' Work of a Mountain Observatory: a brief account of the Mount Wilson Solar Observatory *by G. E. Hale, 1915.*

"It is perhaps not an undue exaggeration to refer to the period between 1890 and 1930 as the Hale-Mount Wilson era in solar physics."

Introduction to The Solar System – *Volume I,* The Sun, *by L. Goldberg, 1953.*

There are three tasks for which big telescopes of traditional design are unsuitable: they are too flexible to measure the position of the stars with high precision: they can "see" only a minute portion of the sky at any one time, and therefore make very poor survey instruments; and, like people, they are unable to look directly at the Sun without risk of damage.

The first of these limitations to assume importance was that concerned with position measurement. The problems were emphasized by the range of discrepant values of the Earth–Sun distance obtained during the transits of Venus in 1874 and 1882. In response, a series of special measuring devices was developed for their solution. One approach to measurement of the distance to the Sun that did not require the measurement of small planetary parallaxes was based on the measurement of Bradley's constant of aberration, which relates the Earth's speed in its orbit to the velocity of light. As the latter was now known from laboratory measurements by Fizeau and Foucault, aberration measurements could give the velocity of the Earth in its orbit and from this the orbit's dimensions could be calculated. In the first determination of the constant of aberration Bradley had used a vertical telescope (the zenith tube) because it was, of all types of telescope, the least affected by problems of flexure and distortion and also gave data that did not require correction for atmospheric refraction. The first new instrument was an improved zenith tube, developed by Airy at Greenwich. The chief improvement was to use a dish of mercury to define a horizontal surface, the liquid mirror being fitted at the bottom of a vertical telescope so that the light was reflected back to form a star image just above the telescope lens. This device, the reflex zenith tube, was insensitive to almost all errors of alignment, but the mercury mirror proved very sensitive to vibration and the original instrument had to be remounted on a new site. A long series of measurements with this telescope, especially those taken between 1882 and 1886, showed an effect that badly upset the original aim of its designer. The latitude of the observatory was changing, confirming an analysis made in Boston by S. C. Chandler.

The instrument that Chandler had developed also used mercury to define a horizontal plane, not as a mirror but by floating the entire telescope in a mercury trough. Measurements with this instrument, again designed to determine the constant of aberration, showed that the Earth was not spinning perfectly smoothly but the axis wandered very slightly in position, an effect now called the "Chandler wobble". The extent of the motion is

Albert Einstein, 1879–1955, occupies the same dominant position in modern science as did Newton in the seventeenth century. Einstein's theories, like those of Newton, were both stimulated and confirmed by astronomical measurement.

minute: the North Pole moves around a roughly circular path, with a radius of about 6 metres (20 feet). About one-third of the motion follows a regular annual pattern; the remainder repeats every 436 days, and is caused by the fact that the Earth is neither exactly spherical nor rigid. The floating telescope idea was further developed by Bryan Cookson at Cambridge, who built a 16-centimetre ($6\frac{1}{4}$-inch) aperture photographic refractor that weighed some 160 kilograms (350 pounds) and could be floated in an annular trough, again filled with mercury. It was, however, a version of Airy's instrument that proved the most accurate device – the photographic zenith tube – and these are still in use today.

A more subtle approach to measuring the Earth's velocity was made in 1881 by a young American naval officer, Albert Michelson, when on leave of absence as a visitor to Hermann von Helmholtz's laboratory in Berlin. Michelson invented an instrument – now known as the Michelson interferometer – that could compare the time taken for two halves of a light beam to travel in two different directions that were at right angles to one another. If one of these directions were along, and the other across, the direction of the Earth's motion, then a time difference should be observable. Michelson's first interferometer was built in Berlin with the aid of a grant from Alexander Graham Bell, but the vibration due to the trams outside von Helmholtz's laboratory made work impossible. He moved the instrument to Vogel's observatory at Potsdam and reassembled it between the piers of the large telescope. At this site vibration was much less severe, but none the less Michelson was totally unable to detect any sign of the Earth's velocity. He resolved to pursue the matter further and, after leaving the navy to become the first Professor of Physics at the Case School for Applied Science in Cleveland, Ohio, Michelson built a new and larger interferometer. The first set of observations, made by Michelson with the help of Edward Morley, the professor of chemistry at the nearby Western Reserve University, confirmed the Potsdam measurements: the interferometer showed no sign of any effect due to the motion of the Earth. This unexpected result, described by J. D. Bernal as "the greatest of all negative results in the history of science", later became a foundation stone of Einstein's theory of relativity.

At the turn of the century a new approach to the size of the solar system superseded the older measurements. One of the hundreds of minor planets, Eros, discovered in 1898, proved to have an orbit that brought it very close to the Earth. The closeness of approach meant that the parallax of Eros as seen from observatories across the Earth should be unusually large and so could be measured with precision. An international campaign was organized to determine the parallax of Eros when near the Earth in 1901, and the data were reduced both by the individual observatories and by A. R. Hinks at Cambridge. The result gave the astronomical unit as 149·5 million kilometres (92·9 million miles). The next close approach of Eros was in 1931 and a new campaign was organized, this time by Harold Spencer-Jones, the Director of the Cape Observatory. The revised result was 149·7 million kilometres (93·0 million miles). Modern measurement of interplanetary distances, using radar techniques, gives a result of 149,597,870 kilometres (92,955,580 miles).

The interval between the two Eros parallax measurements saw, among other things, the publication of Albert Einstein's two great theories of relativity. The simpler of the two, "special" relativity, came out in 1905. It explained the negative result of Michelson and Morley's experiments by stating that light beams always travel at the same speed (in a vacuum) regardless of the speed of the light source. In addition the theory presented the famous equation $E = mc^2$, relating energy (E) to mass (m) multiplied by the square of the speed of light (c). This at last offered a possible solution to the problem of the origin of the energy that kept the stars shining, though the mechanism was not apparent. If the Sun's radiation were to come from a process that converted its mass into energy, then, although it would need to lose 5 million tonnes a second, the total mass available is so large that this rate can be maintained for more than 5 billion years. The more sophisticated of Einstein's two theories, "general" relativity, was, broadly speaking, a new theory of gravitation. It relied very heavily on astronomical measurement to test its validity. In return the theory, in the hands of Willem de Sitter, Alexander Friedmann and Hermann Weyl, predicted that the universe could not be static but must expand. It was several years after this prediction that Hubble's studies of distant galaxies provided experimental evidence that this was the case. The three predictions of general relativity that astronomers could test quantitatively rather than qualitatively were the precession of the perihelion of Mercury and small changes in the direction and wavelength of light in the presence of a strong gravitational field.

The data on the orbit of Mercury were already available. The discrepancy that had concerned Le Verrier and Newcomb proved to match the new theory with gratifying accuracy. The other two tests took rather more effort, as the number of test cases was severely limited – two for the wavelength shift and one for the deflection. The limitation arises from the smallness of the effects: to deflect the path of starlight by one arc second needs a mass comparable to that of the Sun – even the 2 million million million million tonnes (2×10^{24}) of the most massive planet, Jupiter, is not enough. The wavelength change is not easily seen even for the Sun: a more compact heavy object is needed. Fortunately one of the classes of stars that are not in the main sequence, the white dwarfs, have exactly this property. The brightest of these objects is the companion to Sirius, called Sirius B and discovered by Alvan G. Clark in 1862. In 1915 Walter Adams at Mount Wilson showed that Sirius B was a faint object not because it was cool and dim, but that it was white hot like Sirius A, and must therefore be very small. As it was one of a pair of stars its mass could be found. The result was the surprising discovery that Sirius B is heavier than the Sun but roughly the size of the Earth. This incredibly dense object, averaging 150

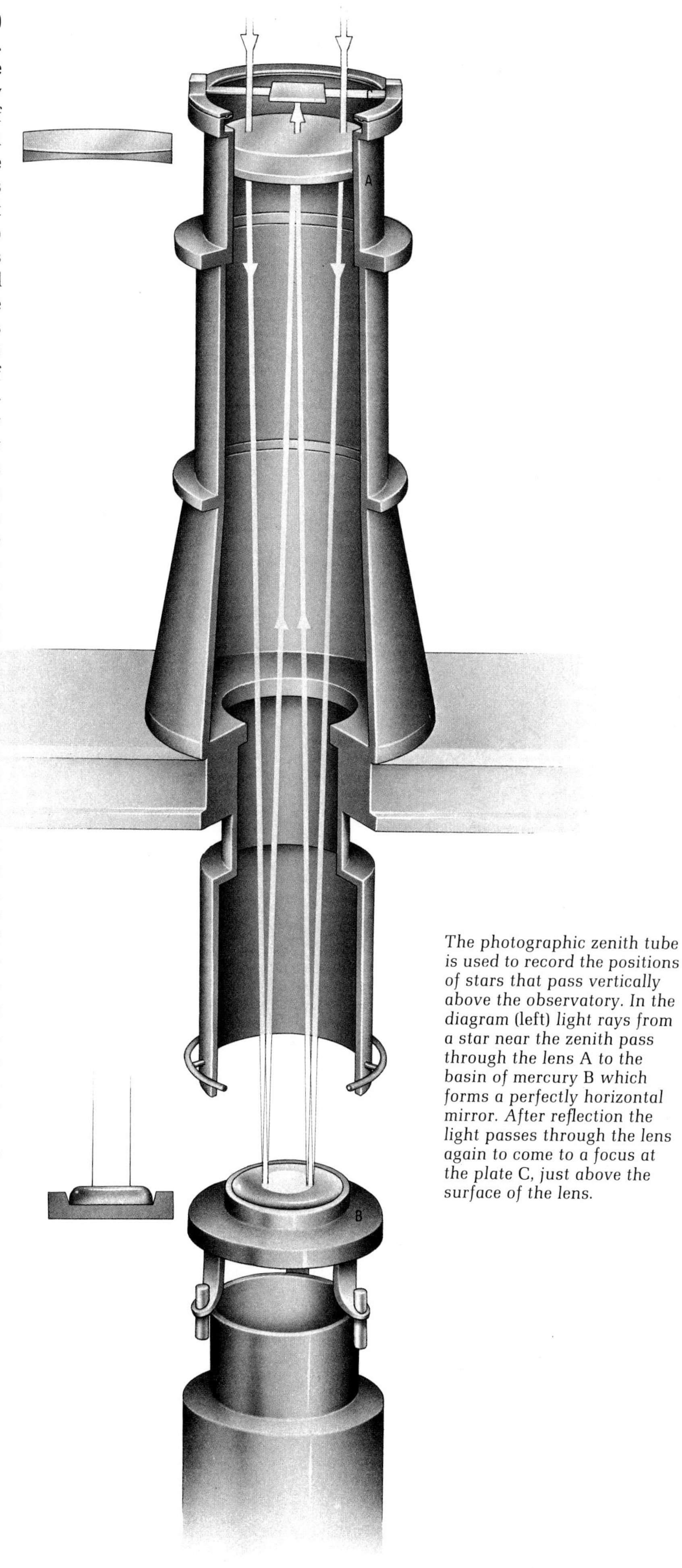

The photographic zenith tube is used to record the positions of stars that pass vertically above the observatory. In the diagram (left) light rays from a star near the zenith pass through the lens A to the basin of mercury B which forms a perfectly horizontal mirror. After reflection the light passes through the lens again to come to a focus at the plate C, just above the surface of the lens.

tonnes a litre, makes a useful gravitational test bed, and Adams attempted the difficult task of getting a spectrum of the little star that was not swamped by light from its vastly brighter companion. He found that all the wavelengths were slightly shifted to longer wavelengths, and by an amount in close accord with Einstein's theory. Later work has shown that the agreement was fortuitous as both the radius of the star and Adams's measured wavelength change were wrong: Sirius B is smaller and the gravitational effect larger than Adams's figures suggested.

The third test of general relativity, the deflection of light from distant stars by the Sun, required the comparison of two photographic plates: one of a field of stars, and the second of the same stars but with the Sun in the centre of the picture. If the Sun's mass were sufficient to deflect the starlight then the stars would appear to be in slightly different places on the two plates. The predicted shift was of the order of an arc second, an amount easily measurable with precision telescopes. However, the star field behind the Sun is only visible during a total eclipse, and eclipses obstinately refuse to happen over established observatories. Due to the Great War, news of Einstein's prediction reached England by indirect routes and Sir Arthur Eddington and F. W. Dyson organized two expeditions, one to the island of Principe in the Gulf of Guinea, and the other to Sobral in Brazil, to exploit the May 1919 eclipse as a test of Einstein's theory. The test was difficult, as the plates taken at the eclipse had to be compared with other plates which could only be taken when the stars were visible at night. This meant a comparison between plates taken at Greenwich in January with those obtained at the eclipse sites in May, with further trouble due to the differences in temperature between night and day. In spite of this and the other difficulties of working at remote and ill-equipped sites, the results were fairly successful and the general theory of relativity received a much more substantial experimental endorsement. The agreement was not found without a preliminary scare. The observations at Principe were made with one of the 33-centimetre (13-inch) "Carte du Ciel" telescopes, but were troubled by cloud and very few stars appeared on the plates. These showed displacements in agreement with Einstein's theory. The Brazil expedition had better weather, but the heat of the Sun before the eclipse distorted the optics of an identical telescope and the results agreed better with Newton's theory than with relativity. Fortunately the Brazil team had taken a second telescope of smaller aperture, and its results supported the meagre data from Principe, so that Sir Arthur Eddington and Dyson were able to announce confirmation of Einstein's prediction.

Observation of the Sun, at times other than at eclipse, also required the development of alternative instruments to the established large reflecting telescopes. The difficulty that arises in making observations of the Sun is largely one of mechanical distortion, often complicated by problems caused by heating. Because there is so much sunshine, the observer's aim is to spread the light out so that the finest detail can be seen. This necessarily involves large apparatus, either to give a large solar image or to form a spectrum of unprecedented quality. Such apparatus is too unwieldy to be swung across the sky to follow the Sun, so the majority of solar telescopes use mirror systems that feed the light to stationary instruments. A classical reflecting telescope with a coudé focus is not, however, usable; the Sun has an angular diameter of 0·5°, which is too large to fit comfortably into a system designed for star images three or four thousand times smaller. In addition, the secondary mirror of such a Cassegrain telescope is positioned near the focus of the primary mirror and receives all the radiation collected by the latter. This places it near the focus of what is, in effect, a powerful solar furnace.

The first satisfactory solar telescope was based on the heliostat developed by Jean Foucault. The ingenuity of the device lies in the invention of a linkage that will convert the steady motion of a clock into the unsteady motion needed to correct for the variation in the Sun's position during the day. The heliostat mirror is flat and does not focus the light. It simply reflects sunlight into a fixed horizontal telescope. During the nineteenth century, solar astronomers used both heliostats and large refractors. Lockyer, for instance, worked with the 60-centimetre (24-inch) Newall refractor, and Hale with the 1-metre (40-inch) Yerkes telescope. The final move to the heliostat came when Hale found the Yerkes refractor had too short a focal length to give the image size he needed. If 19 metres (63 feet) was too short, and Cassegrain reflectors unusable, only the heliostat could serve. After some pioneering work at Chicago, where Hale and Ritchey built a horizontal telescope that burned down almost as soon as it was finished, Hale persuaded Miss Helen Snow to fund a horizontal telescope with an aperture of 60 centimetres and a focal length of 43 metres (140 feet). It was completed in 1903 and later became the first big instrument at what started as the Mount Wilson Solar Observatory.

The Snow telescope was scarcely reassembled on Mount Wilson before it was

(Above) *The pier of the Snow telescope, just after its completion in 1904. The top of the pier supported a heliostat which reflected light into a long, horizontal telescope, set in a low shed running across the top of Mount Wilson.*

(Right) *The 45-metre (150-foot) tower telescope on Mount Wilson. The tower is a more subtle structure than is apparent. An outer hollow shell, on separate foundations, encloses the main frame and shields it from buffeting by the wind.*

(Below) *The coolest layers of the visible disc of the Sun can be studied spectroscopically. If light from the extreme edge of the Sun's disc is examined, the top layers are seen alone and their spectrum appears as emission lines. The brief appearance of this effect during an eclipse leads to a "flash" spectrum.*

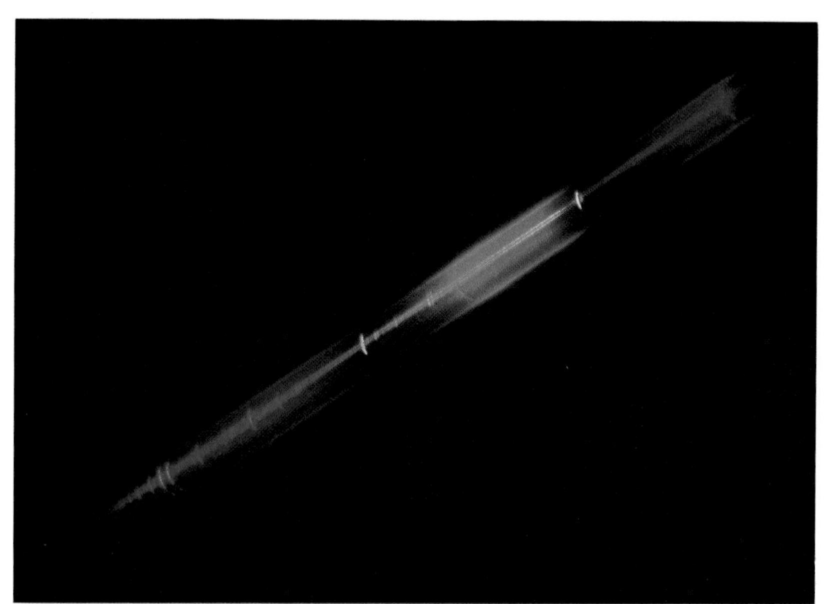

superseded. The heat of the Sun distorted the heliostat mirror and warmed the air in the long telescope, causing it to well up across the light beam. Both effects spoiled the quality of the image. Thicker mirrors could help with distortion, but the internally generated seeing problem could only be solved by using a vertical telescope. In 1907 Hale completed the first of two tower telescopes. Each had a two-mirror guidance system at the top of a vertical tower, feeding a refracting telescope of great focal length attached to a large spectrograph, which was mounted in a pit dug vertically beneath the telescope focus. The balance between telescope and spectrograph was pushed strongly towards the latter. The focal length of the first vertical telescope was 18 metres (60 feet) and that of its spectrograph 9 metres (30 feet). The second tower telescope, completed in 1912, matched a 45-metre (150-foot) telescope to a 22-metre (72-foot) spectrograph.

With these instruments Hale and his co-workers studied the Sun in enormous detail. At the least exciting level, the solar spectrum was remeasured and a more accurate list of the Fraunhofer lines published. Hale's own work showed that sunspots were associated with magnetic fields and that close pairs of sunspots had the opposite polarity. The eleven-year sunspot cycle, discovered in the mid-nineteenth century, proved to be in reality a twenty-two-year cycle: in each eleven-year surge in sunspot numbers the magnetic behaviour was opposite to its predecessor. Sunspots proved to be cooler regions of the Sun's surface, and their spectra showed the presence of small molecules, too frail to exist at the higher temperatures of the rest of the solar surface. As well as studying individual spectrum lines from differing regions of the Sun, Hale invented an instrument, the spectroheliograph, that allowed him to obtain pictures of the whole Sun in light of a single wavelength. These pictures, which show a wealth of detail unobservable in "white" light, are in effect maps of the brightness of the Sun at different levels of the outer surface. Spectroheliograms, as the maps are called, can be made at different wavelengths, and those made in the light of the strongest Fraunhofer lines show the coolest portions of the solar atmosphere. The pictures obtained by Hale, especially those photographed using the calcium line in the violet and the hydrogen line in the red, showed very clearly three classes of solar phenomenon (seen only poorly in white light). As well as dark sunspots, there were bright patches called plages, dark streaks called filaments, and sudden brief explosions now called solar flames. The use of a mountain site and better apparatus also showed the solar prominences, observed earlier by Lockyer and Janssen, in much more detail, and they were identified with the filaments; they appeared dark against the Sun's surface or bright when seen at the edge of the disc.

The only method for solar observation that was developed in America but not by Hale was the use of cinephotography. This was pioneered by Robert McMath at the McMath Hulbert Observatory on Kitt Peak, Arizona, and proved a very powerful method for studying the complex motions of the solar atmosphere.

The last aspect of the Sun's visible appearance to become the subject of regular study was the corona, its outer atmosphere. The Sun's disc is about a million times brighter than the inner corona, and a long series of gifted observers, including Huggins, Bond and Hale, had failed to detect the corona except during a total eclipse. The successful development of an instrument able to see the corona in broad daylight was achieved by a Frenchman, Bernard Lyot, in 1930. Lyot very carefully analysed all the sources of scattered light within a telescope and set about their elimination or reduction. His list included faults in the glass of the lenses, tiny scratches on their polished surfaces, unwanted faint reflections, dust within the telescope and dust in the atmosphere above the observatory. A special polishing technique on near-perfect glass discs was developed, and finally a special-purpose telescope was erected in the clean, clear air on the Pic du Midi in the Pyrenees. The inside of the telescope was coated with sticky grease to catch the dust, and

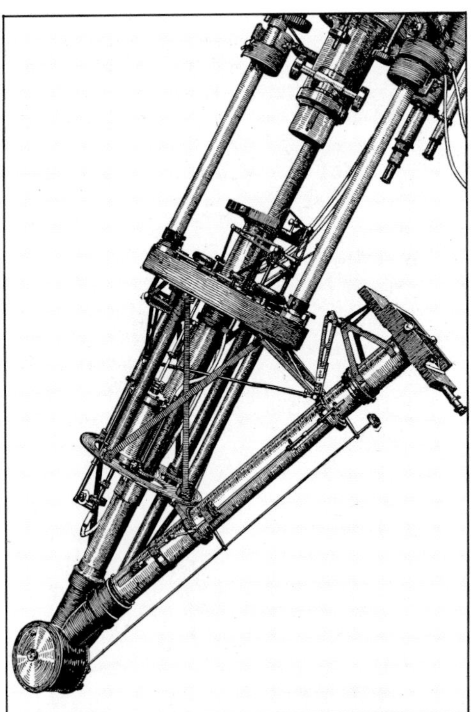

G. E. Hale's spectroheliograph, mounted on the 30-centimetre (12-inch) refractor bought by Hale's father. The telescope formed an image of the Sun on the entrance slit of the spectrograph. A system of levers, cranks and linkages then moved the entrance slit across the solar image and the plate across an exit slit. The result is a picture of the Sun's disc in light of only one wavelength.

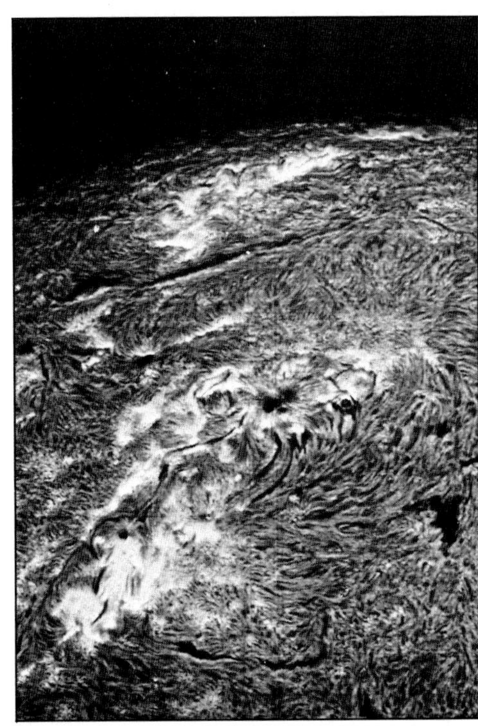

A "spectroheliogram", taken with a spectroheliograph, of part of the solar surface. This picture was taken in 1958 using light from a narrow wavelength range set at the centre of the very strong red Fraunhofer line due to hydrogen (the "C" line). The view is therefore of the uppermost layer of the Sun's atmosphere. The atmosphere was turbulent in 1958, a year of maximum solar activity.

The observatory founded by Bernard Lyot in 1930 on Pic du Midi de Bigorre, a 2865-metre (9400-foot) peak in the French Pyrenées. The clear mountain air above the peak greatly reduces the seeing effect.

the lens rubbed with "nose oil" – a mixture of sweat and human grease which Lyot had found filled the tiny polishing scratches and further reduced the scattered light.

The surprise that came from much improved data on the corona was that it turned out to be much hotter than the Sun's surface, the photosphere. Bengt Edlén, a Swedish spectroscopist, identified nineteen out of the twenty-three spectrum lines in the corona as being due not to some unknown element "coronium" but to atoms that had been heated to the point where ten to fifteen electrons had been detached from their parent nuclei rather than the one or two detached from atoms in the photosphere. This implied that the corona was at a temperature of between 1 million and 2 million degrees. It must also have a very low pressure, though not as tenuous as the gaseous nebulae. The question left hanging once "coronium" had been identified was how could heat flow from the solar surface at 5000 or 6000 degrees to a region many times hotter? The laws of physics lead to an expectation that energy flows from hot to cold and not vice versa.

While specialists in solar and positional astronomy pursued their own aims with their purpose-built instruments, the general run of astronomer was much concerned with the problem of how to find

CORONAGRAPHS

The coronagraph, shown schematically below, produces its own miniature eclipse and then records the light of the corona, a millionth of the brightness of the Sun's visible disc. The first lens, A, produces an image of the Sun at S. A metal disc very slightly larger than the solar image is placed at S to block out the vast majority of the sunlight. Only the minute fraction of the light from the corona is allowed past the edges of the disc and this is re-imaged by the right-hand lens, C, to give the final image on the plate at D. The purpose of the lens at B, a "field" lens, is to guide the rays from the edge of the image through the centre of lens C.

On the right is a nineteenth-century engraving of the Sun's corona as seen during a total eclipse. The corona is the outermost layer of the Sun's atmosphere extending millions of kilometres above the solar surface. During the eleven-year solar cycle the shape of the corona is changing continually: it is almost spherical when sunspot activity is maximum and elliptical when spot activity is minimum.

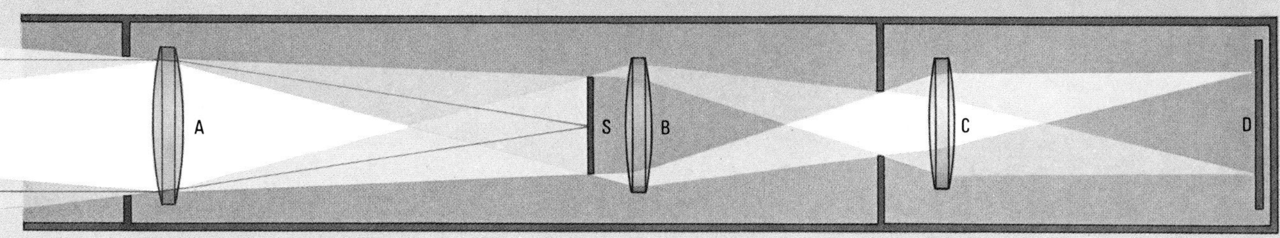

(Far right) *The 1-metre (40-inch) Ritchey-Chrétien telescope of the US Naval Observatory. The telescope was built in Washington after Ritchey's return from France in 1930, where he had worked with Chrétien on the first Ritchey-Chrétien telescope. In 1955 it was moved to a better site in Flagstaff, Arizona.*

the stars to which his big telescope should be applied. The late nineteenth-century atlas of the skies, the photographic "Carte du Ciel", was made using 33-centimetre (13-inch) aperture telescopes, so the new generation of 2-metre (78-inch) reflectors could see objects very much fainter than those in the atlas. A star chosen at random would probably be one of the millions of middle-sized, middle-aged unexciting main sequence stars and little that was new would result from its study. Astronomers' needed a survey instrument to locate the interesting objects in the sky, but the big reflectors themselves were totally useless for survey work. The field of view of the 1·5- and 2·5-metre (60- and 100-inch) Mount Wilson telescopes was tiny – it would take between fifty and sixty plates to photograph the full Moon, yet that covers an area that is only a few millionths of the whole sky. A complete survey with such a telescope would take about 10,000 years, so instruments that combined a good field of view and a large aperture became a matter of priority.

The prototype for one class of survey instruments was the high-quality camera lens developed for portrait photography which used additional components so that the designer could increase the field of view. In 1892 Miss C. W. Bruce provided the funds for a telescope lens based on a design developed for cameras but enlarged to be 60 centimetres (24 inches) in aperture and still only 3·3 metres (11 feet) in focal length. The compact telescope gave a large angular field on plates of reasonable size and was built for Harvard by Alvan Clark in 1893. In 1896, it was moved to Peru, where it provided the data on Cepheid variable stars, which were to become so important in estimating the size of the universe. The Bruce telescope, however, still suffered the disadvantages of lensed instruments, especially the absorption of ultraviolet radiation, and it did not lead to a general development of refracting survey telescopes. One group, that of Percival Lowell at Flagstaff in Arizona, did persevere with the refractor, although Lowell's own highly controversial data on the planet Mars were obtained with an orthodox 60-centimetre refractor built by Clark and not with a wide field instrument. Lowell himself made a survey of the Zodiac in an unsuccessful search for a planet beyond Neptune, using a 28-centimetre (11-inch) refractor. The planet – later named Pluto – was found by Clyde Tombaugh of the Lowell Observatory in 1930, after a second survey of much of the Zodiac. Trombaugh used a 33-centimetre aperture photographic refractor, but this time with a triplet lens and a field of view more than 5° across, a major step forward.

The first step towards a reflecting telescope with a larger field than the classical Cassegrain was taken by George Ritchey, who had established himself as an independent telescope-maker after completing the Mount Wilson 2·5-metre. A Frenchman, Henri Chrétien pointed out that if the mirrors of a Cassegrain were slightly modified, then the secondary mirror could correct the poor image given by the primary. In the Cassegrain the primary mirror gives a perfect image on axis and the secondary merely re-images this perfection at the final focus. Away from the telescope axis the poor primary image is refocused with all its imperfections, so the Cassegrain field is not significantly better than that at the prime focus. The difficulty in Chrétien's design was that the shape of the primary mirror had to be a hyperboloid, not the usual paraboloid. Such a mirror was difficult to make and even more difficult to test. Ritchey's contribution was to show that it could be done, and the first Ritchey-Chrétien telescope, of 50 centimetres (20 inches) aperture, was built in 1923. The telescope had a field of view of roughly 1°, a vast improvement over the superficially similar Cassegrain. Ritchey returned to the United States and began the construction of a 1-metre (40-inch) Ritchey-Chrétien telescope for the US Naval Observatory in Washington. It was the last big telescope he was to build before retiring at the age of seventy-two, more than fifty years after making his first reflector.

A field of view 1° across is still only a tiny fraction of the sky and the next step to better performance involved using the components of both reflector and refractor. While the Ritchey-Chrétien design philosophy was to use one mirror to correct the errors of another, the alternative, developed by Bernard Schmidt in Germany, used a lens to correct the faults of a mirror. Unfortunately, the correcting lens of the Schmidt telescope is even more difficult to make than Ritchey's hyperbolic mirror. The crucial point in the design is that the "lens" of the telescope is very nearly a flat, parallel-sided plate of glass. In consequence this "corrector plate" does not suffer from appreciable chromatic aberration, and so does not degrade the ability of the mirror to bring all wavelengths to a common focus. Schmidt, an individualistic and independent worker, devised a method for polishing the corrector plate so that the critical difference between flat and "very nearly" flat could be made with precision. The first Schmidt telescope, completed in 1930, had a corrector plate 36 centimetres (14 inches) in aperture and a field of view 16° across. This was followed by a 60-centimetre (24-inch) aperture system with a more modest

BERNARD SCHMIDT
(1879–1935).

Schmidt was a maker of mirrors and lenses for amateur astronomers. He was persuaded to work at Hamburg Observatory and there developed his telescope, known as the "Schmidt camera".

(Above) Schmidt's "observatory" near Jena, East Germany, in about 1920. On the left is a fork-mounted Cassegrain reflector and on the right part of a horizontal telescope.

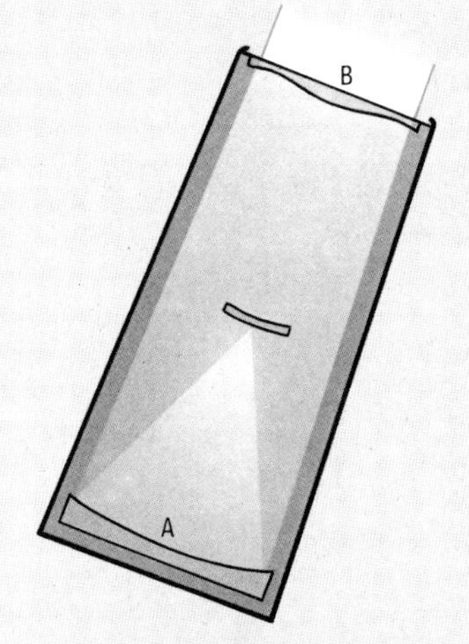

The Schmidt camera, shown schematically above, combines a spherical mirror, A, with a glass "corrector plate", B. The corrector plate compensates for the spherical aberration of the mirror and is placed at the centre of curvature of the mirror. The shaping of the corrector plate surface is much exaggerated in the diagram.

(Left) The 1·2-metre (48-inch) Schmidt camera on Mount Palomar. This instrument was completed in 1948 and first used to survey the entire northern sky. The plate-holder takes plates 34 centimetres (14 inches) square, which cover a region of the sky 6° across, producing single photographs of as many as a million stars.

field, a telescope that was still incomplete at Schmidt's death in 1935.

The Schmidt telescope is not without its disadvantages. The instrument is twice as long as an orthodox reflector of the same focal length because the corrector plate is positioned not at the focus but at one focal length above the focus. As with the lens of a refractor, there is a limit to the size that the corrector plate can be made. The approach to correcting the defects of the primary mirror with lens elements can, however, be applied to orthodox telescopes. The use of a lens system to improve the field of view available at the prime focus of a parabolic mirror was pioneered by F. E. Ross, who was driven by the need to improve the field available at the prime focus of the 5-metre Hale telescope. At the design stage of the 5-metre it was clear that the focal length would need to be kept small, even if only to reduce the heavy cost of the dome. As a result the field at the uncorrected prime focus would be only two arc minutes across. In 1938, ten years before the telescope was complete, Ross designed a special corrector lens that would give a good image for a field 0·5° across – still small, but 200 times the area of the uncorrected system.

There was a considerable delay after Schmidt's death before other opticians learned to polish the difficult corrector plates. The optical shop of Mount Wilson Observatory was one of the first professional teams to master the art, though many amateurs took up the challenge with success. The Mount Wilson group produced a series of small Schmidt systems to use in spectrographs and then larger optics for telescopes with an aperture around 50 centimetres (20 inches). As the optical components grew larger the design settled to a configuration that gave a field 5° or 6° across with a ratio of aperture to focal length of about 1:3. The end of this development was the 1·2-metre (48-inch) Schmidt telescope built at the Hale Observatory on Mount Palomar in the early 1950s. This instrument was first used to carry out two surveys of the sky of the Northern Hemisphere, one through a red filter and the other through a blue so that a comparison of the two black-and-white prints would reveal how cool (red) or hot (blue) a star was. The surveys involved taking 1758 plates and the entire programme took about five years. In comparison, the nineteenth-century Carte du Ciel, using 28-centimetre refractors, had needed 22,154 plates taken by some twenty observatories over twenty years. The Palomar sky survey also recorded stars five hundred times fainter than did the smaller, ultraviolet-absorbing refractors, and not only is it still the standard reference atlas of the sky a quarter century after the start of the programme, but will remain so for many years to come.

The Veil nebula (below) is one of the remnants of a supernova explosion that took place about 50,000 years ago. The debris – the Cygnus Loop – spreads $2\frac{1}{2}°$ across the sky, a field far too large to photograph with an orthodox telescope. The finest filaments in the nebula are about twice as far across as the entire diameter of the solar system.

CHAPTER ELEVEN

A NEW WINDOW ON THE UNIVERSE

The beginnings of radio astronomy 1930–50

"The foregoing observations confirm that radiation in the radio spectrum is apparently coming from the Milky Way. . . . A few bright stars such as Vega, Sirius, Antares, Deneb and the Sun gave negative results. Mars and the Orion nebula also gave no readable indication."

George Reber
212 West Seminary Avenue
Wheaton, Illinois.
December 1939

From Astrophysical Journal, *Volume 91.*

The greatest achievements of nineteenth-century physics were presented in a paper "A Dynamical Theory of the Electromagnetic Field" and a book *A Treatise on Electricity and Magnetism*. Both were written by the same author, the Scottish physicist James Maxwell; the paper appeared in 1865 and the book in 1873. Maxwell's theory linked electrical and magnetic phenomena to the properties of light, interpreting light "waves" as oscillations of electrical and magnetic fields – electromagnetic radiation. Maxwell suggested that such radiation might also be found at wavelengths very different from those of infra-red, visible and ultraviolet light. The German physicist Heinrich Hertz undertook a series of brilliant researches to test this prediction. By the end of the 1880s Hertz had shown that there existed a form of electromagnetic radiation, "radio waves", which could be reflected and refracted but had a wavelength millions of times longer than visible light.

By the beginning of the twentieth century, radio equipment had become more sensitive and two contradictory results emerged which were relevant to the study of radio waves from beyond the Earth. Although the Sun emits large quantities of visible radiation, experiments by Sir Oliver Lodge in England, J. Scheiner and J. Wilsing in Germany and C. Nordman in France failed to detect any radio waves. At the same time, other experimenters, particularly Guglielmo Marconi, showed that long-distance radio communication was easier than expected. It seemed that radio waves could travel over the horizon, and Oliver Heaviside and Arthur E. Kennelly independently suggested an explanation of this in 1902. High above the surface of the Earth the upper atmosphere is ionized (the electrons are stripped from their parent atoms), and such ionized gas reflects radio waves very efficiently. This tenuous mirror, the "ionosphere", would reflect downward any radio waves transmitted on Earth and so aid communication; but it would also reflect back into space any radio waves arriving from outside the Earth's atmosphere from astronomical bodies. It followed that radio waves from space were neither expected nor sought in the thirty years following Nordman's unsuccessful work.

The First World War stimulated the development of radio, and after the end of the war the 1920s saw an explosive development of broadcasting. By 1924 there were well over a thousand radio stations in America alone. Since the early stations were all operating on long wavelengths, radio engineers were forced to build systems of shorter and shorter operating wavelength in order to find a region not already occupied by other transmitters.

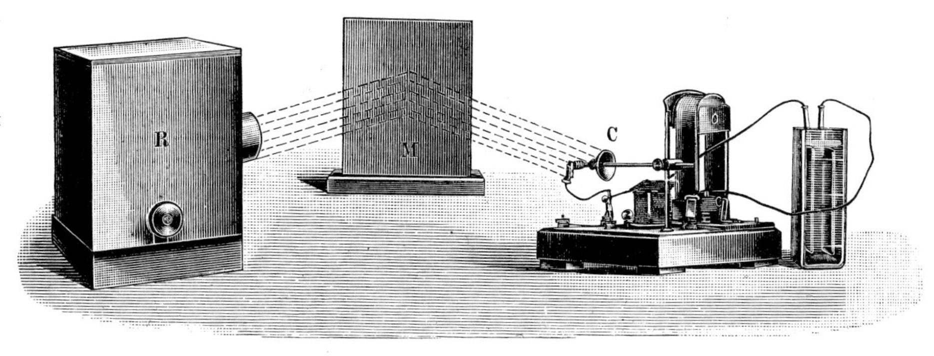

An early illustration of an experiment to show that, like visible light, radio waves obey the law of reflection. The source of radio waves, inside the box R, and the detector circuit at C are both electrical circuits tuned to resonate at the same frequency. Heinrich Hertz showed that the waves emitted by the source could be reflected from the metal surface M, in agreement with a prediction by Maxwell, and that the angles of incidence and reflection were equal. Hertz's original experiments used systems that resonated at about 100 million cycles per second and produced radio waves with a wavelength of a few metres.

ELECTROMAGNETIC SPECTRUM

The range of wavelengths covered by electromagnetic waves – radiation closely related to visible light – extends for at least 20 decades or 70 octaves. In empty space all the different forms travel in straight lines with the same velocity, 300 million metres a second (about 200 million miles per second) or one parsec in $3\frac{1}{4}$ years.

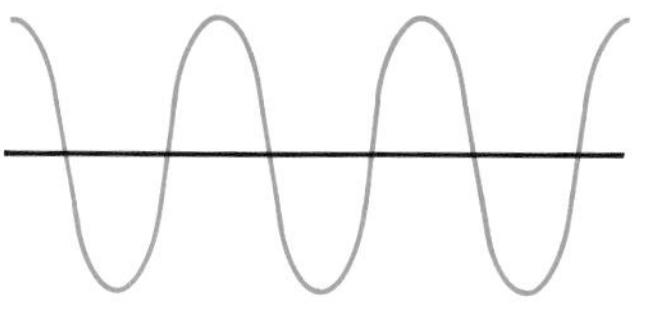

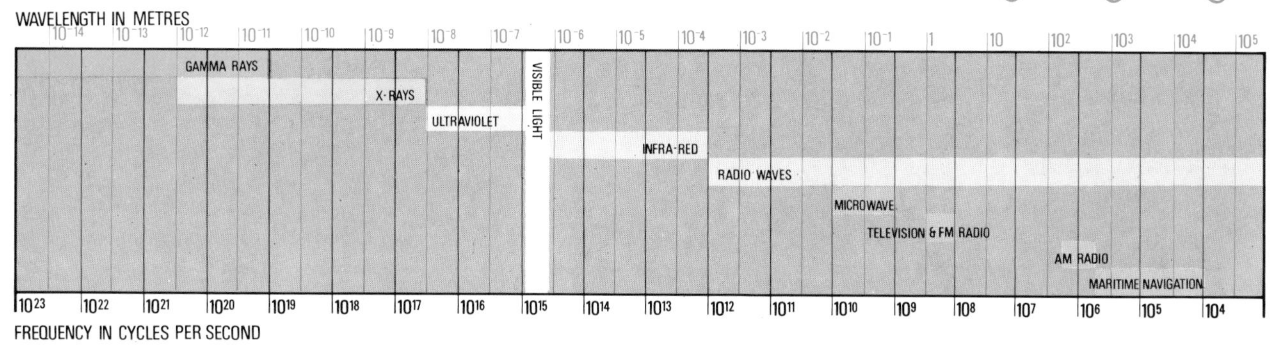

(Right) *A poster for a lecture publicizing Marconi's "wire-less" communication. The range achieved rose rapidly – 2 kilometres in 1895 and 3500 kilometres in 1901.*

Lecture & Practical Demonstration
OF
MARCONI'S
WONDERFUL SYSTEM OF
WIRELESS TELEGRAPHY
BY MR.
WILLIAM LYND.

The Instruments used are supplied by the WIRELESS TELEGRAPH SIGNAL CO.,

And prepared specially for Mr. LYND by Signor MARCONI.

MARCONI AND HIS MARVELLOUS APPARATUS.

How to Send Telegrams through Space and Apparent Obstacles will be demonstrated during the course of the Lecture. Mr. LYND will be assisted by Mr. W. W. BRADFIELD, one of Signor Marconi's Experimental Assistants.

"The Demonstration of Wireless Telegraphy by Mr. William Lynd must have proved to the audience little short of a revelation."
—Belfast Evening Telegraph, *Jan. 21st.*

They also discovered that sensitive radio receivers were limited in performance because they were picking up a background noise due in part to radio emission by lightning strikes and called "atmospherics" or "static". This static determined the maximum performance attainable by a receiver and, in consequence, was a matter of commercial importance. For this reason, a commercial company in the USA, Bell Telephone Laboratories, assigned one of their research engineers to study the origins of this unwanted electrical interference. The engineer, Karl Jansky, assembled a high-quality receiver and antenna system designed to work at short wavelengths, specifically at the wavelength of 15 metres (50 feet), which was coming into use for ship-to-shore communication. The antenna was built up as an array of separate aerials, mounted on a frame 30 metres (100 feet) long and 4 metres (13 feet) high. The array was set on wheels so that it could be rotated to see if the static varied in strength as the antenna was pointed in different directions.

Jansky found that the signals he detected could be ascribed to three types of static, one of which was the well-known large signal associated with nearby thunderstorms. Less noisy than this was a second component due to the net effect of

(Above) *Karl Jansky and his rotating aerial array built at Holmdel, 40 kilometres (25 miles) south of New York. The set of aerials was 30 metres (100 feet) in length and nearly 4 metres (13 feet) in height and thickness – twice and one quarter of his operating wavelength respectively.*

(Right) *Grote Reber and a reconstruction of his 9·4-metre (31-foot) aperture radio telescope erected at the US National Radio Astronomy Observatory in West Virginia. The original – the first radio telescope – was built almost entirely of wood, except for a galvanized iron surface to the dish.*

a vast number of distant thunderstorms. Beneath these two intermittent crackly signals was a steady hiss, similar to that caused by a bad amplifier. Jansky knew his apparatus to be too good for this to be the cause and he attributed the hiss to man-made interference from local electrical equipment. Careful study of the directional properties of the signal showed this to be wrong: the static due to a multitude of electrical motors and other devices would not move steadily round the horizon, as this radiation source did. At the beginning of the measurements the loudest hiss occurred when the aerial faced the Sun, and this seemed a probable source. Over the months, however, the aerial position for the strongest signal moved slowly away from the Sun's direction. By 1933 it was clear to Jansky that his static hiss was coming from a region of the sky that moved not with the Sun but with the stars. The aerial array could not discriminate directions accurately, but Jansky and his partner, A. M. Skellett, realized that the source was in the same direction as the centre of our galaxy.

After a couple of years' further study Jansky was able to show that the entire band of the Milky Way across the sky was a source of "cosmic static", though the signal was strongest from the direction of the constellation Sagittarius, that is from the centre of the galaxy. His discovery also demonstrated that the ionosphere does not reflect short wavelengths back into space, as it does long radio waves, but allows them to penetrate to the Earth's surface. There was now a second region of the electromagnetic spectrum in which it was possible to study the stars: a radio window on the universe.

Although the discovery of radio waves from space was announced in widely read scientific journals and publicized both by newspapers and, very appropriately, on radio, there was no immediate follow-up. Jansky himself proposed to Bell Telephone Laboratories that they build a bigger aerial, but the company felt that they had achieved their aim – to find the sources of static – and further expense could not be justified. It was regrettable, though not entirely surprising, that no university picked up the trail blazed by Jansky. While commerce was struggling with the Depression, American universities were trying to cope not only with shortage of funds but also with a flood of academic refugees from Fascist Europe.

After two years of total neglect, the nascent science of radio astronomy was rescued from oblivion by an enthusiastic amateur. Grote Reber, of Wheaton, Illinois, decided to study "cosmic static" in his spare time. Reber was a professional radio engineer, who earned his living by designing radio receivers, and he set about building the first purpose-built radio telescope. Reber's first assumption was that the radio waves, like those of visible light, would be emitted by the hot stars forming the Milky Way. If that was so, then more radio energy from stars would be detected at shorter wavelengths than Jansky had used: each tenfold decrease in wavelength would raise the power by a factor of a hundred. Working at shorter wavelengths would also allow much more accurate position measurements. Reber thus built his first radio telescope to collect radiation of only 9 centimetres ($3\frac{3}{5}$ inches) wavelength.

In principle a radio telescope is the same as an optical reflecting telescope. A paraboloidal surface is covered with a thin sheet of metal which reflects the radiation to a focus. At the focus of a radio telescope is a short wire aerial (dipole) to collect the radio signal and convert it to a voltage which can then be amplified. In both optical and radio telescopes the mirrors need to be precise to a fraction of a wavelength, and this leads to very great differences in engineering practice. The optical telescope, working at a wavelength around 0·5 micrometres (1/50,000 inch), uses carefully polished mirrors coated with thin metal films to achieve the necessary precision. The radio telescope, if working at 9 centimetres, requires a mirror accurate to only about a centimetre if it is to give essentially perfect performance. Reber's telescope was nearly all built of wood, with a mirror surface made up from forty-five pieces of thin galvanized iron. The larger tolerances also allowed Reber to build a bigger mirror and to use a much "faster" focal ratio than was practicable for visible light, so the mirror had an aperture of 9·4 metres (31 feet) and a focal length roughly equal to the aperture, giving a focal ratio of about 1.

Reber built the telescope in four months, almost entirely within his spare time, though he had part-time help with the foundations and the overall assembly. His first attempts to detect signals at the selected wavelength of 9 centimetres were completely unsuccessful. This upset the idea that the source of Jansky's signals was radiation from hot objects. Reber made further unsuccessful observations in 1938 and early 1939 with a receiver ten times more sensitive, working at 33 centimetres (13 inches) wavelength. He once more rebuilt his receiver to detect yet longer wavelengths, again with an improvement in sensitivity. The new choice, 1·87 metres (74 inches), was more susceptible to terrestrial interference, especially from vehicle ignition systems, but Reber at last detected signals from the Milky Way. The source could not be the stars, as the 1·87-metre wavelength signals were weaker than those at 15 metres (49 feet), the reverse of that expected for a hot source. Reber pointed out that a more likely source of the radiation was collisions occurring in the tenuous interstellar gas: we now know that some of the radio waves are from high-speed electrons moving through interstellar magnetic fields.

Now that he had a signal to study, Reber was able to do more than confirm that the Milky Way was a radio source. His telescope, like Romer's seventeenth-century transit telescope, could only track up and down the meridian, but with the rotation of the Earth the telescope scanned a strip of the sky as it passed overhead. By pointing his telescope to different elevations on successive days, Reber could build up a map of the intensity of radio emission over much of the sky. The resolution was terrible, though significantly better than that achieved by Jansky. The price exacted for working at long wavelengths is that, although the mirror need not be so precise, the sharpness of the image is fundamentally restricted. Jansky's 30-metre aerial array was two wavelengths long, and was unable to resolve any detail less than 30° across. Reber's mirror had a diameter five times his 1·87-metre wavelength so that his telescope achieved a resolution of 12°–two and a half times better. (The much shorter wavelengths of visible light mean that even a 10-centimetre/4-inch aperture is 200,000 wavelengths across, and can resolve a detail an arc second across. Optical telescopes of larger aperture are limited in performance by the atmosphere, not by the wave properties of light.)

Reber's first results were published in 1940. By this time there were other, more pressing reasons for the development of highly sensitive shortwave receivers. The increased positional accuracy obtained with short wavelengths applies to the observation of radar echoes as well as to radio astronomy, and the Second World War saw tremendous strides in shortwave techniques. The usefulness of radar is set by the ability of the receiver to detect faint signals reflected from distant targets, and these signals can be severely disrupted by jamming – the transmission of obstructive high-power signals by enemy transmitters. James Hey, an X-ray crystallographer, was relabelled "radio expert" by the British Army and directed to study German radar jamming techniques. Soon after starting work Hey was confronted with reports of a new and very effective jamming method. British anti-aircraft radar sets, working at different wavelengths and spread over most of the country, were all rendered inoperable by an extraneous source of radio noise. Hey found that the source moved during the day; fortunately, it proved to be the Sun and not a new and unexpectedly powerful enemy transmitter. Although the Sun was at a period of minimum activity, as it had been during Jansky's experiments, there was a large sunspot near the centre of the Sun's disc and this active region was acting as a powerful shortwave radio source.

Hey's discovery of radiation from active regions on the Sun was fortuitous, but the next advance in radio astronomy was deliberate. In June 1942, G. C. Southworth of Bell Telephone Laboratories completed a very short wavelength ("microwave") receiver, working at 3·2 centimetres ($1\frac{1}{4}$ inches) wavelength. As a test of the new system, the antenna connected to this receiver was pointed at the Sun and an increased signal was detected. The amplitude of the signal was roughly consistent with that expected from a hot, opaque body at the temperature of the Sun. It was steady in intensity and was therefore due

INTERFEROMETRY

It is impossibly expensive to build radio telescopes above 100 metres (330 feet) in aperture. Interferometry achieves the effect of a large aperture by the addition of signals from smaller telescopes; the resolution of the two systems is comparable.

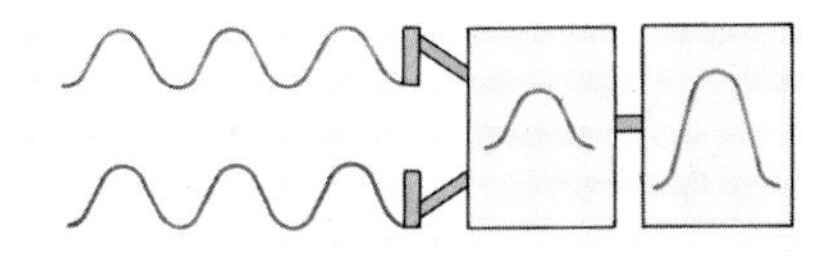

In the radio region the signals from the two mirrors can be recorded separately and then combined in a common amplifier. If the source is precisely on the axis of the interferometer the waves arrive with their motions exactly in step (right). The addition of these two signals in the final stage of the system gives a signal with twice the amplitude. As the rotation of the Earth turns the pair of telescopes, so that the radio source is off-centre, the two waves no longer arrive in step (right, centre) and the amplitude of their sum diminishes. Further rotation can result in the two signals being exactly out of step, so that the sum is reduced to zero (right, below). This alternate addition and cancellation is called "interference" of the two waves. A big swing between large and zero signal occurs only if the source is small in extent. Larger sources give an output in which the contrast is much reduced.

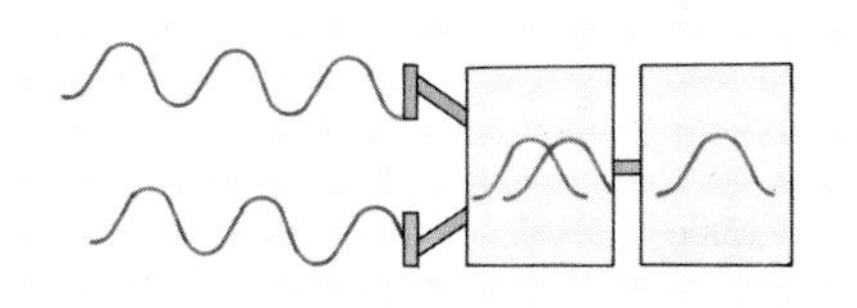

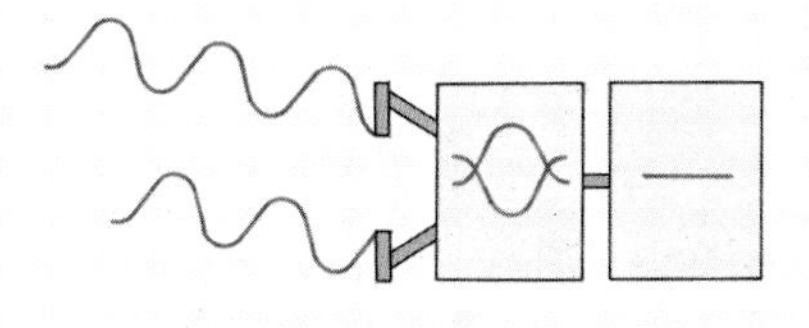

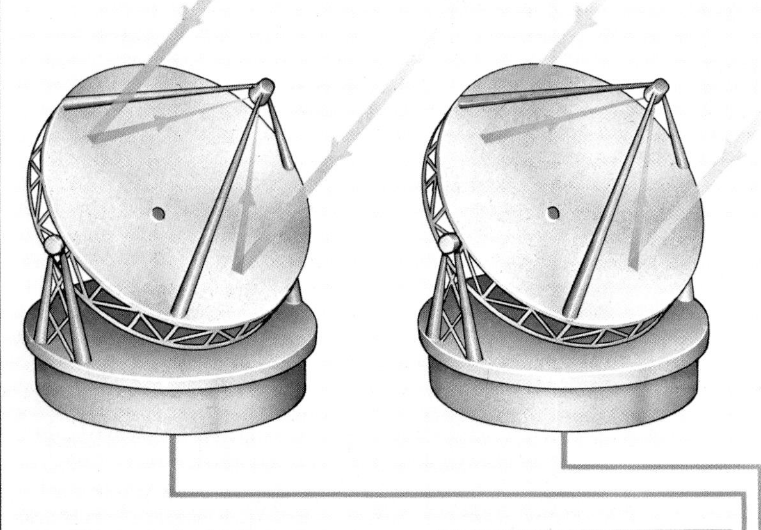

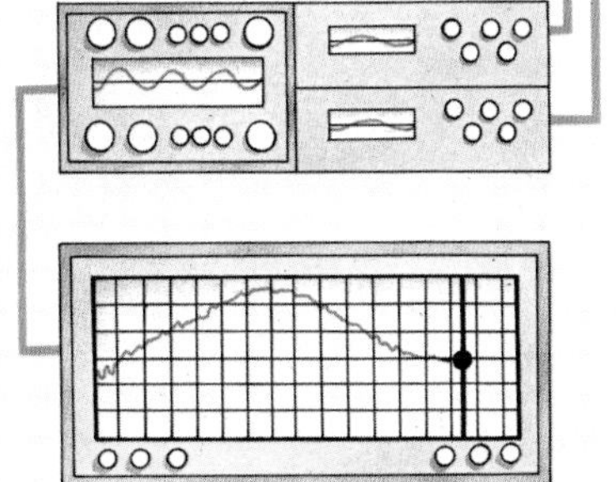

As with an optical telescope, a radio telescope is dominated by the structure needed to support a thin reflecting surface. The reflector concentrates incoming radio waves on to a small aerial mounted at the focus. From this aerial the signal is fed into a receiver and compared with radiation from a standard source. The power received is less than a billionth of a watt.

to the "quiet" Sun rather than some active region on its surface. The result was "confirmed" by Reber in 1944: this took the form of an independent rediscovery, Hey and Southworth being unable to publish findings that were of military relevance.

The next discovery in radio astronomy came as a result of a desperate advance in weapons. The Germans developed a new device to bombard London: the "Vergeltungswaffe-2" – reprisal weapon No. 2 or simply V2. This was the first long-range missile, a rocket with a trajectory which rose 150 kilometres (90 miles) in the air to reach a target 250 kilometres (150 miles) down range. Defence against this weapon required British radar sets to be realigned to observe high above the horizon. The new system proved successful. It not only detected radar echoes from V2 rockets and Jansky's and Reber's cosmic static, but also observed a new class of effect – brief radar echoes, one every five or ten minutes, from altitudes around 100 kilometres (60 miles). The echoes continued after the V2 rocket bombardment ceased and by the end of the war Hey and his colleagues were able to divert several radar sets to the study of these echoes. The cause proved to be the ionized (and thus radar-reflecting) trails of meteors.

Although no further experimental developments were achieved before the end of the war, there was one major development in the theory of radio astronomy. Copies of Reber's papers reached Leiden Observatory in German-occupied Holland and were discussed at a colloquium by the director, Jan Oort. Oort pointed out that the radio waves originated in processes which emitted energy at all wavelengths – there were none of the spectrum lines that proved so useful in the study of visible starlight. One of his colleagues, Hendrik van de Hulst, realized that there was a possible process that would give a line in the radio spectrum. This involved a very rare process, the emission of radio energy as a result of an internal realignment of the proton and electron, which together make up a hydrogen atom. He calculated that the wavelength would be 21 centimetres ($8\frac{1}{4}$ inches), and it was hoped that the vast reaches of our galaxy would contain enough hydrogen atoms for the line to be detectable. One hydrogen atom per cubic centimetre undergoing a "21-centimetre" transition once every million years or so might still add up to a detectable energy, since a radio telescope would be picking up radiation from over thousands of parsecs of the interstellar medium. No experimental work was possible: the problem in Holland in 1944 was one of simple survival.

After the war the radar equipment, either as complete systems or just the new

In Britain, radar tracking stations (above) were used to detect the imminent threat of V2 rockets and this stimulated the development of radio astronomy.

Jan Oort (above) is the father of Dutch radio astronomy and an authority on the structure of our galaxy.

Sir Bernard Lovell (below), founder of Jodrell Bank Observatory. His research has included telescope design and the study of "flare" stars.

and highly sensitive shortwave and microwave receivers, were rapidly turned to peaceful use. The former were used in radar astronomy, bouncing radio signals off astronomical bodies and detecting the echoes, while the receivers alone were ideal for radio astronomy, the detection of natural radio emissions from celestial objects. The improvement in shortwave radio technology during the war was most clearly demonstrated in January 1946, when J. H. De Witt of the US Army Signal Corps Laboratory succeeded in detecting radar echoes from the Moon. He was able to measure the 2½-second delay sufficiently accurately to determine the Earth–Moon distance with a precision of about 1500 kilometres (930 miles). This measurement, and a similar experiment by Z. Bay in Hungary, were the first direct determinations of the distance between two bodies in the solar system.

While De Witt had used an entire radar system, a group at the Massachusetts Institute of Technology, led by Robert Dicke, applied a newly developed microwave receiver to radio astronomy. The receiver was built to study atmospheric absorption and worked at 1·25 centimetres (½ inch), an even shorter wavelength than Southworth had used. Its performance was unprecedented as Dicke had invented a new technique which vastly improved the sensitivity of his apparatus. He demonstrated the effectiveness of his methods by measuring the microwave radio emission of the Moon. The Moon has no atmosphere and so no thunderstorms or other atmospheric radio phenomena. What Dicke was doing was detecting the radiation from the Moon as a "hot" body. All objects emit electromagnetic radiation, the strength at each wavelength depending on their temperatures. Just as Southworth had found microwave radiation from the Sun, emitted by a region at a temperature of several thousand degrees, so Dicke detected radiation from the Moon's surface and measured the temperature as 19°C.

The difficulty that Dicke had overcome in his measurements arises from the fact that the weaker a signal, the longer it must be observed to be certain that the signal is really there. This is impossible if there is any slow and unpredictable drift in the response of the system. Indeed, such an instrument problem had led Reber to believe that he had detected radio waves from M31, the Andromeda galaxy. Dicke's new method involved switching the input to the amplifier to and fro between two sources. One was the radio telescope itself, and the other a standard and stable calibration source. Slow drifts in the receiver response would affect both inputs equally, and the spurious electronic signals, called noise, that are generated by an amplifier of even the highest quality, would also be the same for both channels. The difference between the two channels was amplified and recorded and was a reliable record of the signal from the radio telescope, free of amplifier vagaries. The gain allowed by this new technique was enormous. Dicke was able to detect signals that were about one-ten thousandth of the "noise" level of his apparatus, and he measured the lunar temperature with an antenna only 45 centimetres (18 inches) in diameter.

Although the two American observations – De Witt's radar determination of the Moon's distance and Dicke's radio measurement of its temperature – were the high points of the immediate post-war period, developments elsewhere were to provide a more enduring base for the infant subject of radio astronomy. Four laboratories took up the subject with enthusiasm, all of them established by individuals who had been concerned with wartime radar development. Hey continued with radar and then radio astronomy in an essentially military situation at the Army Operational Research Group (AORG). Also in Britain, Bernard Lovell returned to Manchester University and Martin Ryle took up research at Cambridge, as he had planned to do in 1939. In the Southern Hemisphere E. G. Bowen and J. L. Pawsey, who had both been working on radar systems for the war in the Pacific, guided the Australian National Radiophysics Laboratory into solar radio astronomy.

All the new groups begged, borrowed or otherwise acquired surplus radar equipment. Lovell set up a copy of Hey's 5-metre-wavelength radar to seek radar reflections from the high-energy particles arriving from outer space – the cosmic rays. Cosmic rays are single atomic nuclei, coming to us from interstellar space and travelling at velocities very close to the velocity of light, 300,000 kilometres (186,000 miles) a second. Although the cosmic rays move faster, the heavier but slower meteors detected by Hey are still a million or more times more energetic than the fastest cosmic rays. Lovell's attempts to detect radar echoes from the cosmic rays were not successful, and he rapidly turned to the radar study of meteors. The unsuccessful search was not without effect. It forced Lovell to move his radar set from the physics department on the university campus in Manchester to a site with less electric interference. He chose the remote Cheshire hamlet of Jodrell Bank, where the botany department had already established a field station. On this site Lovell, with the help of John Clegg, built the first of his big radio telescopes – a paraboloid 66 metres (220 feet) in diameter. The telescope consisted of a mesh of wire forming a fixed dish pointing vertically upwards, a design chosen to allow maximum size since the main priority was to gain maximum sensitivity.

While Lovell and Hey pursued radar observations of meteors, the other two groups, Pawsey in Australia and Ryle at Cambridge, set out to study the radio physics of solar activity. Both groups faced the same problem: solar radio noise appeared to be associated with sunspots, but the smallest detail that could be resolved, even by Lovell's big dish, was still several times the size of the entire Sun. If a single radio telescope was to resolve details comparable in size to a large sunspot, it would need to be three or four thousand wavelengths across. This was clearly impossible at the metre wavelengths at which the active Sun was known to radiate. Ryle and Pawsey turned instead to devices known as interferometers. An interferometer consists of two separate small telescopes whose outputs are connected electrically. Although two suitably connected aerials a thousand wavelengths apart could not give the sensitivity of a big dish, they would give the angular resolution needed.

The first system that demonstrated an angular resolution able to show detail smaller than the Sun's disc was set up by Pawsey in Australia. His method was ingenious and used a single aerial mounted at the top of a cliff looking out to sea towards the rising Sun. The sea acted as a flat and efficient mirror for 1·5-metre (60-inch) wavelength radio waves and the instrument performed as if there were two aerials, one on the cliff top and the other its reflection in the sea. Two different cliffs, one 85 metres (280 feet) and the other 120 metres (400 feet) high were used to study the great sunspot of February 1946. The two interferometers were just sufficient to show that the area of the Sun that was emitting radio waves was comparable in size to the sunspot.

The Cambridge interferometer, which was operating by June 1946, was simpler and more versatile. Ryle employed two

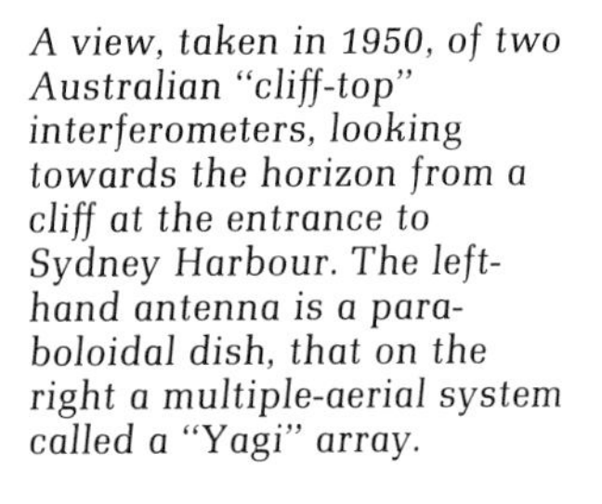
A view, taken in 1950, of two Australian "cliff-top" interferometers, looking towards the horizon from a cliff at the entrance to Sydney Harbour. The left-hand antenna is a paraboloidal dish, that on the right a multiple-aerial system called a "Yagi" array.

aerials separated horizontally, a geometry which allowed observation away from the horizon and made changing the separation of the aerials a simple task. Like Pawsey, Ryle was able to show that active radio regions on the Sun were about the right size to be sunspots, but both British and Australian groups were overtaken by a discovery from Hey's unit, the AORG.

One part of Hey's group was concerned with the construction of a new radio map of the sky, working at 5 metres ($16\frac{1}{2}$ feet), the wavelength of most of the group's radar equipment. J. W. Phillips discovered that there was a powerful radio source in the constellation Cygnus, whose power fluctuated rapidly, changing by fifty per cent in a few seconds. This implied a small angular diameter for the source, and the position, far from any member of the solar system, meant that the source must be distant and thus very powerful.

J. G. Bolton and G. J. Stanley applied the Australian cliff-top interferometer to the new source and showed that it was less than eight arc minutes across. The discovery also spurred both interferometer groups to look for celestial radio sources other than the Sun. Several were found and a new notation developed – the strongest source in Cygnus was called Cygnus A and the next Cygnus B, in a Roman equivalent of Bayer's seventeenth-century Greek letter labels for the bright stars. At first none of these radio sources could be identified with any visible astronomical object. This was in part because an eight-arc-minute circle in the sky still includes a lot of objects, and in part due to the difficulty of estimating the corrections due to atmospheric refraction. Refraction affects radio waves as strongly as it does visible light, and was particularly significant for the cliff-top interferometers which looked almost horizontally through the atmosphere. The main clue for the identification of the new radio sources was that the sources had to be rare objects. Nearby stars like Sirius and Vega gave no signals, and neither did the stars of the Milky Way. In 1949 Bolton and his colleagues identified three radio sources – M1, the Crab nebula; M87, a distant galaxy; and NGC 5128, a bright galaxy too far south to be included in Messier's list. The Crab nebula – radio name Taurus A – was known to be the debris from a supernova in our galaxy, while both the galaxies were exceptional in structure. M87 – Virgo A – has a streak of light projecting from an almost spherical profile, suggesting a cataclysmic event in the interior of the galaxy. NGC 5128 – Centaurus A – is a dusty and turbulent galaxy with a peculiar spectrum in visible light.

These identifications completely altered the direction of radio astronomy research. Radio astronomy ceased to be just a study of objects in the solar system and now offered a new technique for the study of the rest of the universe. This outward-looking aspect of the subject was further emphasized by the discovery in 1950 of the predicted 21-centimetre ($8\frac{1}{4}$-inch) line of interstellar hydrogen gas. The discovery was made virtually simultaneously by three groups, including Oort's team at Leiden. The same year saw the identification of radio emission from the Andromeda galaxy by Ryle's group. The radiation proved to be comparable in type with that from the Milky Way, though much weaker, since the source is seventy times farther away than the centre of our own galaxy.

The year 1950 was also notable for developments in the technology of radio astronomy. A new group at the US Naval Research Laboratories started work with a 15-metre (50-foot) aperture radio telescope. The telescope's importance lay not only in its ability to point at any radio source but in its also unusually precise construction. The surface finish and mechanical stiffness were such that the telescope was able to operate at wavelengths as short as 1 centimetre ($\frac{2}{5}$ inch). The second development was the invention of a new type of interferometer by Hanbury Brown at Jodrell Bank. This was a controversial device which later proved to be more use for visible region interferometry than for radio studies.

The third new project, and one that was to prove more stormy than even Hanbury Brown's interferometer, was the start of

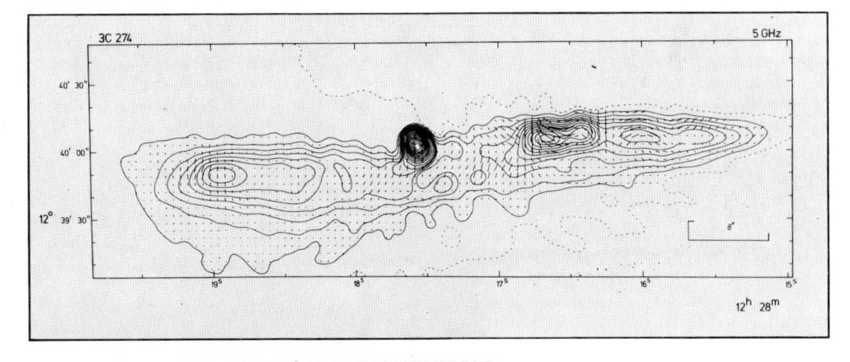

The giant galaxy M87, the most massive known object in the universe, seen in visible light (left) *and in the radio region* (above). *It is a member of a cluster of galaxies on the northern border of the constellations in Virgo. The jet, which radiates blue light and radio waves, contains enough matter to make a million suns, but the source of its energy is not understood. The entire galaxy, 13,000 parsecs across and 10,000 billion times more massive than the Sun, is receding from us at a velocity of 1220 kilometres (760 miles) a second.*

the 76-metre (250-foot) steerable dish at Jodrell Bank. Lovell drove through the construction of this giant radio telescope in the face of a series of crises, mostly due to escalating cost. Reber had proposed in 1948 that a 67-metre (220-foot) aperture telescope should be built, but the proposal was not so much rejected as politely ignored. Lovell was not a man to be ignored and he had the considerable support of the head of the physics department at Manchester, P. M. S. Blackett. Blackett was one of the most influential of his generation of British physicists and he believed that progress in post-war research necessarily required big instruments. The telescope was completed seven years later at a final cost that was roughly ten times the original cost estimate: indeed, it appeared at one time that Lovell was likely to finish in jail for building the instrument without having sufficient funds to pay for it.

The interval during which the Jodrell Bank dish was being designed and built was chiefly noteworthy for the development of a series of larger and more sophisticated interferometers and for the rise of American radio astronomy. The United States had not been a major contributor in the early days of radio astronomy, partly because the majority of early apparatus was based on anti-aircraft defence radar. Defended by oceans too wide to be crossed by hostile bombers, the Americans did not need to deploy such equipment. In addition, the new optical telescopes and the clear sky above Mount Palomar guaranteed that the western United States would continue to be the world centre for optical astronomy. Other countries gained from the doubtful advantage of cloudy skies. Radio waves can penetrate cloud, and students eager to do astronomy in such climes are more easily guided to radio astronomy.

Three groups were building interferometers at this time: Hanbury Brown at Jodrell Bank, Ryle and Graham Smith at Cambridge and B. Y. Mills and W. N. Christiansen in Australia. Hanbury Brown made his own interferometer obsolete by showing that it was possible to combine two radio telescopes as a conventional interferometer by connecting them with a radio link when the separation is too great for a cable to stretch. Lovell and Clegg's fixed 66-metre (217-foot) dish was used as one of the two telescopes. The other was a small mobile unit 6 metres (20 feet) square. The radio link was complicated, as it had to be exceptionally stable. The interferometer was used to study a set of five "radio stars" which were well away from the plane of the Milky Way and thus not likely to be objects within our galaxy such as the Crab nebula and other supernova remnants. (In addition to Virgo A (M87) and Centaurus A (NGC 5128), the first radio star Cygnus A had by now been identified with a very distant galaxy.)

The first trial was with the telescopes separated by 910 metres (1000 yards) and took place in mid-1954. The interferometer spacing was then increased again and again in order to give ever better angular resolution: 1·2 kilometres (1320 yards) in March, 1955, 1·8 kilometres ($1\frac{1}{8}$ miles) in April, 3·9 ($2\frac{1}{2}$ miles) in June, 12·8 (8 miles) in August and 20 kilometres ($12\frac{1}{2}$ miles) by the beginning of 1956. This last separation, 10,600 wavelengths, achieved an angular resolution of twelve arc seconds. Even so, three of the five sources still showed no structure and must be smaller than the telescope resolution. The small size and high intensity suggested that the three sources were similar to Cygnus A, one of the few sources that had been identified with a distant galaxy.

The Australian development of radio interferometers pursued a different design approach. The limitation of a two-telescope interferometer is that, while it can match the resolution of a dish with a diameter equal to the aerial separation, it gives only one component of the overall picture. To build a complete picture it is necessary to use many pairs of aerial spacings. One set, with the pair of aerials at different separations, gives information on a picture as a set of slices of different widths. A second set, observed with the aerials set up in a direction at right angles to that used for the first, is needed to complete the picture. The alternative is to use an array of aerials which includes all the separations simultaneously. This concept led to two multi-aerial interferometers. In the first of these, the Mills Cross, Australian radio astronomer B. Y. Mills used two rows of aerials, forming a cross with arms 450 metres (500 yards) long. The signal strength was increased by mounting a trough-shaped reflector behind the aerials and the entire cross was roughly equivalent in resolution and signal strength to a telescope made by selecting two perpendicular strips from a (hypothetical) 1500-metre ($\frac{7}{8}$-mile) single dish. The interferometer could not be continuously guided: it was used as a transit telescope in the same way as Reber's and Clegg's dishes. Its importance was that it achieved a high angular resolution at low cost.

Neither the two-element interferometer nor the Mills Cross was ideal for studying the Sun. The rotation of the Sun and the short lifetime of sunspots and other active regions on the solar surface prevent a sequence of observations with a series of two-aerial spacings. By the time a new aerial separation has been established the

Sun's surface has changed and the new observations are unrelated to the earlier data. Transit instruments can only observe the Sun for a couple of minutes a day and a combination of the angular resolution of the Mills Cross and the ability to track given by a steerable radio telescope was needed.

This next step was taken by W. N. Christiansen, one of Mills's colleagues at the Australian Radio Physics Laboratory, who replaced the aerials of Mills's cross with lines of separate, small, radio telescopes. The first trial employed thirty-two telescopes 1·8 metres (6 feet) in aperture in a line 220 metres (240 yards) long. The resolution was much better than that of the Mills Cross as the array – the "Chris-Cross" – worked at a much shorter wavelength (21 centimetres/8¼ inches) and so was much larger in relation to the wavelength of the observations. From this array Christiansen proceeded to a cross of radio telescopes with thirty-two dishes in each arm, each 5·8 metres (19 feet) in diameter.

In addition to the study of active regions of the Sun, Christiansen's array was able to map the radio emission of the quiet Sun. The result is very different from the appearance of the Sun in visible light.

The 76-metre (250-foot) aperture telescope at Jodrell Bank (above). *The whole 2000-tonne telescope turns on a circular railway 110 metres (350 feet) in diameter.*

B. Y. Mills's cross-shaped interferometer (below) *built near Sydney, Australia, in 1952. The low cost of Mills's system was partly due to the fact that radio waves will reflect from wire mesh, provided the holes are much smaller than the wavelength.*

Because the solar corona emits radio waves the "radio" Sun is larger than the visible disc. It is also brightest at the edges of this larger image, in contrast to the visible appearance, where the edge of the Sun is less bright than is the centre of the solar disc.

The third interferometer group, that at Cambridge, was chiefly concerned with the vital task of surveying the sky to map an increasing number of radio sources. A preliminary survey, using a two-aerial interferometer, had shown some fifty sources, and a new interferometer was built in the early 1950s. The new device used four trough-shaped reflectors at the corners of a rectangle 580 by 50 metres (630 by 55 yards). A survey at a wavelength of 3·7 metres (12 feet) was not very successful as the long wavelength meant a poor resolution and consequently the analysis was difficult when the signal contained radiation from several overlapping sources. Halving the wavelength doubled the resolution, and a new survey and catalogue resulted. This, the third Cambridge catalogue, abbreviated to 3C, is the standard list of nearly 500 bright radio sources visible in the skies of the Northern Hemisphere.

The 25-metre (83-foot) reflector at Dwingeloo in the Netherlands, completed in 1955, at which time it was the largest in the world. This telescope, light in weight (only 127 tonnes) and therefore of modest cost, was deliberately designed to study the interstellar (21-centimetre) hydrogen line.

The interferometers at Manchester, Cambridge and Sydney were all operating by 1955 and during the rest of the decade radio astronomers were more concerned with the construction of single-dish steerable radio telescopes. A large collecting area and the ability to track one object for hours at a time give the telescope a much greater sensitivity than the interferometer, though without the ability to see fine detail. Three large radio telescopes were completed in 1956. One, of aperture 25 metres (82-feet), was built by the Dutch to study the 21-centimetre hydrogen wavelength. The others, two 26-metre (85-foot) aperture instruments, were funded by the United States Naval Research Laboratory, one for their own use and the other for the University of Michigan. In 1957 Lovell's 76-metre (250-foot) telescope was completed at Jodrell Bank and the first of a new class of instrument was built in Owen's Valley, California.

The latter instrument comprised a pair of telescopes, each of aperture 27 metres (90 feet), connected together as an interferometer and with one of the pair mounted on a 1600-metre (1-mile) long railway track. The rail-mounted system combined the enormous sensitivity of steerable telescopes with the resolution and flexibility of variable-spacing interferometers. A third 26-metre telescope marked the establishment of the United States National Radio Astronomy Observatory, while the Russians built a 22-metre (72-foot) aperture instrument designed to reach the unusually short wavelength of 8 millimetres ($\frac{1}{3}$ inch).

The new large-aperture telescopes quickly made significant contributions to radio astronomy. The Naval Research Laboratories group extended Dicke's work to measure the surface temperatures of Venus, Mars and Jupiter.

Both interferometers and "dishes" proved able to detect complex structures within the 21-centimetre line that was a component of the radiation from our own galaxy. This work, by Oort, van de Hulst and others at Leiden and by Christiansen and Kerr in Australia, gave a third dimension to radio maps of the galaxy. Optical astronomers had used observations of nearby hot stars – Baade's Population I stars – to map out the local star distribution in our own galaxy and, in the early 1950s, had shown that there were indications of spiral structure. The Leiden/Sydney map, based on 21-centimetre radio observation, showed the spiral structure for eighty per cent of the entire galaxy. For the first time astronomers could leaf through the pictures of other spiral galaxies in *The Hubble Atlas of Galaxies* and say, with some confidence, "Our galaxy looks like that one."

CHAPTER TWELVE

A BETTER DETECTOR IS A SHORT CUT TO A BIGGER TELESCOPE

The reflecting telescope 1950–75

"My qualification is that I have observed for forty years, probably longer than anyone here, with telescopes from 0·6 to 5·0 metre apertures, starting with photography but currently using all existing electro-optical auxiliaries, e.g. guiding in a brightly lit data room (using an intensified TV or SIT viewer), using Oke's 32-channel spectrophotometer, or the Gunn-Oke intensified-silicon-vidicon digitised spectrograph, or Boksenberg's image-photon-counting system at high and low resolution, or Shectman's linear array of diodes in the coudé."

Jesse Greenstein at the European Southern Observatory Conference on Telescopes of the Future, Geneva, December 1977.

The effect of the Second World War on optical astronomy was, to a large extent, to put the subject into cold storage. Scientists were transferred to the war effort, and the pace of work at observatories, including the building of the 5-metre (200-inch) Hale telescope on Mount Palomar, slowed considerably. One of the few gains was the work of Walter Baade at Mount Wilson. The German-born Baade was not involved in the American war effort and was able to take advantage of both the large amount of time available on the big telescopes and the reduced sky brightness that resulted from the black-out of nearby Los Angeles. His aim, attempted unsuccessfully many times before by other astronomers, was to resolve the inner region of the Andromeda galaxy into its individual stars. The spectrum of the galaxy showed that it was made up of stars, but they were so closely packed that all previous photographs had shown only a continuous hazy blur. The first requirement for a new attempt was to have the telescope in absolutely first-class condition. The mirror had to be freshly coated and its supports checked and adjusted to give an image as near perfection as possible. On nights when the air was still, the seeing unusually good and the temperature steady, the plate-holder was adjusted to be in perfect focus. In freezing cold conditions Baade had to watch through the guiding eyepiece to make sure that whenever the image drifted it was immediately re-centred. Many plates were required to find variable stars by their changing brightness from plate to plate and to allow the use of colour filters to discriminate between stars of different colour and, hence, temperature.

Baade's skill and concentration were rewarded by the discovery that the stars in the central region of the Andromeda galaxy were fundamentally different from those in the spiral arms. The two types of star could be described as different populations. The spiral arm stars, which Baade called Population I, included hot blue stars as the brightest members; these were similar to the stars near the Sun. In contrast the Population II stars, found in the nucleus, were dominated by cool red giants similar to those observed in the globular clusters of our own galaxy. Cepheid variables occurred in all parts of the Andromeda galaxy, but Baade found that the two populations showed different relations between brightness and period for these important stars. Population I Cepheids were three times as bright as Population II of the same period. The Cepheid variables were assumed to be identical when Hubble used them to find astronomical distances, so much work was needed to re-establish a reliable scale.

The new studies could not be done at Mount Wilson. The end of the war was followed by a surge in outdoor lighting, and this ended the forty-year supremacy of the observatory. In Europe, many of the great observatories were in ruins. The hope for the future lay in the more remote parts of California, where on Mount Palomar the Hale telescope was nearing completion and on Mount Hamilton a 3-metre (120-inch) was being designed for the Lick Observatory.

The pre-war design of the Palomar 5-metre set a new pattern for the post-war telescopes. The enthusiasm of George Hale for a bigger telescope had required the design team to solve three problems. The first was the need to find a support frame that was sufficiently stiff but, unlike the Mount Wilson 2·5-metre, would allow the telescope to study the whole sky, including the region near the Pole Star. Second, the bearings would have to provide smooth motion for a telescope that would weigh many hundreds of tonnes. Third, the secondary mirror, itself over a metre in diameter, must be held in alignment with the primary as the telescope swung across the sky. All these problems

The Andromeda galaxy (left) *is the brightest and most massive marker of the "local group" – an isolated cluster of eighteen galaxies, all within a million parsecs of the Earth.*

Although Greenwich Observatory was damaged during the Second World War (below), *observations were maintained.*

could be eased if the telescope tube could be short and therefore stiff and the main mirror kept as light as possible. A lightweight mirror offers a big advantage as it leads to a lighter support system and a lighter overall structure, so the 5-metre Pyrex mirror blank was cast not as a solid disc but as a hexagonal cellular structure, rendering it only about half the normal weight. The reduction in overall length was achieved by making the ratio of mirror aperture to focal length (the focal ratio) much larger than had been used in previous telescopes – for the new mirror the ratio was 1:3·3, giving a focal length of only 16·5 metres (54 feet). This reduction in overall length greatly reduced the size and cost of the dome, but it also upset the alignment tolerances for the secondary mirror. The solution, invented by Mark Serrurier, was to deliberately allow the telescope to bend. The art was to make both ends of the tube droop by the same amount, and in such a way that they remained parallel and kept the primary and secondary mirrors in alignment. The final design has a simple telescope tube carried in a modified yoke mounting and supported on a hydraulic support called an oil pad, in which the load is carried by a thin film of oil at high pressure. The giant structure was assembled on Mount Palomar and testing began in early 1948. Within three years the observatory was able to publish pictures of star images only half an arc second in diameter.

With the new telescope Baade was able to quantify the behaviour of his two different types of Cepheid variable and as a result revise the astronomical distance scale. All distances beyond our own galaxy were immediately doubled, a change which showed all the other galaxies, as "extra galactic nebulae" were now re-named, to be much larger than was previously believed. This had the advantage that our galaxy, the Milky Way, was no longer ostentatiously bigger than all the others, a result that supported the underlying Copernican view that the Earth does not occupy an exceptional site in the universe. Other facts fell into place – the brightest stars in the Andromeda galaxy were now of much the same brightness as those in our own environs and no longer inexplicably faint. It also became clear that, in our own galaxy, the 1·5-magnitude difference in brightness of the spiral arm and the globular cluster Cepheids was offset by absorption due to interstellar dust. The dust is only found near the Population I (spiral arm) stars and it was a most unfortunate coincidence that the dimming due to the dust precisely balanced the extra brightness, so that from Earth all "local" Cepheids looked the same.

The critical part of Baade's observations involved the detection and study of brightness variation of faint stars and was done by measuring the changes in blackness of photographic images. This

The 5-metre (200-inch) telescope at Palomar was the first instrument big enough to carry the observer at the prime focus. The telescope is pointed to the horizon to enable the observer to climb into the observing "cage". When built in the late 1940s, the Hale telescope incorporated two important innovations. The "Serrurier" truss, which comprised a built-in flexibility in the telescope tube, permitted a high focal ratio and close alignment of the primary and secondary mirrors: these were set 13 metres (43 feet) apart at opposite ends of a 140-tonne tube and remained in alignment to within 0·25 millimetres (1/100 inch). Second, a new yoke system allowed the telescope to reach the whole sky; this changed the shape of the instrument dramatically from traditional reflectors. The north end of what, in the Mount Wilson 2·5-metre (100-inch), had been a simple oblong frame, was enlarged on to a 14-metre (46-foot) diameter disc with a notch cut on it to allow the telescope tube to swing right over to the Pole Star.

method suffers from many limitations: photographic emulsions are inefficient detectors of light and their response does not show any simple relation between the quantity of light incident on the plate and the resultant blackening. By the early 1950s a new electronic device, the photomultiplier, was coming to the fore, as it was both more efficient and had a direct relation between input (light) and output (an electric current). The origins of the photomultiplier can be traced back to early experiments by Heinrich Hertz, the discoverer of radio waves, which showed that ultraviolet light caused the ejection of electrons from metal surfaces. The phenomenon, called the photo-electric effect, was explained by Albert Einstein in 1905 in a paper that won him the Nobel Prize for physics. Einstein drew heavily on an earlier controversial idea put forward by Max Planck in 1900, that a beam of light could be regarded as a vast number of little packets of light energy called photons, and the energy of each individual photon depended on the wavelength. Ultraviolet photons were more energetic than those of longer-wavelength visible light and so were more able to eject electrons or blacken photographic plates.

The thirty years from 1905 to 1935 saw the development and application of a simple light-sensitive detector based on the photo-electric effect, the photo-cell. In this device the incident light fell on a sensitive surface called the photocathode housed within an evacuated cell, and the ejected electrons travelled across the evacuated space to a collecting electrode. The stream of electrons – an electric current – could be measured using a very sensitive ammeter, and this current was directly proportional to the intensity of the incident light. The longest wavelength to which the photocell would respond and the efficiency of its response depended critically on the detailed composition of the photocathode. The majority of the improvements made in the interval between the two world wars were due to developments in the black art of photocathode preparation. Plates of silver or tungsten overlaid with mixtures of other, more exotic metals, treatment of the surfaces with non-metals – all found a place in the photocathode chef's recipe book.

The photocell was first applied to astronomy in 1924 by Paul Guthnick at Babelsberg in Germany and soon after used by the Danish astronomer Elis Strömgren and his son Bengt to register star transits. The problem was that, although a typical photocathode would produce an electron for every twenty or so photons incident on its surface, the resulting current was (when measured in amperes) infinitesimal for all but the very brightest stars. The best measuring devices were not sensitive enough to register such small currents, and an electronic amplifier was no use, as the amplifier itself generated a spurious signal ("noise") that was much larger than the electrical signal that was being sought. The result was that the photocell was more use when applied indirectly to astronomy. In 1937 A. E. Whitford and G. E. Kron built an automatic telescope guider using a photocell to watch a bright star at the edge of the field of view and moving the plateholder to keep the image steady; S. McCuskey and R. Scott developed a photo-electric device for counting and measuring the intensity of the star images on their plates.

The first intimations of the development needed if the photocell were to be applied directly to astronomy appeared in the years before the Second World War. Physicists had known for many years that if an electron were accelerated in an electric field and thumped into a metal surface, it was not always absorbed. Sometimes it just bounced off, but on other occasions it would chip a few additional electrons out of the surface. As with photocathodes, the process was much affected by the composition of the surface, and more esoteric cookery followed. In 1935 H. Iams and B. Salzburg put together a photo-cell with an intermediate electrode in which the electrons that had been ejected from the photocathode were accelerated towards a metal plate. On impact with the plate each electron knocked five or six more electrons out of its surface, and the whole collection then travelled to a final collector electrode. The net result was an output current six times that of a simple photo-cell. The next year G. Weiss of the German Post Office built a device with eight consecutive stages of acceleration and multiplication, and achieved an overall gain of about ten thousand. Better design, guiding the electrons from stage to stage, was pioneered by Vladimir Zworykin and his colleagues at the Radio Corporation of America (RCA), and by the end of the war the fully developed system, the photomultiplier, was in operation. This device could take the initial electron produced by light hitting the photocathode and amplify it down a cascade of intermediate electrodes to give a million electrons at the output.

After the war the photomultiplier was rapidly adopted by the more forward-looking astronomers. The gain was enormous: one out of every three or four photons incident on the photocathode would eject an electron and give a pulse of a million or more electrons, while it took between 300 and 400 photons to produce just one blackened grain on a photographic plate. The new detector was also

THE PHOTOMULTIPLIER

Light enters through the transparent wall of the tube and impinges on the photocathode (at the top). Electrons are ejected when the light is absorbed, and these are accelerated towards the first intermediate electrode. On impact, the incident electrons eject more electrons and the increased number are then accelerated again to the next electrode. The process of multiplication on impact, followed by further acceleration and impact, is repeated for many stages and the resultant flood of electrons collected at the anode. If, at each stage, there are three times as many electrons leaving than arrived, then a twelve-stage tube will produce 500,000 electrons for each one ejected by the incident starlight.

STELLAR EVOLUTION

(a) If a cloud of dust is very dense it will slowly condense, being drawn together by gravitational attraction.

(c) The hydrogen becomes exhausted and the star expands to form a red giant. It then burns helium – a less efficient process.

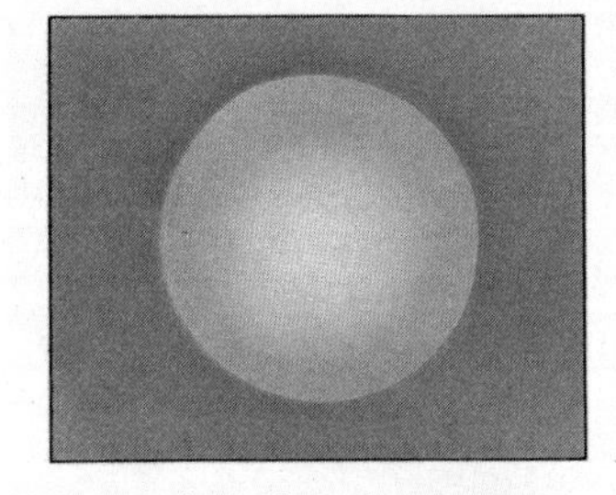

(e) A white dwarf star is unstable and will collapse in a cataclysmic explosion (a supernova) to force the iron nuclei to merge into one colossal atom – a neutron star.

(f) Very massive stars are unable to stop at the neutron star stage, but collapse further into a black hole, from which not even light can escape.

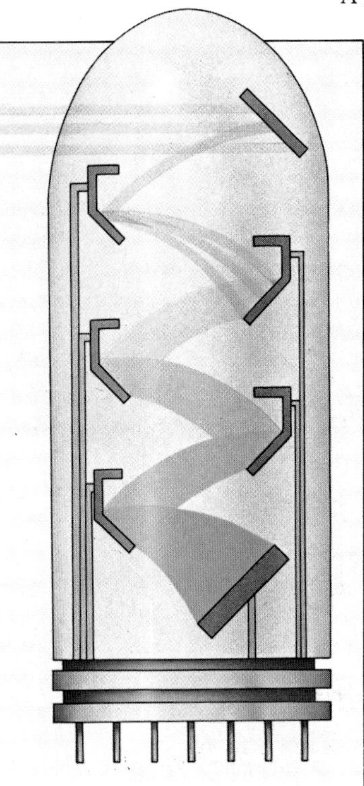

(b) In the centre of the star nuclear reactions begin and the energy released keeps the new star shining for aeons.

(d) As the star's fuel becomes exhausted a shell of gas is ejected, leaving a ball of iron atoms which collapses into a small dense white dwarf.

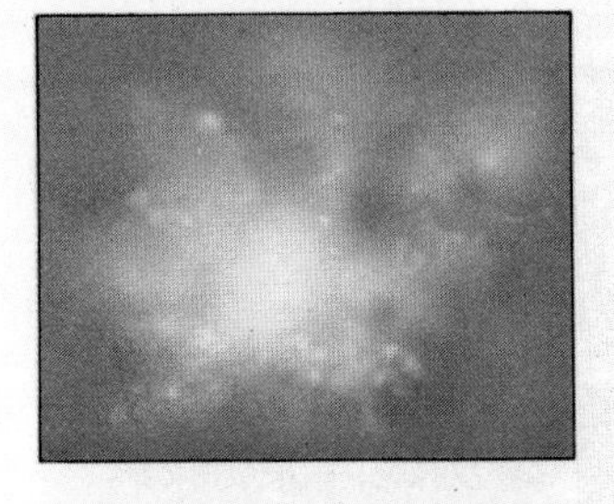

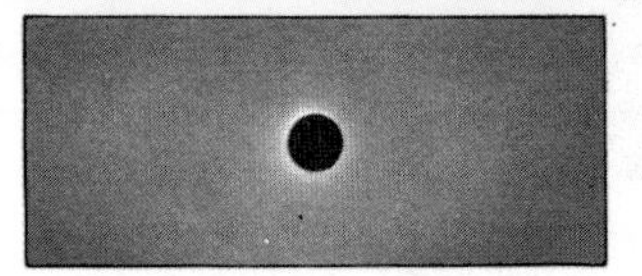

much better at coping with the problem that was becoming of increasing importance – the light of the night sky. Even at the darkest and most isolated sites the natural brightness of the sky, due to light-emitting chemical reactions in the upper atmosphere, was significant compared with the faintest stars and galaxies. The optics between telescope and photomultiplier could be designed to switch to and fro between star plus sky and sky alone. The difference is the brightness of the star itself, and the photomultiplier could go on accumulating signals until the difference was significant. The photographic plate reached a limit once the exposure was long enough for the light of the sky alone to blacken the plate.

Probably the most important application of the photomultiplier was to the measurement of the brightness of the stars in globular clusters with our galaxy. The stars of a cluster have the advantage that they are all at essentially the same distance from us, so that effects due to interstellar and atmospheric absorption of light are the same for the whole of the cluster. It could also be assumed that each cluster had condensed from the primeval dust and gas at a defined time in the formation of the galaxy, so that the stars in the cluster were all the same age and made from the same raw material. The clusters were, as a result, the major test for theories of stellar evolution, where pre-war work by Sir Arthur Eddington, George Gamow and Hans Bethe had been developed by Subrahmanyan Chandrasekar, Fred Hoyle and others. These theories were confronted by observations made by Allan Sandage, Halton Arp and Richard Baum, who determined the brightness and effective temperature of cluster stars as faint as twenty-first magnitude – ten million times fainter than Spica or Antares.

The stars in each cluster were found to fall into three main groups. The stars of low mass, which evolve very slowly, were still part of Russell and Hertzsprung's main sequence. More massive stars, which evolve faster, had left the main sequence and had inflated into red giants much larger and brighter than the main sequence stars. The third group, the most massive stars, had completed both the main sequence and red giant phases of their evolution and, after an unstable interval as Cepheid variables, had become much smaller and hotter again. They appear as a minority of blue stars scattered among the red stars which are members of the main sequence and the giant phases. These three stages were interpreted as showing the "combustion" of different fuels in the nuclear reactions taking place within the stars. The main sequence star "burns" hydrogen to form helium, at a rate which depends on its mass – the bigger the star, the faster the reaction. The giant phase, during which the outer atmosphere of the star is enormously expanded, draws its energy from the fusion of helium nuclei in its central core into carbon and oxygen. The third phase is sustained by a series of ever less-efficient reactions, occurring at higher and higher temperatures, which convert the star's centre into a ball of white-hot iron nuclei. With all its nuclear energy sources used up, the star then collapses to form a white dwarf. Such stars are far too small and faint to be seen in the distant clusters.

The telescopes built in the twenty-five years after the completion of the 5-metre at Mount Palomar were influenced only indirectly by the precision and speed of the photomultiplier. Even the most obvious application, the building of automatic guiders, was slow to be adopted, perhaps because the need to sit in the cold and dark to guide the telescope manually was part of the background to being a "real" astronomer; an observer who felt in close contact with his data. The main effect was to restrict the size of new telescopes. As it was possible to use the new detectors to do world-class astronomy on an instrument of aperture less than 5 metres, it was not necessary to face the appalling task of funding an even larger telescope. It was twenty years before the Hale telescope was even approached in size. For two telescopes, the 3-metre (120-inch) at Lick Observatory and the 2·5-metre (100-inch) at the Royal Greenwich Observatory, the main design input was the existence of a mirror blank. The Lick 3-metre was built round a flat mirror that had been used to test the optics of the Palomar 5-metre, while the British instrument was based on a mirror blank cast in 1936 for Michigan Observatory but never built into a telescope. The mirror for the Lick Observatory telescope was too thin to allow a deeply concave surface to be fabricated, and as a result the designers were forced to accept a longer focal length than would have been ideal. The lighter telescope tube, 50 tonnes compared with the 140 tonnes for the 5-metre, meant that they could use an ordinary fork mounting rather than a complete yoke, but the tines (prongs) of the fork needed to be very long (7·3 metres; 24 feet) to accommodate the long telescope, even after 5 tonnes of ballast were added to the primary mirror cell to bring the balance point well below the mid-point of the tube. Serrurier's balanced deflection truss was developed to work with arms of unequal length above and below the telescope declination axle. The final design is a very elegant telescope, of slim, almost willowy appearance. The slender design is accompanied by an embarrassing flexibility – the flexure of the fork explores how far it is possible to go rather than how far is desirable.

When free of the constraints that limited the Lick telescope (which was none the less the second largest telescope in the world for many years) the post-war telescope settled into a well-defined pattern. The majority of the twenty or more instruments of 1·5- to 3-metre aperture completed in the 1950s and 1960s used the Ritchey-Chrétien optical layout. This gave a good-sized photographic field at the "Cassegrain" focus behind the primary mirror; relatively short focal lengths ("fast" focal ratios) to fit the telescope inside small and economical domes; unequal-arm Serrurier trusses to support primary and secondary mirrors; and a fork mounting and oil pad bearings to support the polar axle. The pace-setter for this generation of telescopes was the 2·1-metre (84-inch) aperture instrument at the new observatory at Kitt Peak in Arizona. The telescope pushed towards the limit of high-quality optical work by requiring a primary mirror with a focal ratio of 1:2·6. A "Serrurier" truss was used, or at least its apparent presence dominated the design of the telescope. It is, however, an aesthetically, rather than a technically, balanced truss, as the components do not precisely satisfy Serrurier's conditions. Also typical of the period is the fact that the telescope was equipped with an elaborate range of spectrographs, and is rarely used for anything but spectroscopic observations.

The background to the Kitt Peak 2·1-metre also set a series of precedents. First and foremost it was the product of a new type of organization – an association of universities, rather than a single institute. The advantage of the association was that many groups would have access to the telescope and that it could be on a site chosen for astronomical excellence with a newly built headquarters nearby, rather than being controlled from, say, Boston, Cleveland or Chicago, none of which offers clear and unpolluted skies. The search for a site was one of the most thorough undertaken. Weather statistics immediately guided the survey to the southwestern corner of the United States, and experience at Mount Wilson, Mount Hamilton and Mount Palomar showed the advantage of a high-altitude site. The designers surveyed several peaks, all well away from major cities but not so inaccessible as to make it impossible to transport major instruments to the mountain top. Special small telescopes were built to measure the seeing at each site so that the astronomers could be sure that the degradation of stellar images by atmospheric turbulence was the minimum attainable.

The final choice fell on Kitt Peak, a mountain about 80 kilometres (50 miles) from Tucson in southern Arizona. The success of the concept of an observatory as a facility for visiting astronomers who reduce their observations elsewhere may be measured by the fact that the peak now supports the biggest collection of working telescopes in the world.

By the end of this interval, around the mid-1970s, it was possible to summarize the experience gained. First, a telescope of given aperture cost five or six times as much as one of half the size while it collected only four times as much light. This added weight to the argument that a large number of medium-sized telescopes was better than a few big ones. It also appeared that, in peacetime, a telescope of aperture D metres took 2D years to build. There were exceptions to this rule. The European Southern Observatory, handicapped by the difficulty of multinational collaboration, took twenty-three years to get from the original concept to a working telescope. The British, without the same excuse, struggled valiantly but were only able to keep going for twenty-one years before completing the 2·5-metre (100-inch) Isaac Newton telescope in 1967.

In parallel with the growth in the number of telescopes larger than 1-metre aperture, roughly doubling in number every twenty years – three by 1920, seven by 1940, sixteen by 1960 and thirty-five by 1980 – there was a major research effort to combine the best features of both photographic plate and photomultiplier. The high efficiency of the photocathode compared with the plate could only be fully exploited if just one measurement was being made. While the plate could record the images of all the stars in the field, the photomultiplier could only give a current that was proportional to the total light incident on the photocathode, with no distinction between one star or many.

The first device that combined the efficiency of a photocathode with the ability to record a two-dimensional image was proposed in 1936 by the Frenchman A. Lallemand. The basic concept was simple: let the image fall on to a large photocathode, accelerate the electrons ejected and then focus them on to a photographic plate. A range of instruments developed in various branches of physics had led to the existence of "photographic" emulsions that would respond directly to electrons rather than light, and if the energy were high enough (25,000 or 30,000 volts) many by each electron. But it was twenty years before Lallemand had a practical "camera electronique" working on a telescope.

The main difficulty comes from the constitution of the photocathode, which is only efficient if it includes the element caesium, one of the most chemically active metals known. A stack of eight electron-sensitive plates had to be introduced into the camera's evacuated chamber and then frozen by a surrounding flask of liquid nitrogen. A sealed glass cell containing the photocathode was then broken open and the photocathode moved, using a magnet outside the vacuum, until it was correctly placed in the "camera". Even with the plates as cold as possible the minute quantities of gas and water vapour that they gave off were enough to ruin the photocathode in about forty-eight hours. Removal of the exposed plates inevitably broke the vacuum, so a new camera had to be prepared after every eight exposures. The struggle was worthwhile in that the gain in sensitivity of a factor of about forty over orthodox photography allowed observations that would not have been possible by other methods. Typical applications, using the Lick Observatory 3-metre telescope, included the observation of the spectra of rapidly changing variable stars as well as very faint stars. In both cases the gain in speed was essential – the spectrum of a variable star must be taken in a small fraction of the time of a complete cycle if the results are not to be confused by superposition on the same plate of the spectra of the different phases. When observing faint stars, the limit was an exposure of roughly half a night to give spectra of sixteenth magnitude stars with a resolution that allowed the simultaneous recording of many points in the spectrum.

The next step was to improve the durability of the image tube by separating the photocathode from the plate by a window that would prevent the gas emitted from the plate reaching the photocathode but would transmit the accelerated electrons travelling in the opposite direction. Two methods were tried to achieve this separation. W. Hiltner and his colleagues at Yerkes Observatory used a very thin (0·1-micrometre) film of sapphire to separate two evacuated chambers, one of which included an air lock to admit the plates and the other contained the photocathode. J. D. McGee at Imperial College, London, used a thicker (10-micrometre) and stronger film of mica and was able to demonstrate an image tube in which the electron-transmitting window was tough enough to withstand atmospheric pressure, so that the plate could be exposed without evacuation or freezing being necessary. This was a major gain for a device that was mounted at the Cassegrain focus of the telescope and used in total darkness.

The Lallemand-McGee style of image tube relies on increasing the energy of the electrons from the photocathode rather than increasing their number. The alternative approach, an imaging device related in concept to the photomultiplier, has developed in parallel to Lallemand's system. In fact it too was first proposed in 1936, by M. von Ardenne. A significant gain can be achieved by accelerating the original electron into a phosphor screen, similar to that used in a television tube. The light output from the phosphor then ejects more electrons from a second photocathode and these are accelerated for another stage of amplification. After three, four or five stages the final phosphor screen is then photographed with an ordinary camera. Although the last process wastes about 99·99 per cent of the light, this loss is far offset by an overall gain within the image tube of a million or more. Tubes of this type were first applied to astronomy at the Crimean Observatory and at Pulkovo, which had been rebuilt after the end of the war. The astronomical aim was the detection of faint galaxies, a difficult problem for direct photography as the faint outer halo of the galaxy, the feature that most directly distinguishes a distant galaxy from a star, is easily swamped by the light of the night sky. By the early 1960s such "cascade" image tubes built with aid from the Carnegie Institution, were being tested at Mount Palomar Observatory.

All the image tubes available by the early 1960s ended up with a photographic plate from which the information had to be extracted in a form suitable for further analysis, usually by computer. The more recent developments of the image tube have concentrated on the removal of the plate to give an all-electronic device with the results presented in a more direct form. One approach, used at Lick Observatory, was to replace the photographic plate with a modest "silicon chip" – a solid-state device comprising a microcircuit with many small diodes arranged in a regular array on the surface. In this system, the digicon, the accelerated electrons from the photocathode are focused on to the chip and produce pulses of current from the diodes that they hit. These pulses can be counted directly and analysed by computer. The early digicons had only a few tens of diodes, each with its own tiny amplifier, but the number has grown as microelectronic technology has developed, enabling finer detail to be distinguished over a larger field of view. The alternative method for doing without the final photographic plate was to replace the camera behind a cascade tube with a television camera and record the electronic output signal from that. The normal television camera could not be applied directly to astronomy: it is designed to give many pictures a second from a well-lit scene – news reader, football match or advertisement – and not to produce one picture

The Russian 6-metre (240-inch) telescope under construction in 1969. One of the pillars that support the elevation axle is being hoisted into the dome. The instrument is sited on Mount Pastukov near Zelenchukskaya in the Caucasus.

Grinding the 4-metre (160-inch) mirror for the Mayall telescope at Kitt Peak Observatory. The grinding tool is used to give the mirror surface the hyperboloidal profile needed for the Ritchey-Chrétien optical configuration.

every few hours from a scene that is a close approach to total darkness. If, however, the incident light is amplified by a cascade image tube the signal is sufficient to allow the use of Vidicon-type television cameras. The early experiments along these lines were undertaken in the early 1960s by McGee and by L. B. Robinson and E. J. Wampler at Lick. The concept was further developed by Alec Boksenberg at University College, London, who added a computer to the output system to give a device called the "image photon counting detector". This device was getting near to the ideal detector – about a quarter of the photons in the incident starlight are recorded, and the results are immediately presented as a television picture to the astronomer. It is possible to sit in a warm control room to watch the slow accumulation of the image and decide precisely when the exposure is sufficient. In contrast, in the 1940s Baade sat at the guiding eyepiece in the freezing cold for hours at a time and still did not know when he took the resulting plate to the darkroom whether he had under- or over-exposed the vital few square millimetres of photographic emulsion.

The one aspect of the photographic plate that the image tube could not match was sheer size. Plates 35 centimetres (14 inches) across were easily available, while few image tube photocathodes were as large as 35 millimetres (1·4 inches) in diameter. The immediate result was to bring the prime focus back as a major observing point. Star images at the prime focus were smaller and closer together than at the magnified Ritchey-Chrétien focus behind the primary mirror and so fitted better on the small photocathodes. The light of the night sky, which would blacken a plate at the Hale telescope prime focus in twenty minutes, could now be subtracted in the computer attached to the electron detector output. The primary mirror of a Ritchey-Chrétien telescope does not give a good image at the prime focus and Charles Wynne and others developed greatly improved corrector lens systems to cure this defect. After some while it was recognized that scarcely anyone was using the Ritchey-Chrétien focus to photograph the sky, and Wynne pointed out that the prime focus of a classic Cassegrain telescope with a parabolic primary mirror could be corrected to give good images just as easily as a Ritchey-Chrétien. This nudge effectively ended the fifty-year career of the Ritchey-Chrétien design.

The mid-1970s saw a renewed burst of large telescopes in the 3·6- to 4-metre (140- to 160-inch) category along with a very large (6-metre; 236-inch) telescope completed in Russia in 1976. These are split between Ritchey-Chrétien and Cassegrain designs, the earlier examples using the more sophisticated system. The 4-metre-class telescopes, starting with the Mayall telescope at Kitt Peak, also mark the end of an era in telescope design. Three of the six telescopes are closely similar, the original at Kitt Peak, a second at its sister institution, the Inter-American Observatory at Cerro Tololo in Chile, and a third at the Anglo-Australian Observatory at Siding Spring in New South Wales. All three are Ritchey-Chrétien telescopes in a yoke that is a development of the Palomar design. Apart from the optical design, the other main change from the 5-metre was the use of mirrors made from new materials of very low coefficient of expansion: quartz and cervit in place of Pyrex. The telescope is pivoted in the horseshoe, not on a girder bridge between north and south bearings. The most striking development was perhaps the recognition that the site in Chile set new standards for seeing, images less than an arc second across being relatively common. The concentration of star-light into small images greatly improves the contrast between faint stars and the night sky and is the main reason for the transfer of the 2·5-metre Isaac Newton telescope from the post-war headquarters of the Royal Greenwich Observatory at Herstmonceux in Sussex, England, to a new mountain site on La Palma in the Canary Islands.

The first large telescope to look to the future, however, was the 6-metre instrument at the Zelenchukskaya Astrophysical Observatory in the mountains of southern Russia. This telescope was designed from the start with the assumption that a computer could be used to drive the telescope as well as to process the data it collected. This major step frees the designer from the trouble and strife associated with the tilted polar axis of equatorially mounted telescopes. The advantage of the equatorial mounting is that one axis (the north-south "polar" axis of the yoke parallel to the Earth's rotation axis) is driven at constant speed, while the other axis (the declination axis on the telescope tube) is kept fixed in position during an observation. Early observers had used an alt-azimuth mounting in which one axis (altitude) was horizontal and the other (azimuth) vertical; its problem is that both axes must be driven, and at rates that vary with the position. Given that a computer can easily handle the variable drive rates, the gain in reverting to an alt-azimuth mounting is enormous and becomes more important as the telescope gets bigger. Rotation about the vertical azimuth axis does not change the orientation of the telescope tube with respect to gravity, so that this motion does not change any aspect of the flexure of the support. In effect an alt-azimuth mounting is a fork mounting with the axis vertical, so that the tines have no transverse load at all. Not only is the system much better able to support the load, but it obviates the twisting of the fork tines that makes it so difficult to design the declination axle bearings of the equatorially mounted telescope. Longer tines are now practicable and the horizontal elevation (altitude) axle can be nearer the mid-point of the telescope tube.

The Russian 6-metre is a mixture of new and traditional design. The alt-azimuth mount and the use of a much thinner than normal blank for the primary mirror are both new developments, although the thin mirror proved a mixed blessing. One blank – the biggest disc of glass in the world – snapped when being lifted and the second was a great deal more difficult to polish than had been expected. On the more conservative side, the focal length was relatively long and the engineering was on a massive scale, the heavy design making a considerable contrast with the graceful elegance of the Lick 3-metre.

Even without the birth pangs of the Russian telescope – it required a lengthy period of development after its first assembly, as had the Hale telescope nearly thirty years before – astronomers began to realize that the end of possible developments in light detectors was approaching. An ideal detector would register every individual photon from a distant star or galaxy and real detectors were getting to within a factor of two or three of that ideal.

At the time of writing, the emphasis is very heavily on all solid-state silicon chip detectors, which can now be produced in arrays containing a quarter of a million elements. The crock of gold at the end of the silicon rainbow is that the new detector will respond to low-energy (red light) photons that the image tube cannot detect and that it responds to two out of every three photons hitting it. The disadvantages of the silicon chips are that they are very non-uniform in response, and this non-uniformity is different at different wavelengths. In addition, the detectors are very sensitive to changes in temperature, and have a wavelength response that makes it difficult to combine the array with a cascade image tube. None the less, the potential of the device as the detector array in an all-solid-state home television system means that an enormous amount of money is being invested in its development for this market. As the mass market for the home television camera replaces that for the photographic camera, so the television detector may replace the photographic plate as the "standard" astronomical system.

EQUATORIAL MOUNTINGS

The fork is the most convenient mount for a telescope of medium size. D = declination axis, P = polar axis.

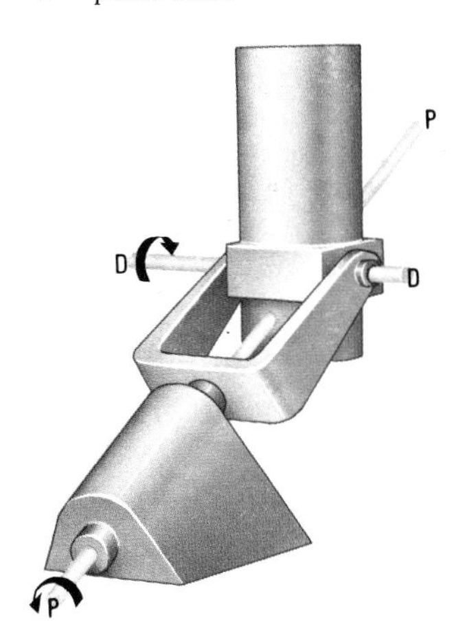

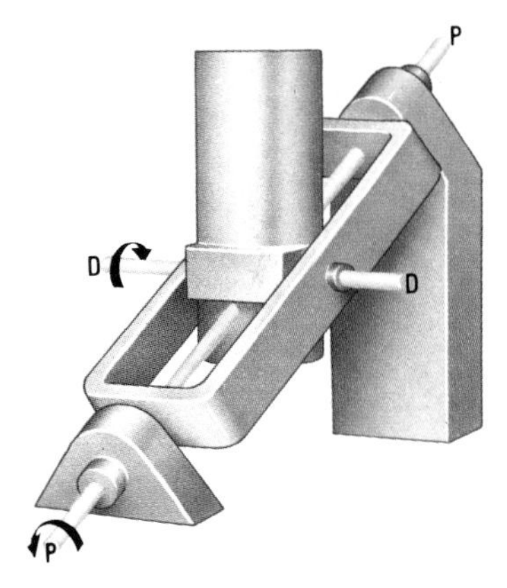

In the yoke mounting (above) supports are north and south of the telescope, which make the yoke frame more rigid and stiffer than in the fork mounting. It is the most economical mount for large telescopes and has been used for the 2·8-metre (145-inch) UK Infra-red Telescope and the 3-metre (120-inch) NASA Hawaii telescope.

The cross-axis mounting (below) carries a counterweight to gain rigidity. The mount is popular with amateur astronomers because two telescopes, one photographic and the other visual, can be used to counterweight one another.

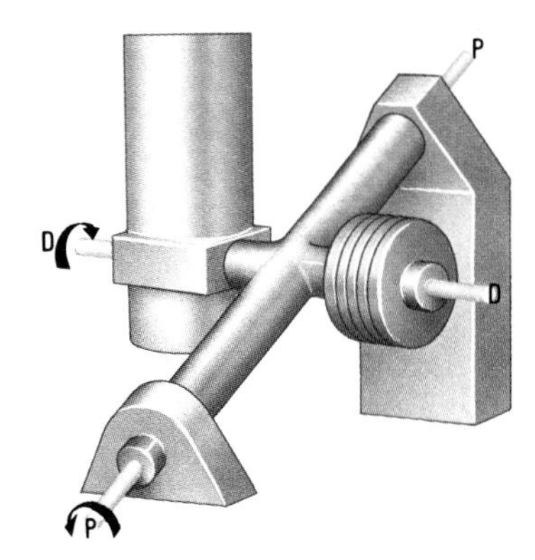

CHAPTER THIRTEEN

TWO RADIO TELESCOPES ARE BETTER THAN ONE

The use of arrays, aperture synthesis, and long-base interferometry 1950 to the present-day

"There is an extremely precise test of the conventional interpretation of galaxy redshifts which consists of comparing the redshifts measured at radio and optical wavelengths. According to the conventional (Doppler) interpretations of redshift the measured shift must be independent of wavelength. . . . Redshifts of the neutral hydrogen 21-centimetre line determined by radio techniques are in excellent agreement with redshifts measured with optical lines. The agreement is better than one part in a hundred over a (velocity) range of 6000 km/sec and a wavelength factor of 500,000. Note also that the 130 galaxies included include galaxies of all types."

John N. Bahcall in The Redshift Controversy. *1973.*

The use of a connected pair of radio telescopes to form an interferometer had been established in the early 1950s, particularly by Graham Smith at Cambridge. The majority of these high-resolution systems were not, however, based on steerable "dish" telescopes. Hanbury Brown's radio link system at Jodrell Bank, England, the Mills Cross in Australia and the interferometer used for Ryle's "3C" third Cambridge catalogue were all passive devices which relied on the rotation of the Earth to scan across the sky. The first large-aperture interferometer which combined two large steerable dishes was built at Owen's Valley in California in 1957. The Owen's Valley interferometer had a maximum telescope separation of 1600 metres (1 mile), and the separation could be adjusted to any smaller value by moving one of the pair of 27-metre (90-foot) telescopes along a railway track.

Using the Owen's Valley instrument, T. H. Williams and his colleagues were able to accurately fix the position of one of the radio sources that had been studied at Cambridge and Jodrell Bank. H. P. Palmer at Jodrell Bank had already shown that the source, number 48 in the third Cambridge catalogue and therefore called "3C 48", was one of a small number of radio sources less than three arc seconds across. The new interferometer gave a position that was accurate to a few arc seconds, which was sufficiently precise to encourage a new attempt with an optical telescope to find a visible source for the radio signals. The search was carried out in September 1960 by an American astronomer, Allan Sandage, using the 5-metre (200-inch) Hale telescope. In the predicted position he found a sixteenth magnitude star that was unusually blue.

Soon after this identification, two more of the very small radio sources, 3C 196 and 3C 286, were also found to coincide with very blue stars. The three stars were unusual objects. The optical spectrum of 3C 48 showed emission lines instead of the usual absorption lines, and the wavelengths did not match those emitted by any known element. 3C 48 was studied with optical telescopes for most of 1961 and it was found that the brightness fluctuated significantly from month to month. This implied that the source could not be more than a fraction of a parsec across, less than one ten thousandth the size of a typical galaxy. The long-expected radio stars had at last been found.

The next step was to return once more to the radio telescopes. At Jodrell Bank, Cyril Hazard developed a new technique which gave accurate radio star positions

The 76-metre (250-foot) steerable dish at Jodrell Bank near Manchester, England. This was the world's first giant radio telescope.

without using interferometers. Hazard made observations of the precise time when radio signals from a star were cut off and then restored as the Moon moved in front of and then away from the source. From this it was possible to calculate the source position with high accuracy – about two arc seconds. The method was tested on the source 3C 212 and then, after Hazard moved to Australia, on 3C 273. The second of these was immediately identified as another very blue star, this time with the added peculiarity of a faint jet projecting from it. Once again the Hale telescope was employed to study the newly identified object, and Dutch-American astronomer Maarten Schmidt observed the visible and ultraviolet spectrum of the star. As with 3C 48, the spectrum of

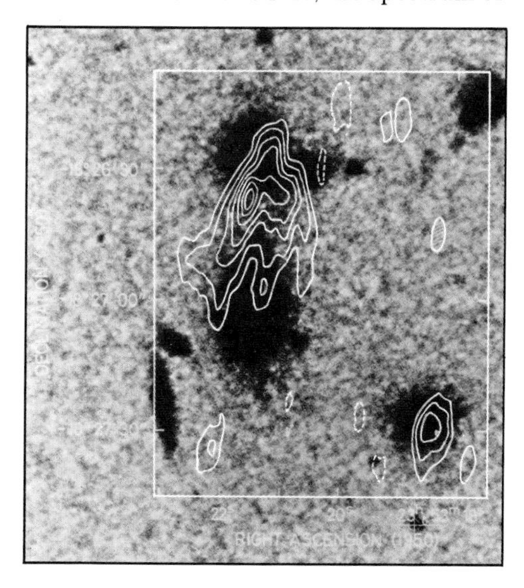

An optical image of a southern galaxy taken by the Anglo-Australian 3·9-metre (150-inch) telescope on which is superimposed a map of the radio emission received from the object.

3C 273 showed emission lines at odd wavelengths. Schmidt realized that these emission lines could be attributed to the most abundant element of all, hydrogen, if it were assumed that the star was moving away from us at nearly 50,000 kilometres (31,000 miles) a second. The Doppler effect would stretch all the wavelengths, moving the spectral lines to unfamiliar wavelengths. This unlikely explanation was confirmed by a new look at Sandage's spectrum of 3C 48 by another astronomer at the Hale Observatory, Jesse Greenstein. He found, as had Schmidt, that the emission lines could be due to hydrogen, provided that the velocity with which 3C 48 was receding was just over 110,000 kilometres (70,000 miles) a second.

These velocities were completely unreasonable for stars in our own galaxy. The nearby stars are in motion in complex orbits round the centre of the galaxy, and the Sun is fairly typical, moving at 250 kilometres (160 miles) a second. If the Sun's velocity were raised by only fifty per cent then it would be moving fast enough to escape altogether from the galaxy's gravitational attraction. The only objects that were known to have high velocities were very distant galaxies, carried away by expansion of the universe. In the thirty years since Edwin Hubble had discovered the law relating the velocity of recession of galaxies to their distance from us, the relationship had been extended to include galaxies moving even faster than 3C 48. But the photographs of the new objects showed sharp-edged circular images, with no sign of either elliptical shape or the fuzzy edges that are the distinguishing features of distant galaxies. They were christened quasi-stellar objects, then QSOs and finally quasars. If Hubble's law applied to the quasars, then they were among the most distant objects in the universe. Yet the variation in brightness from month to month meant that they had to be small. The actual brightness could be calculated from the observed magnitude and the distance, and required the quasars to be completely unprecedented objects, radiating fifty times the power of the brightest galaxies from a volume less than a thousandth that of even the smallest galaxies.

The debate over the quasars continued for a decade. The critical question was whether Hubble's law for galaxies could be applied to quasars to convert their observed velocities into distances. As the high velocity is inferred from the displacement of all the spectrum lines to longer wavelengths – that is, ultraviolet and visible lines are shifted towards the red – the debate was christened "the redshift controversy". The two sides each had one advantage and one difficulty. Those in favour of the quasars being at great distance were able to keep Hubble's law but were unable to explain the source of the vast energies radiated by the object. The opposite camp, who preferred "local" quasars, had no difficulty explaining the apparent brightness, as the quasars were much nearer and therefore less energetic. On the other hand, they could not explain the observed redshifts, partly because the most obvious alternative, a gravitational red shift, could not possibly produce a large enough effect and still give an observable line spectrum.

Radio astronomy in the early 1960s was not only concerned with what were to prove to be the most distant objects in the universe. As with the astronomy of the visible region, the subject was enormously aided by the development of new and more sensitive detectors. Two new types of amplifier, each with a performance vastly superior to previous equipment, arrived almost simultaneously. These were the maser and the parametric ampli-

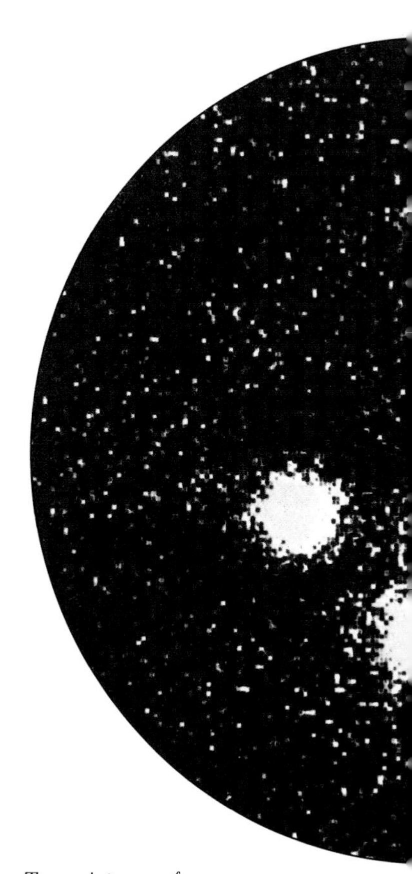

Two pictures of a quasar (PON 256) taken using the 5-metre (200-inch) Hale telescope and the image photon counting detector. The upper frame shows the central star-like nucleus of the quasar (at the centre of the field). The lower frame is a longer exposure and shows

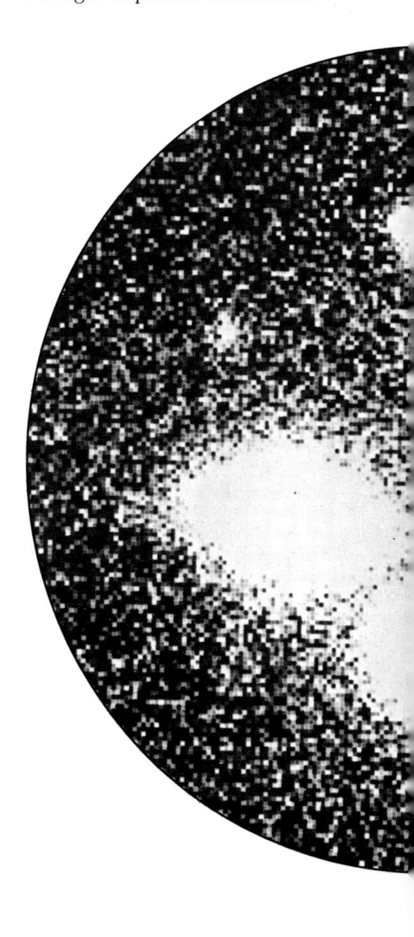

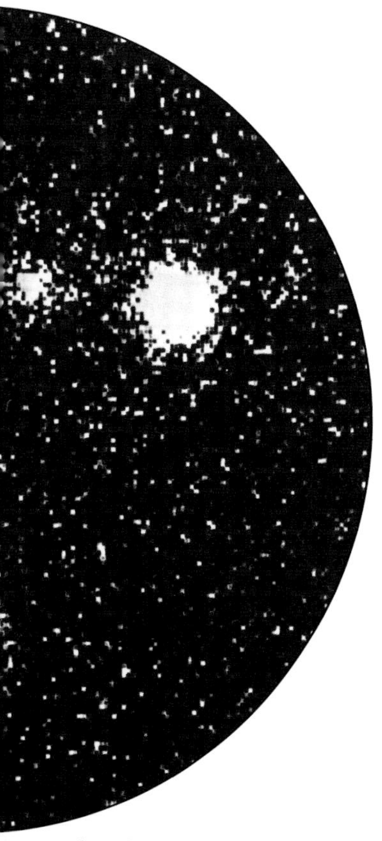

that the quasar has an elliptical halo, similar in shape to the adjacent galaxy (below and to the right of the quasar). In both frames the uniformly bright sky light has been subtracted in the detector system computer—the sky is considerably brighter than the quasar halo.

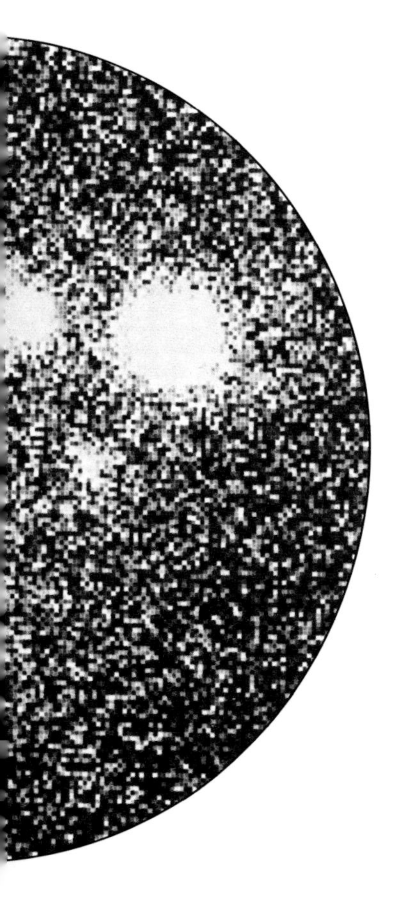

fier. The maser was the microwave predecessor of the better-known laser and was the more sensitive, but the parametric amplifier was easier to use and found wider application.

The new parametric amplifiers were fitted to existing radio telescopes, and their application coincided with a surge in new telescopes completed in the early 1960s. The period from 1961 to 1964 began and ended with the completion of two 64-metre (210-foot) fully steerable telescopes, one at Parkes in Australia and the second for the Jet Propulsion Laboratories (JPL) at Goldstone in California. In the relatively brief interval between these two instruments, a third large steerable dish, three transit telescopes, two steerable interferometers and a large shortwave radar telescope were all completed, and two large array interferometers were started. These successes were accompanied by two failures. The United States National Radio Astronomy Observatory undertook the construction of a 180-metre (600-foot) steerable telescope, an instrument nearly two and a half times the size of Lovell's telescope at Jodrell Bank. The project proved beyond both the engineering and financial resources available and was abandoned soon after the completion of the foundations. On the other side of the Atlantic, at Jodrell Bank, a telescope with an elliptical aperture 36 by 24 metres (120 by 80 feet) was built as a prototype for another giant instrument. Lovell hoped to build a 300-metre (1000-foot) aperture telescope and had chosen an elliptical geometry to avoid the problems (especially wind pressure) connected with having a telescope 300 metres high. Unfortunately, the giant dish was never funded, though a near copy of the 36-metre prototype was built in 1966 to work in association with the original telescope.

One of the new interferometers was a pair of 25-metre (82-foot) telescopes at the Royal Radar Establishment, near Malvern in Worcestershire. Both the dishes were mounted on rails in a layout similar to the Owen's Valley instrument, but more versatile in operation. This interferometer was also noteworthy because it was linked, using a microwave radio relay, to the big telescope at Jodrell Bank to form an interferometer with a baseline, at its working wavelength of 21 centimetres (8¼ inches), of 600,000 wavelengths. The high resolution of this Malvern-Jodrell pair showed some of the controversial quasars to be about a tenth of an arc second across. This result not only added to the facts and the problems of the quasars, but also was of immediate application to the problem of the structure of the universe. Fred Hoyle at Cambridge showed that for some theoretically possible universes very distant objects would not appear small, but would always appear to be at least a few arc seconds across whatever their distance.

The most influential interferometer, however, was the "One Mile" telescope at Cambridge. This system used three 18-metre (60-foot) aperture telescopes, two of them immobile and set 800 metres (½ mile) apart and the third movable on an 800-metre (½-mile) long railway track. The array gave an angular resolution of twenty-three arc seconds and was used in a new mode pioneered at Cambridge using simpler systems. This was aperture synthesis, a technique in which very long runs of data collection from the interferometer pairs were fed into a computer. After it had processed this data, the computer was able to construct the view that would have been obtained if a complete 1600-metre (1-mile) diameter telescope had been available.

The "One Mile" telescope was both very sensitive and able to separate faint sources too close together to be resolved by large single dishes. Martin Ryle used it to supply the data for a twentieth-century equivalent of William Herschel's star counts of two centuries earlier. The objects counted this time, however, were radio galaxies, not stars, and the intention was to find the structure of the universe, not that of our galaxy. Just as Herschel had assumed that "taken one with another" all stars were similar, so Ryle assumed that all distant radio galaxies were roughly the same. The raw data is the number of radio galaxies of differing apparent brightness in a random area of the sky viewed by the One Mile telescope. In a uniform universe the number of galaxies of a given apparent brightness (and so at a given distance) is related to the number that appear brighter or fainter.

The results of Ryle's radio galaxy counts showed that the distribution of galaxies with distance was not uniform. There were far too many faint galaxies and far too few very faint galaxies. This result was not compatible with a uniform universe and favoured a model which evolved with time. In the 1950s cosmologists were split into two camps: the proponents of a steady state (and steadily expanding) universe, which had to be uniform, and those who favoured a model that started from a single point, exploding outward and evolving as it travelled. The debate now became somewhat stormy, with a great deal of argument about how Ryle's data should be interpreted. It was abruptly ended in 1965 by further radio astronomical measurements.

The measurements, by A. A. Penzias and R. W. Wilson at Bell Telephone Laboratories, were almost a replay of those carried out by Jansky in the 1930s.

The 1600-metre long ("One Mile") radio telescope array at Cambridge, England, employs three dishes that will work together as an interferometer; one of the dishes moves on rails. The almost complete mobile dish is shown with, in the distance, the central dish.

The 38-metre (125-foot) steerable dish of the Mark 2 radio telescope at Jodrell Bank. Completed in 1964, it is a more accurate instrument than was the 76-metre (250-foot) Mark 1 telescope (seen in the background) and can be used at much shorter wavelengths.

The company had developed a very sensitive antenna and receiver working at 7·35 centimetres ($2\frac{9}{10}$ inches) for communications links using the Echo satellite. Penzias and Wilson modified this system to do radio astronomy, specifically to seek the very weak radiation from the Milky Way that was expected at short wavelength. To their surprise, they found a signal that was some sixty times the expected strength and which was the same no matter where their antenna was pointing. The first and most obvious explanation was that the apparatus was not as good as its users had hoped – a noisy amplifier would give a spurious signal that would be independent of the aerial's position.

Quite by chance, Robert Dicke's research group at Princeton, not far from the Bell Telephone Laboratories at Holmdel, New Jersey, was setting up apparatus to look for just the sort of signal that Penzias and Wilson had found. Collaboration between the groups and a new measurement at a wavelength of 3·2 centimetres ($1\frac{1}{4}$ inches) confirmed the signal was real: in the spaces between the radio galaxies, the quasars and other sources, there is a faint background of radio waves that is the same in all directions. The radiation corresponds in wavelength and intensity with that from an extremely cold body, one with a temperature only three degrees above absolute zero (−273°C). This weak uniform radiation had been predicted by the Russian-born American astronomer George Gamow and Dicke himself as a direct consequence of an explosively expanding universe. Just as the air escaping from a punctured tyre cools as it expands, so the hot energetic radiation from the earliest days of the universe has cooled as the universe itself expanded. The observation of this cool radiation – called the "three degree background" – effectively ended the career of the elegant, but incorrect, steady state theories of the universe. Within a few years, cosmology – the study of the history of the universe – had evolved from a field for unlimited theoretical speculation into a subject disciplined by direct observation.

The study of the most distant reaches of the universe was only one of the subdisciplines of astronomy that were reshaped by radio astronomy. Much closer at hand was an increased knowledge of the planets due to advances in radar astronomy.

One of the first achievements of the new generation of sensitive detectors had been the extension of the earlier radar observation of the Moon to the detection of the planets. In 1961 five different research groups reported radar observations of Venus, a target requiring a gain in sensitivity of ten million over that needed to

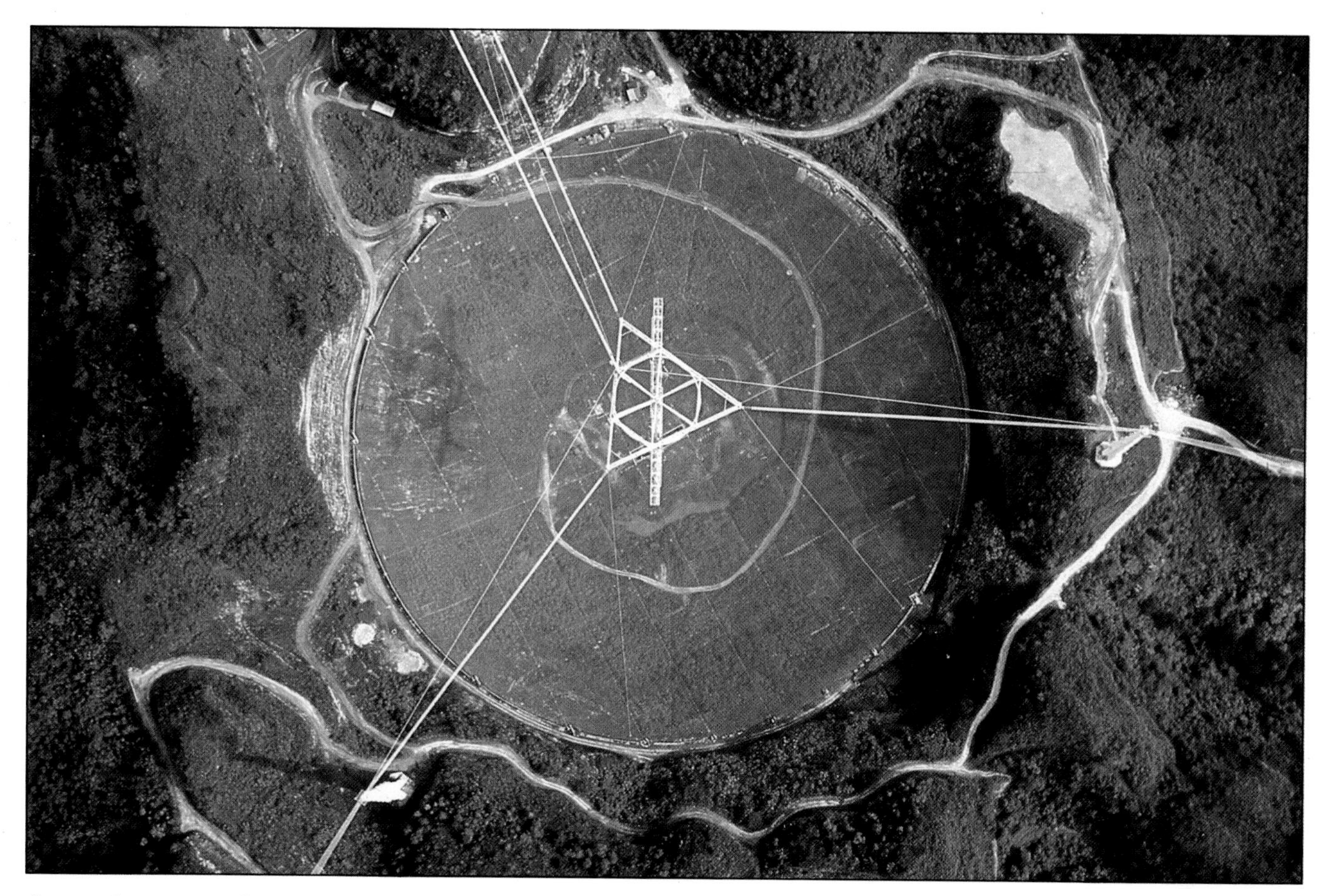

A view from directly above the 305-metre (1000-foot) aperture radio telescope at Arecibo. The telescope dish is fitted into a bowl-shaped valley and works as a transit instrument.

detect the Moon. The radar echoes from Venus gave a highly accurate value of the distance to the planet, enabling astronomers at long last to fix the size of the solar system with confidence and precision. The following year the Russian group detected radar echoes from Mercury, while astronomers from the Massachusetts Institute of Technology (MIT) detected radar echoes from the Sun. They used a system built in Texas and of truly Texan scale – a 500,000-watt transmitter and an aerial array covering 3·2 hectares (8 acres) to pick up weak reflected signals.

These successes led, by the mid-1960s, to three new planetary radar systems, the specially built Haystack radar at MIT and two systems using large radio telescopes, one at JPL in California and the other at Arecibo in Puerto Rico.

The three systems were to some extent complementary. The 37-metre (120-foot) Haystack telescope was designed to work at centimetre wavelengths and is one of the few radio telescopes in a dome. The JPL Goldstone dish is an orthodox large steerable radio telescope. When used as a radar it operates at a longer wavelength than the MIT dish – 12·5 centimetres compared with 3·8 (5 and $1\frac{1}{2}$ inches). The third system is based on the world's largest radio telescope, the 305-metre (1000-foot) aperture fixed dish built in Puerto Rico but operated by Cornell University. The giant telescope has, as a consequence of its size, a narrow beam even at relatively long wavelength, and was used as a radar set at 70 centimetres ($27\frac{1}{2}$ inches) and 7·5 metres (25 feet).

All the new systems demonstrated a dramatic improvement over the radar observations made in 1961. The older systems had only detected simple reflections from the point on the planet's surface directly facing the radar transmitter. The new equipment was more than a thousand times more sensitive. The much improved performance made it possible to study structure in the returning radar echo. The structure included differing reflections from different parts of the surface of the planet – causing the pulse duration to lengthen – and changes in the frequency of the transmitted radar signal due to the rotation of the target planet. As the radar waves could penetrate cloud, the reflection from Venus was from the planet's surface and not from the top of the sheet of clouds, which are all that can be seen in visible light. The echo showed that Venus is rotating about an axis roughly perpendicular to its orbit. This seemed quite reasonable, but it was the only prediction borne out by the new observations. The surprises were that Venus goes round very slowly and, compared with the rest of the planets, it turns backwards. The clockwise rotation takes 244·3 days, rather longer than the Venus year of 224·7 days, and is very close to a period of 243·16 days, which would mean that Venus always has the same face turned towards Earth when the two planets are closest to one another.

The radar measurement of Mercury's rotation also proved to be a surprise. Astronomers had believed for almost a century that the planet's "day" and "year" were the same, so that the same face of the planet was always towards the Sun. The new observations showed that Mercury turns in 59 days, exactly two-thirds of the planet's "year". The revised data showed how subtle were the pitfalls in the interpretation of "simple" visual observations. The classical telescopic studies of Mercury necessarily concentrated on those times when the planet was in the darkest sky, that is, when Mercury was both at its greatest distance from the Sun and farthest from the Sun as seen from Earth. On these occasions the faint markings on the planet look the same because the same face is towards the Earth. It is hard indeed that this interval between favourable appearances turned out to be precisely every six complete rotations of the planet.

Other details were extracted from the radar echoes – the surface of Venus, in spite of the clouds, reflected radar with an intensity that showed it to be dry rock, and Mars proved to have mountains higher than any on Earth. The accuracy with which the distances to the nearer planets was known improved from one-tenth of one per cent of classical measurement to one part in a hundred million, and planetary orbits had to be completely recalculated to fit the new precision. The accuracy was enough to add a fourth successful test of Einstein's general relativity to the classical measurements. Radar reflections from Mercury when the signal was passing very close to the Sun proved to be delayed by about a hundred millionth of a second, as required by the theory.

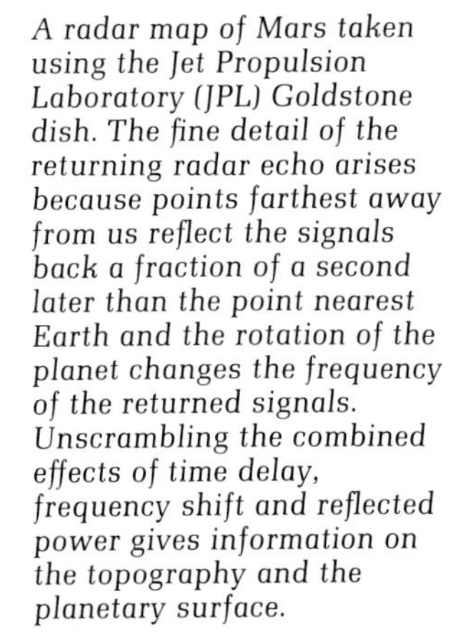

A radar map of Mars taken using the Jet Propulsion Laboratory (JPL) Goldstone dish. The fine detail of the returning radar echo arises because points farthest away from us reflect the signals back a fraction of a second later than the point nearest Earth and the rotation of the planet changes the frequency of the returned signals. Unscrambling the combined effects of time delay, frequency shift and reflected power gives information on the topography and the planetary surface.

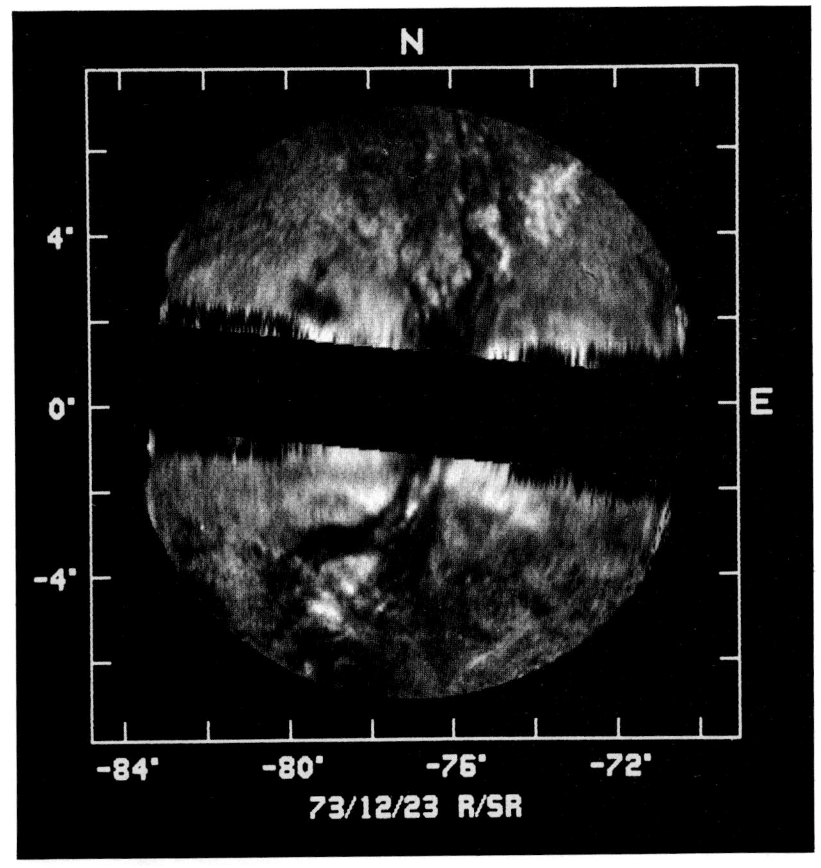

Further information on the solar system came from the distant and controversial quasars. It had been known for a long time that the outermost layer of the Sun, the corona, extends into an outward-streaming flow of ionized gas, the solar wind. Turbulence in this very tenuous wind can cause twinkling, or scintillation, of radio stars as their radiation passes through the wind on its way to Earth. The effect is similar to the way turbulence in the Earth's atmosphere affects light, and makes the stars twinkle. Just as the tiny images of stars twinkle while the larger images of planets do not, only very small radio sources scintillate. The smallness of the quasars led astronomers to expect their radio signals would twinkle, and that a study of the process would give information on both the turbulence of the solar wind and the angular size of the quasars.

The twinkling of the quasars is a process that cannot be studied by standard radio-astronomy techniques. The scintillation involves rapid changes in signal strength so it is not possible to obtain clear data by adding up weak signals for a long period. The only solution is to build a big enough collector to accumulate a useful signal strength in the fraction of a second of one twinkle. A group at Cambridge, led by Antony Hewish, built a 1·8-hectare ($4\frac{1}{2}$-acre) aerial array, using the low-cost stretched-wire method pioneered by Clegg and Lovell.

Built to study quasars, the aerial system unexpectedly discovered pulsars – radio stars which emit pulses of radiation with clock-like regularity. The fast repetition rate for the pulses – one every 1·33 seconds for the first discovered pulsar – was extremely difficult to explain. The first explanation was that the signals were man-made interference. Even now, when hundreds of pulsars are known, the majority of new discoveries turn out to be due to the regular pulsing of electric fences. For a brief interval after the discovery of the first pulsar, in 1967, there was the tantalizing thought that the signal might be coming from an extra-terrestrial intelligence. This idea faded rapidly when a second and then a third pulsar were found. One source of intelligent signals might just be imaginable: to find several in a few months is not.

It was not easy to find an explanation of the pulsars, as no known astronomical object could move fast enough to produce a signal once a second. The orbital motion of a double star system could be ruled out: there is a natural limit to the closeness of a pair of stars, inside which tidal forces will break up the components. For two main-sequence stars this makes all periods less than about two and a half hours impossible. There is also a natural limit for the speed of rotation of a star before it flies apart. Cool stars are not observed to spin rapidly, and the fastest possible rotation period for a hot star is still a few hours – 10,000 times the pulsar period. Single stars which pulsate regularly are well known, but even the fastest of these – the hottest dwarf Cepheids – take over an hour to complete a single cycle.

The solution of the pulsar problem required an extension of the theory dealing with the last stages of a star's life, how it evolves after its nuclear fuel is exhausted. While a star like the Sun finally evolves into a white dwarf star similar to the companion to Sirius, this is not the final step for a star which is a bit more massive. The added mass increases further the gravitational load on the iron nuclei of which the white dwarf is made. At a critical point, at a mass about 1·4 times that of the Sun, the nuclei collapse. The pressure forces protons and electrons to combine to make neutrons and all the electrical forces helping to support the star disappear. As a result the star collapses catastrophically, ending as an incredibly dense ball of neutrons more massive than the Sun but only 10 to 20 kilometres ($6\frac{1}{4}$ to $12\frac{1}{2}$ miles) in radius.

This incredibly small, dense object, a neutron star, has a density of about 200,000 tonnes per cubic millimetre (3000 million tonnes per cubic inch). It is small enough and strong enough to spin once a second without breaking up. More evidence for pulsars being neutron stars came once again from the combination of radio telescopes, radio interferometers and optical telescopes. Detailed study of the radio pulses indicated that the pulsars must be within our own galaxy. This was supported by evidence from one of the interferometers, the giant Mills Cross at Molonglo, which has discovered the majority of the pulsars now known. The Molonglo surveys showed that pulsars are distributed in much the same way as the stars of the Milky Way. The One Mile interferometer at Cambridge obtained precise positions for some of the pulsars and these were examined with optical telescopes, which found there were no obvious stars or galaxies at the places indicated – a negative result but compatible with the small size expected of neutron stars.

Then in 1968 a pulsar was identified with a known visible star. The giant radio telescopes at Greenbank and Arecibo found a pulsar within the Crab nebula. The next year W. J. Cocke and M. J. Disney at the University of Arizona found a star at the centre of the nebula that was flashing in synchronism with the radio pulses. The discovery fitted the theoretical picture of a neutron star very well. The Crab nebula is the result of a supernova observed by Chinese and Japanese astronomers in AD 1054. This dramatic event, in which a star collapsed into a neutron star, gave out a burst of light so intense that the star was visible in daylight for three weeks. At its peak, it blazed out to appear fifty times as bright as Sirius, the brightest star, although it lay six hundred times farther away from us.

The importance of neutron stars was not only that they could explain the behaviour of the pulsars. The demonstration of a new, more compact stage of collapse for a dying star too massive to end as a white dwarf gave impetus to theoretical studies of another and possibly final stage of stellar evolution.

If a sufficient mass is concentrated in a small volume, the force of gravity becomes so large that even the neutrons are mashed together and the result is an even smaller and denser body than the neutron star. Such an object has so intense a gravitational field that even objects moving at the velocity of light, including light itself, cannot escape. This total lack of radiation gives the object its name – a black hole.

What happens inside a black hole is, by definition, unobservable and not susceptible to experimental test, so such phenomena have been a fertile source of theoretical speculation. This includes the problem of what happens to the collapsing object at the centre of a black hole. The only thing that can be said with confidence about this aspect of black hole theory is that it is wrong, or at the very least inadequate. Their strong gravitational pull does however mean that black holes have their uses. In particular they are a remarkable source of energy and have been put forward as an explanation of the enormous energy radiated by the quasars.

To see why this is so, consider a simple experiment, that of the fall of a 1-kilogramme mass a distance of 1 metre (roughly a 2-pound mass a distance of 3 feet): in more prosaic terms, the dropping of half a brick from waist height. On the surface of the Earth the fall takes a little less than a quarter of a second and releases enough energy to warm the half-brick up by a hundredth of a degree. Now repeat the experiment at a few other places in the universe. At the surface of the Sun the force of gravity is twenty-eight times greater than on Earth. The fall takes a twenty-fifth of a second and the energy

Two images of the stars at the centre of the Crab nebula. The upper frame is built up of exposures taken during the pulsar flashes, the lower frame is between flashes. Normal stars show in both frames, the pulsar only in one.

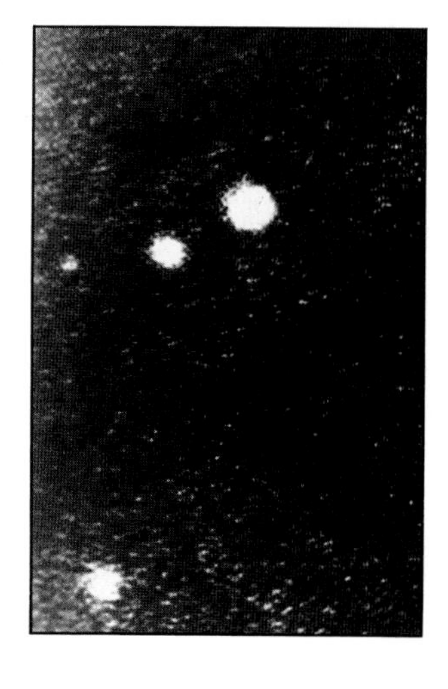

released could give a rise in temperature of about a quarter of a degree. The next step is to go to a white dwarf star – say the companion to Sirius. This star has almost exactly the same mass as the Sun, but is only one-fortieth of the radius. Its gravity is therefore more powerful. The astronaut holding up the brick will need to be fairly strong as the brick will appear to weigh 30 tonnes. When released it will fall in a thousandth of a second and release enough energy to heat up by over 800 degrees, hot enough to melt lead. This was the most extreme situation that classical stellar evolution could offer.

The new stars, pulsars and black holes, are in a different league. The force of gravity on the surface of a pulsar is 100,000 times that on Earth. The half-brick now falls in a millionth of a second and heats up into a blinding ball of dissociated atoms, at a temperature of 100 million degrees. This temperature is much the same as the highest temperature expected in stellar physics, that of the hottest stellar interior. The black hole is even more drastic. If the brick is carefully dropped just outside the rim of the black hole, so that we can still see what happens, the temperature rises to more than ten times the previous record and the astronaut would find his simple half-brick about as devastating as a modest atomic bomb.

All this energy comes from 1 kilogramme of matter, dropped 1 metre. The Earth alone receives 400 tonnes a day of assorted interplanetary material, mostly dust, and this falls from a height very much greater than 1 metre. Given enough matter concentrated in a small enough volume to produce an intense gravitational field, simply letting more material fall in will produce the enormous energy output of the quasars.

The 1960s were an heroic age in astronomy, with vast distances, enormous energies and new and exotic objects like quasars and pulsars. The 1970s have been quieter and less dramatic and interest has swung from the very powerful to the very obscure. In place of pulsars, the last stage of many dying stars, attention has swung to the dark nebulae, opaque clouds of interstellar material in which stars are being formed. Just as radio waves can see through terrestrial clouds, so they proved able to penetrate the clouds in our galaxy, especially those towards the centre of the galaxy. Radio emission from bright nebulae, such as the Orion nebula, had been known since the mid-1950s. This emission was similar to that from the interstellar gas of the galaxy, except that the presence of hot stars inside the nebula raised its temperature quite considerably and greatly enhanced the radiation from ionized gas collisions.

The cool clouds are very different. The temperature is only about a hundredth that of the bright nebulae, about 150°C below zero (100°K). In spite of the low temperature, the dark clouds proved to show emission lines in the shortwave region of their radio spectrum. With only a few exceptions the lines are at centimetre and millimetre wavelengths. Hydrogen was the only atom expected to be observable in the radio region at its 21-centimetre wavelength, and these new lines all proved to be radiation from molecules. Some of the molecules are small, stable molecules of the kind that exist in the atmospheres of the planets, such as water, carbon monoxide and ammonia.

The Horsehead nebula is a dense cloud of cool gas and dust that obscures the stars that lie behind it from everything but radio telescopes. The heavily overexposed bright star to the left is Zeta Orionis, at the eastern end of Orion's Belt. The bright, diffuse light is an emission nebula, excited by Sigma Orionis, the star on the upper right of the picture.

The radio telescope array at Westerbork, Holland (above). *One telescope, at the end of the array, can be moved on rails to give eleven sets of different separations simultaneously.*

The Y-shaped Very Large Array (VLA), on the Plains of San Augustin, Soccoro, New Mexico. The VLA is the largest astronomical instrument ever built and combines the signals from twenty-seven radio telescopes, each 25 metres (80 feet) in aperture. The entire array can be adjusted in size and aerial configuration by moving the dishes on special railway tracks.

Other simple, well-known molecules probably exist in interstellar space but are not detected because the rules governing the interaction of molecules and radiation prevent them from emitting radio waves. Hydrogen, oxygen, nitrogen and carbon dioxide molecules are all invisible at radio wavelengths. Among the molecules which are found are some much nastier chemicals – hydrogen cyanide, formic acid and formaldehyde are all observed. On a more cheerful side, so is ethyl alcohol – some interstellar clouds contain enough of it to intoxicate the entire population of Earth.

Over fifty molecules are now known and the effect of their discovery over the last decade has been to start a new branch of astronomy – astro-chemistry. In the past the discoveries of astronomy and radio astronomy have been dominated by high temperatures, and so are related to atomic and nuclear physics. The new data in the field of low temperatures and molecular physics are giving us the first view of a very different face of the universe.

The developments in quasars, molecules and other astronomical objects best studied in the radio region has justified the funding and construction of the largest single piece of astronomical equipment ever assembled. Its background was a development of aperture synthesis by Ryle's group at Cambridge, which led to the construction of the "One Mile" telescope in 1964. The success of the concept led to the construction of a telescope array at Westerbork, in Holland, using twelve 25-metre (82-foot) steerable dishes in a 1600-metre (1-mile) long line. Ryle then extended the wavelength range and resolution available, again with an array, this time of eight 13-metre (42-foot) dishes in a 5-kilometre (3-mile) long line. This system was able to work down to a wavelength of 2 centimetres ($\frac{4}{5}$ inch), with a resolution just under an arc second, comparable to optical telescopes. The latter half of the 1970s has seen the building of the Very Large Array – the "VLA" – near Socorro in New Mexico. This is an aperture synthesis system laid out as a Y-shape with arms 21 kilometres (13 miles) long. Each arm is a railway, so that the twenty-seven separate radio telescopes, each 26 metres (85 feet) in aperture, can be moved to give the aerial pattern best suited to the observations required. The scale of the enterprise is enormous. The site survey found only three sites in the whole of the United States large enough and flat enough to support the array. The cost was equally enormous – $100 million. It was built in hope that the new instrument will make discoveries such as those that have followed the constructicn of "the biggest telescope in the world" at almost every step taken for the last 350 years.

CHAPTER FOURTEEN

ONWARD AND UPWARD

The arrival of space astronomy – lunar and planetary fly-by and landing; ultraviolet space telescopes, 1950 to the present-day

"The first artificial satellite in the world launched."

Headline in Pravda, *6 October, 1957.*

"The achievement is impressive. It is the first time the Russians have been first in the field with a major revolutionary technical advance."

The Times, *London, 7 October, 1957.*

"New voyage to the cosmos – Soyuz 12 in orbit."

Headline in Pravda, *28 September, 1973.*

"Observers believe that the main objective of Soyuz 12 is to prove to the world in general and the Americans in particular, that Soviet space technology has not fallen behind."

The Times, *London, 28 September, 1973.*

The Earth's atmosphere imposes a whole range of limitations on the astronomical observations it is possible to make using a telescope that is immersed in that atmosphere. Some of the restrictions are temporary nuisances: wind shake on radio telescopes and cloud above optical telescopes are in this class. A different set of troubles makes observation less accurate than is desirable: turbulence in the ionosphere causes fading of signals arriving at radio telescopes, and turbulence lower in the atmosphere leads to "seeing" problems in the visible part of the spectrum – twinkling and blurring of optical images. In addition, the atmosphere itself gives out light – the airglow – which makes the observation of very faint stars (those dimmer than twentieth magnitude) increasingly difficult. More important than any of these difficulties, however, is the complete opacity of the atmosphere at virtually all wavelengths. The only radiations reaching the ground from space are light and shortwave radio signals.

By the end of the war in 1945 physicists had studied the electromagnetic spectrum over an enormous range in wavelength. At longer wavelengths than the 0·4 to 0·75 micrometres of visible light were the near and far infra-red, followed by microwave, short-, medium- and long-wave radio waves, the latter extending to several thousand metres wavelength. In the opposite direction the spectrum runs through the near and far ultraviolet, then soft and hard X-rays and finally to the gamma-rays with wavelengths less than 2·4 millionths of a micrometre. Radiation of virtually all this vast range of wavelengths, from a billion times to one-billionth that of visible light, can travel freely through the vacuum of space but cannot penetrate the Earth's atmosphere.

The first opportunity to make measurements from above the atmosphere was provided by captured German V2 rockets in 1946. The V2 was able to lift a payload of 1000 kilogrammes (2200 pounds) to an altitude of 160 kilometres (100 miles), which gave it about five minutes of observations above the ultraviolet-absorbing ozone layer some 30 kilometres (20 miles) above the ground. The first attempt to do astronomy using the V2 rocket was made in June 1946 and was not a success. The rocket, carrying a spectrograph in place of a warhead, reached an altitude of 118 kilometres (73 miles) and then turned over and returned to Earth. The impact of the 9-tonne rocket made a crater about 10 metres (33 feet) across, and only a few fragments – enough to fill a wastepaper basket – were recovered after much digging.

The first success was achieved a few months later by a group from the Naval Research Laboratories in Washington led

by Richard Tousey. On 10 October, 1946, they launched a rocket carrying a solar spectrograph mounted in one of the tail fins. An armoured cassette of film was recovered, with the first astronomical measurements from "space". The peak altitude of the rocket flight, 174 kilometres (108 miles), was outside all but two-billionths of our atmosphere.

For some years progress was slow and was restricted entirely to studies of the Sun. In the five years following the first successful rocket flight the Aerobee replaced the V2 and about 150 "scientific" rockets were launched, chiefly from White Sands in New Mexico, the proving ground of the prototype atomic bomb. Almost all of these experiments were concerned with the Earth's atmosphere itself. The main astronomical discoveries of these early years were that the active Sun emits X-rays and that its ultraviolet spectrum differs from its visible spectrum. The familiar solar spectrum with dark Fraunhofer lines in absorption gives way to a bright emission line spectrum at shorter wavelengths. In particular, the Sun was found to emit a great deal of radiation at a wavelength around 0·12 micrometres, suggesting that the strongest line in the spectrum of hydrogen, called Lyman

A V2 rocket displayed on its launcher. Developed by the Germans in the early 1940s as a long-range missile, the V2 was later used by the Americans to lift astronomical equipment above the atmosphere.

Wernher von Braun, creator of the V2, later joined the US rocket research team. He is nicknamed "the father of the Apollo Missions".

alpha, was a major feature of the Sun's ultraviolet spectrum. Following up this deduction, W. Rense of the University of Colorado actually photographed the Lyman alpha ultraviolet radiation in 1953. Three years later Rense obtained pictures of the whole Sun in Lyman alpha light.

The pace of the research was modest. A few teams, principally from the Naval Research Laboratories (NRL) in Washington, and academic groups at Johns Hopkins University in Baltimore and the University of Colorado in Boulder, each launched a rocket or two a year. All this changed when the Soviet Union launched the first satellite, Sputnik 1, on 4 October, 1957. A satellite in Earth orbit can study the sky continuously for prolonged periods and is evidently a better platform for astronomy than a short rocket flight. The immediate impact of the Sputnik launch was, however, felt most strongly in the planetary sciences. The quickly developing generation of rockets able to lift heavy payloads into Earth orbit were powerful enough to launch lighter spaceprobes to the Moon, and later to the planets. Their cameras and other equipment could study at close quarters the other worlds of the solar system.

Provided an Earth-based astronomer uses a moderate-sized telescope, his observations of the planets are essentially set by the seeing conditions. Twentieth-century telescope improvements had therefore had relatively little effect on planetary studies, and very little additional data had been gathered in the period from 1900 to 1960. This situation was transformed by the space programme, which brought in an era of exploration unparalleled since the discovery of the New World. The pattern was similar in the two cases. The first successful voyage was followed by a series of visits to the most accessible and best-known destination. Within a few years of Columbus's voyage, Cabot went north and Vespucci south of his route: on much the same timescale, voyages to Venus and Mars followed the first visits to the Moon. There is even rough agreement between the number of ships that sailed the Atlantic between 1492 and 1502 and the number of spacecraft launched between 1958 and 1968. This time, however, in place of a three-horse race for the riches of the Indies, there was a two-horse race for a lot of propaganda and a little science. The result, over the first decade, was roughly speaking a draw. The Russians, by and large, did everything first and the Americans did everything better. For planetary studies, the astronomical telescope was replaced as the primary research tool by the remote-controlled television camera. Even the maps changed. Up to the 1960s, maps of the planets had been drawn with north at the bottom, as seen in the inverted view of an astronomical telescope. Since then the camera view, with north at the top, has prevailed.

The first spaceprobes to leave the immediate vicinity of the Earth and approach the Moon were Luna 1, 2 and 3, all launched by the Russians in 1959. Of these Luna 2 was the first man-made object to reach the Moon, which it hit on September 14. Luna 3 attracted more interest, as it carried a camera and sent back the first picture of the face of the Moon that is never turned towards the Earth. The quality of the picture was disappointing, considering that it was taken only 6200 kilometres (3850 miles) from the Moon's surface.

After this brief salvo the Russians paused to improve their spaceprobe tech-

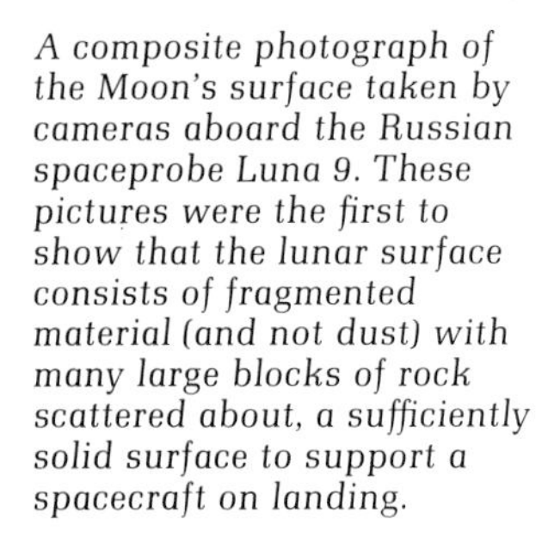
A composite photograph of the Moon's surface taken by cameras aboard the Russian spaceprobe Luna 9. These pictures were the first to show that the lunar surface consists of fragmented material (and not dust) with many large blocks of rock scattered about, a sufficiently solid surface to support a spacecraft on landing.

nology, especially the navigation systems, while the Americans strove to perfect the bigger rockets they needed to accomplish similar feats. Undoubtedly, the chief gain for astronomy in general came from the need for a giant radio telescope to pick up weak signals from the new probes. Bernard Lovell's 76-metre (250-foot) radio telescope at Jodrell Bank proved ideal and the new task, and the resulting new funds, arrived just as the financial problems surrounding the building of the telescope had reached the edge of catastrophe.

Five years after Luna 3, a veritable bombardment of the Moon began. The United States spacecrafts Ranger 7, 8 and 9 sent back high-quality pictures of the Moon as they crashed on to the lunar surface. This programme was followed by a series of Surveyor soft-landing craft and by five Lunar Orbiter satellites. The Lunar Orbiter missions achieved a 100 per cent success record and sent back almost 2000 pictures, covering the entire lunar surface, front and back. The finest detail shown was a metre or so across, though not for the whole Moon. One shot, the epitome of the high level of activity, was a picture taken by Lunar Orbiter 3, showing one of the soft landing spacecraft, Surveyor 1, on the surface of the Oceanus Procellarum.

The equivalent Russian programme, although it had been the first to begin, resumed with a number of poor results. Luna 4, 5, 6, 7 and 8 all failed to achieve the planned soft landing. Luna 9, launched on 31 January, 1966, arrived intact on the Moon on 3 February, four months before Surveyor 1. The $1\frac{1}{2}$-tonne spacecraft, four times the size of the American Lunar Orbiters, carried a television camera and instruments to analyse the soil on which it had landed. One conclusion needed no elaborate analysis: the surface was not, as had been suggested, a thick layer of loose dust into which the spacecraft would disappear. Two months later, and well before Lunar Orbiter 1, Luna 10 arrived to become the first satellite of Earth's satellite, the Moon. The main conclusion from its eight weeks of observation was that the Moon did have a magnetic field, albeit a very weak one.

Luna 11, 12, 13 and 14 – three lunar satellites and a "soft lander" – followed the successes of Luna 9 and 10. Luna 15 was a failure, crash landing on 21 July, 1969, the same day as the first men on the Moon – Aldrin and Armstrong of the US Apollo 11 mission – landed safely.

Since that historic day the pace has slackened: there were six more manned Apollo missions (including the near disaster of Apollo 13) and eight further unmanned Luna spacecraft, the last in 1976. Three of the Luna spacecraft landed, collected soil and rock samples and returned to Earth, while two Russian and three American flights used lunar vehicles.

The results from these experiments have completely rewritten the textbooks on the Moon. At one time, about 1965, the amount of data coming in was such that supporters of virtually any theory could find some observation to add plausibility to their speculation, but this phase is now past. The Moon has been recorded in visible, infra-red and ultraviolet light and prodded with seismic, chemical, radioactive, geological and other probes. The net result, as the jigsaw of new facts has been slowly fitted together to make a picture, is that the history and structure of the Moon are now understood.

As with the Moon, so with the planets. All but Uranus, Neptune and Pluto have become subjects for the television camera rather than the telescope. The study of

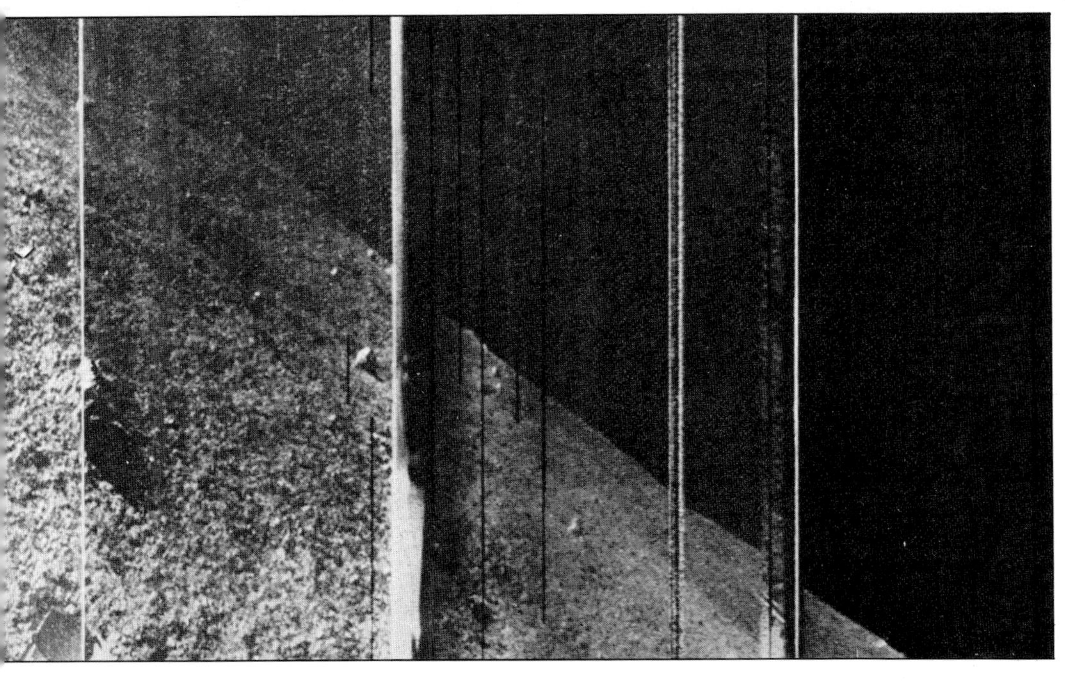

the planets by spaceprobe has involved technical ingenuity of the highest order. The control requirements, especially the sophistication of the on-board computers, are enormous. The manoeuvres of Voyager 1 and 2 as they took their superb and unprecedented pictures of Saturn's rings had to be controlled when the machine was so far away that it took an hour for a radio message to reach the spacecraft and another hour to hear whether anything was achieved as a result. Other types of "impossible" experiments have been done. One intellectual challenge that has been met was the task of designing an experiment to test the Martian soil for a form of life that was necessarily totally unknown. Four experiments soft-landed by Viking 1 and 2 in 1976 have convinced most scientists that Mars has no indigenous life. The Russian Venus landers, Venera 9 and 10, had to satisfy such a tough specification that any one aspect would give pause for thought. It was necessary to design a television camera that could be launched into space, travel for four months in the interplanetary vacuum, slow down and then descend through a fog of strong sulphuric acid to land gently on an unseen surface, far hotter than any domestic oven and at a pressure usually only found inside heavy steel cylinders of compressed air.

The impossible is not only done as a matter of routine, it is also done on time.

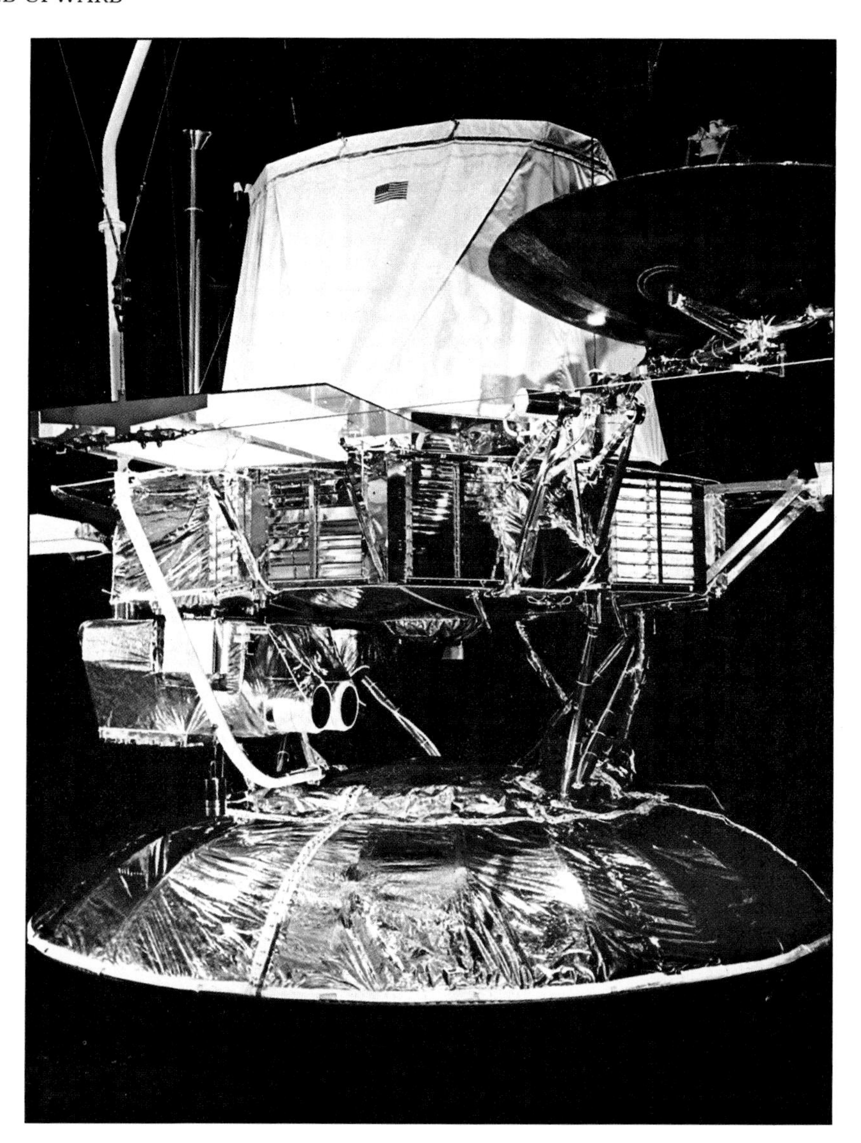

Maintaining the temperature balance of a spacecraft in the vacuum of space is critical. Above, the thermal properties of a test model of the US Viking Orbiter are being determined.

Voyager 1 (below) *being launched from Kennedy Space Center on 5 September, 1977. Its journey will take it to the outer planets.*

A global mosaic of Mars (below) *composed of some 1500 television pictures taken from the US Mariner 9 spacecraft, which went into orbit around Mars in November 1971.*

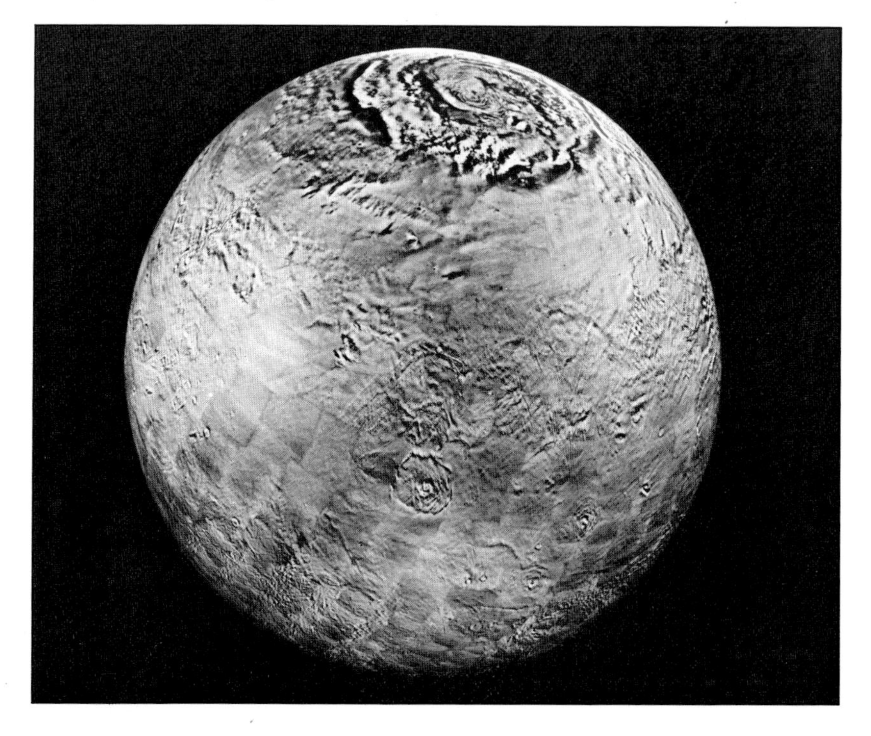

These vastly complex systems have to be designed, built, tested and calibrated for a launch date specified years in advance. The relative positions of the planets are always changing, and a delayed launch means an inaccessible target.

What then have been the gains from the hundred or so lunar and planetary spacecraft launched in the last twenty-three years? First of all, fact has replaced ignorance or error. There are "canali" (channels) on Mars, though few of them are straight and none is water-filled or artificial. Nitrogen is a major component of the atmospheres of Earth and Titan, Saturn's largest moon, but not of Venus and Mars. Mercury, Mars and some, but not all, of Jupiter's moons are cratered. The list of new facts is endless. The biggest gains have been for the sciences that try to understand the Earth. Meteorologists concerned with the global circulation of the atmosphere can now test their theories against data on four other atmospheres: Venus, Mars, Jupiter and Saturn. From similar comparisons, studies of the interior of the Earth can benefit. Not so long ago only the diameter, mass and rotation rate were available for most planets and in many cases they were wrong. For the first seventy years of this century, for example, the value quoted for the mass of Mercury was little better than the personal whim of one influential astronomer, the American Simon Newcomb.

The spacecraft data have become fundamental to the science of cosmogony, the study of the origin of the solar system.

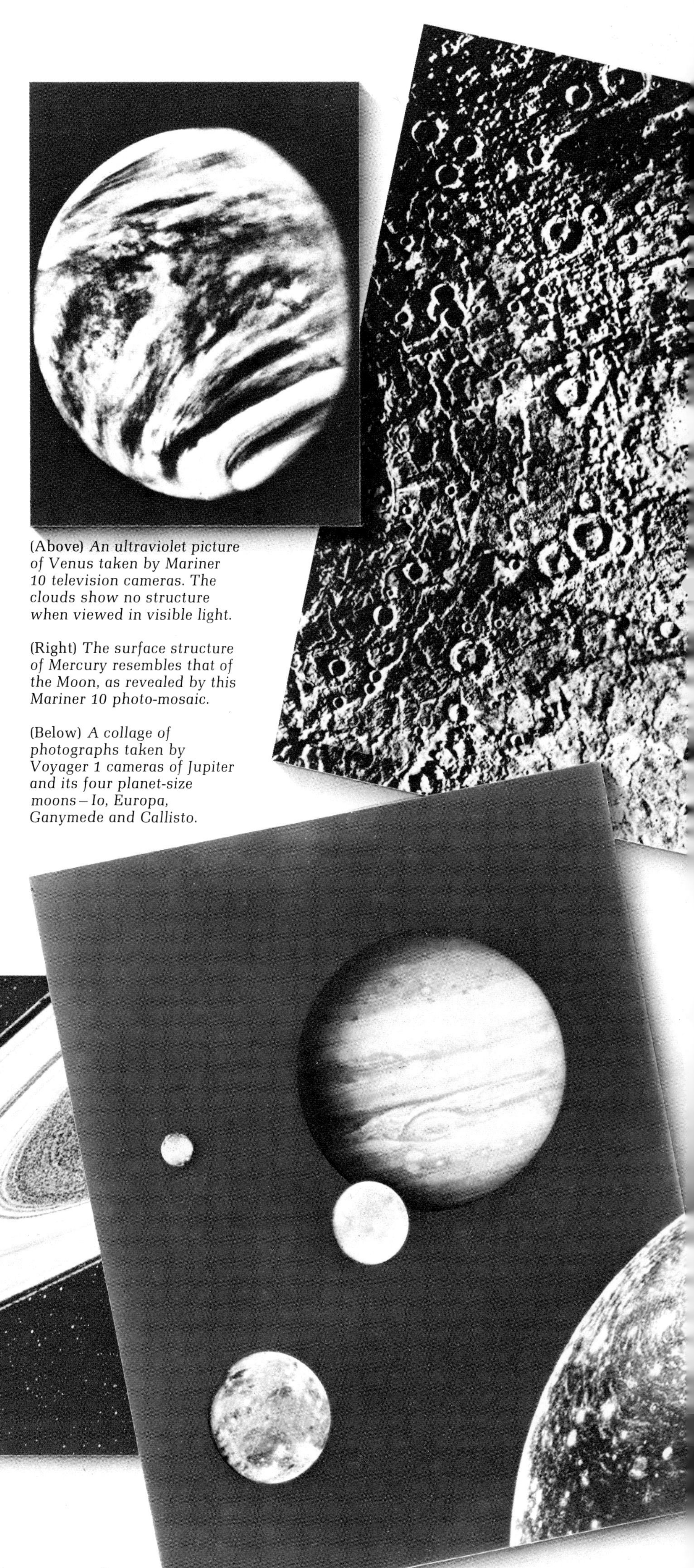

(Above) *An ultraviolet picture of Venus taken by Mariner 10 television cameras. The clouds show no structure when viewed in visible light.*

(Right) *The surface structure of Mercury resembles that of the Moon, as revealed by this Mariner 10 photo-mosaic.*

(Below) *A collage of photographs taken by Voyager 1 cameras of Jupiter and its four planet-size moons – Io, Europa, Ganymede and Callisto.*

(Below) *Another photograph taken from Voyager 1, of the shaded side of Saturn's rings. From this aspect, never seen from Earth, the normally dark "Cassini division" appears as the brightest feature of the rings.*

When Sputnik 1 was launched, the subject was little more than a heap of untested and untestable speculation. Now there is a consistent and generally accepted picture: a process of accretion and fractional distillation in the dust cloud surrounding the infant Sun, in which the planets gradually built up. The differences between the small dense inner planets and the large and less dense outer ones are now matters of calculation not guesswork. To take an example, the decision between the groups of theories that suggest that the Earth and Moon were formed as near neighbours and those that look to a capture of our satellite after formation elsewhere can now be made. The former group are preferred. Only they can explain the details of the abundance of oxygen in the lunar rocks that have been brought back to Earth for analysis. Less fundamental topics have also become matters of fact, particularly the properties of the interplanetary medium. This is a mixture of the hydrogen of the outward-flowing solar wind and a weak magnetic field and it cannot be studied near the Earth as the terrestrial magnetic field disturbs the flow.

The branch of space science concerned with objects more distant than the Moon and planets extended in parallel with all this, as detectors were flown above the Earth's atmosphere to study radiation from the Sun and from astronomical objects beyond the solar system. The impetus that came from Sputnik 1, the opening shot of the space race, did not however immediately transform X-ray and ultraviolet astronomy into a satellite-borne science, although Sputnik 2 did carry a small bank of ultraviolet sensitive photomultipliers. It took a further five years to build and launch the first specifically astronomical satellite, Orbiting Solar Observatory Number One, or OSO 1.

The interval between Sputnik 1 and OSO 1 saw a surge forward in rocket technology. Instruments flown on American Aerobee and British Skylark rockets extended knowledge of the solar spectrum to ever shorter wavelengths. The gap between ultraviolet and X-ray wavelengths (0·1 to 0·001 micrometres) was filled in and this region of the solar spectrum proved to be dominated by emission from the corona, the tenuous outermost layer of the Sun which has a temperature of a million degrees. The first crude studies of ultraviolet radiation from stars came in the early 1960s. By 1963 some sixty stars had been observed in the far ultraviolet at wavelengths down to about 0·1 micrometres. Three times as many had been

Launch of a British Skylark rocket. The rocket carried equipment for astronomy and to study the Earth's upper atmosphere.

recorded at longer wavelengths between 0·2 and 0·3 micrometres. These data were collected using 10- or 15-centimetre (4- or 6-inch) aperture telescopes mounted on Aerobee rockets. The most significant discoveries were however made by X-ray astronomy groups at MIT and at NRL. In 1962 they found X-ray emission from sources other than the Sun. Simple sums, based on how bright the Sun would appear in X-rays at interstellar distances and on theoretical estimates of X-ray emission from hot stars like Vega and Spica, predicted that there would be no detectable X-rays from any stars. The discovery of the first cosmic X-ray source, Scorpius X-1, came as a by-product of an unsuccessful search for X-rays from the Moon caused by the impact of the solar wind on its surface. Within a few months, ten sources were found, one of which was the ubiquitous Crab nebula, producing copius X-rays as well as every other form of radiation.

Such "observations" involved about a dozen rockets and a total observing time of well under an hour and, in all these cases, were confined to the merest glimpses of the star concerned. The rockets were not stable, but rolled, yawed and tumbled through the upper atmosphere, so that the field of view of the ultraviolet telescopes or X-ray counters swept across the sky. An early published record of X-rays from the Crab nebula showed the detection of a mere twenty X-ray photons in the 0·12 seconds during which the source was in the field of view.

By the mid-1960s, a significant step forward had been made in the capability of sounding rockets. New developments in rocket stability gave the experimenter the ability to point his apparatus at a specified target for the whole of the brief flight. The solar radiation could now be studied at different positions on the Sun's disc: ultraviolet observation of the brightness variation between the centre and the edge of the disc, differences between the inner and outer corona or between quiet and active regions all became possible. The ability to point the rocket payload also meant a massive gain in the time spent looking at a particular star.

The first detailed ultraviolet observations of stars, by D. Morton and L. Spitzer of Princeton University, came in June 1965. Two earlier flights had failed; faults – in one mechanical and in the other electronic – had disrupted the pointing control. The third flight was almost a failure: the payload recovery parachute failed and the telescopes were smashed to pieces. The film from the telescope was not recovered until an hour after sunrise. It was found fogged and assumed to be useless. Two weeks later, back at Princeton, careful development of the longest exposure showed the ultraviolet spectra of Delta and Pi Scorpii, two hot second-magnitude stars.

This pioneering effort, using 5-centimetre (2-inch) aperture telescopes, was rapidly followed by two others. G. R. Carruthers of NRL flew a 15-centimetre (6-inch) telescope which was integrated into its own image intensifier, while T. P. Stecher, at a new institution, the Goddard Space Flight Center, developed a 33-centimetre (13-inch) telescope, attached to a photoelectric spectrometer. Survey work, by A. Boggess at Goddard, T. Chubb at NRL and R. Wilson at Culham, England, using 10-, 15- and 23-centimetre (4-, 6- and 9-inch) telescopes, extended observations down to a magnitude of 6·5 and detected several hundred stars. The stars themselves proved to be less bright in the ultraviolet than expected, as their spectra were found to be composed almost entirely of "dark" Fraunhofer lines that masked the predicted bright spectrum. Much more interesting and important were the spectrum lines due to atoms of the interstellar medium through which the radiation travels on its way to us. This material, a gas at a pressure far below any vacuum attainable in the laboratory, is the material from which new stars are made: its constitution is fundamental to studies of stellar evolution. The rocket measurements gave the first indications of the ratio in the interstellar gas of elements such as carbon, nitrogen and oxygen.

The interstellar gas also acts to divide the ultraviolet from the X-ray region of the astronomical spectrum. The hydrogen atoms in space absorb all wavelengths between 0·09 and about 0·01 micrometres, so it is impossible to see distant stars at all in this wavelength range. The stabilized rockets were used both sides of this divide, the main X-ray work being carried out at wavelengths around one nanometre (0·001 micrometre). The difficulty of X-ray studies is that it is not possible to use orthodox refracting or reflecting telescopes: no lens material can focus X-rays, and in a conventional reflecting system the primary mirror would absorb the radiation. Until very recently the only way to limit the field of view of an X-ray detector was to mount a screen in front of it. This screen – a "mechanical collimator" – acts only as a mask to block out X-rays from the rest of the sky and does not form an image of the source being observed.

A great deal of ingenuity went into the design of collimators, but the first analysis of the size and shape of an X-ray source used an alternative approach. In 1964 H. Friedman of NRL launched a rocket with a payload designed to study the Crab nebula as it was eclipsed by the

Lyman Spitzer, Professor of Astronomy at Princeton, has been one of the most influential figures in the development of US ultraviolet astronomy. The Princeton group of scientists designed the satellite Copernicus. Launched in 1972, this is still the largest telescope ever placed in orbit.

Moon. The results showed that the X-rays came from the expanding gas of the nebula and not from the central star. Just as rocket-borne ultraviolet telescopes grew larger, so did X-ray detectors. The two sources known in 1962 increased to thirty-two by 1969 as a result of a hundred-fold increase in sensitivity. Of these sources all but four were in the plane of the Milky Way and so presumably within our galaxy. Two of the remainder were definitely not local – M87, a galaxy already noted for its strong radio emission, and the Large Magellanic Cloud, an irregular clump of stars forming a small galaxy close to our own. The brightest X-ray source, Scorpius X-1, was a temperamental object, sometimes bright and sometimes not. Allan Sandage at the Hale Observatories identified it as a faint blue variable star. Solar flares give out X-rays, and Scorpius X-1 appeared to be a star in a state of almost perpetual violent flaring.

The modest rocket payloads were not only employed to study the X-ray and ultraviolet radiations. A search for gamma rays, the shortest wavelength component of the spectrum, was attempted unsuccessfully as early as 1951, using a V2. Nothing was found, largely because there are so few gamma rays arriving from space that longer exposure times are needed than the rockets could provide. Groups from Cornell University and NRL flew 16-centimetre (6-inch) infra-red telescopes in the late 1960s, chiefly to study the far infra-red cosmic background radiation. Short radio wavelengths are best studied from the ground, but radiation with wavelengths longer than 20 metres (66 feet) cannot penetrate the ionosphere and requires space observations. The long wavelengths were first observed by satellites (the Canadian Transit 2A and Alouette 1) and in this case follow-up using rockets proved difficult. The detection of long wavelengths needs a large antenna and complete isolation from the ionosphere. Various groups tried rockets that unreeled long aerials, spinning slowly to keep the wire straight, but these showed that an altitude of 100 or 200 kilometres (60 or 125 miles) was not enough.

Even when astronomy satellites became reality, the rocket programmes were not replaced overnight. Long after the launch of OSO 1, the Skylark and Aerobee teams were still busy. One duty they assumed was the calibration of the satellites. Experiments in orbit often decline slowly in sensitivity, and a carefully calibrated series of rocket-borne experiments can determine this loss of performance. Equally, there may not be a satellite in the right place at the right time. Such a situation led to the most intensive fusillade of rockets ever fired for peaceful purposes. It occurred at the solar eclipse of 7 March, 1970, the only eclipse to pass directly over an established rocket launch facility – that at Wallops Island, on the east coast of the United States. The total eclipse over the island lasted four minutes and in a brief interval centred on this event thirty-two rockets were launched. One of the two largest, the NRL Aerobee, went into the sea, and there was a very tense atmosphere while Robert Speer of the Imperial College-Culham-Harvard-Toronto team developed the film recovered from the other. The results justified all the effort, and yielded a unique series of ultraviolet spectra of the eclipsed Sun.

Despite the success of rocket experiments, it is beyond dispute that astronomy satellites have dominated space science for the past decade. Although each is far more expensive than a rocket flight, their protracted observing time – often amounting to years – gives modern satellites an unapproachable superiority. The early days of satellite astronomy were dominated by studies of the long-wavelength radio waves. In 1964 Benediktov of the Gorkii Institute in Leningrad made observations at wavelengths up to 400 metres (1300 feet) using the satellite Elektron 2, while British and Canadian groups studied the same range with receivers on Ariel 2 and Alouette 1. The need for such experiments to be well away from even the outermost fringes of the ionosphere led to an interval when the experiments were carried on spacecraft with some other long-range goal. Experiments by Slysh and others were mounted on the Russian spacecraft Zond 2 and 3, Luna 10 and 11 and Venera 2. These simple experiments had very little directional sensitivity, but were able to detect radiation over a wavelength range from 30 to 1500 metres (100 to 4900 feet).

The first radio astronomical satellite with the ability to distinguish one region of the sky from another was the American Radio Astronomy Explorer 1 launched in July 1968. This remarkable spacecraft built a simple aerial array 0·5 kilometre ($\frac{1}{4}$ mile) across after being placed in orbit. On board the satellite were long reels of silver-plated copper tape and these were unrolled through dies that shaped the strip into stiff tubes. RAE 1 fabricated four aerials each 229 metres (750 feet) long, two 35 metres (115 feet) long and a pair of stabilizing booms 96 metres (315 feet) long. The result covered 9 hectares (22 acres) and achieved a resolution of roughly twenty degrees at a wavelength of 150 metres (500 feet). This array was able to show that the majority of the long-wave radio emission came from the plane of the galaxy, the peak power being at a wavelength of 300 metres (990 feet). A

(Above) *Spectra produced by the International Ultraviolet Explorer (IUE) satellite showing two different high-temperature sources within the globular cluster NGC 6752: the cluster is old and should only hold cool stars.*

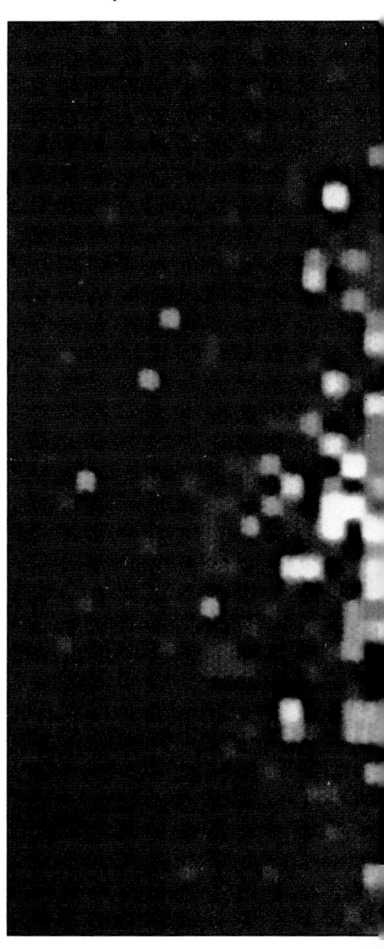

(Below) *The star Eta Ursa Majoris as viewed by the IUE telescope using digital data, with the darkest portion representing the star itself and the large white areas five neighbouring stars invisible to the naked eye.*

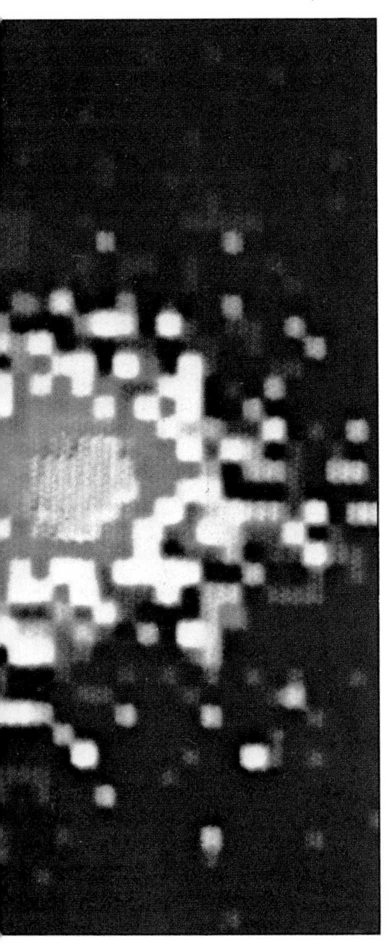

longer wavelength component with a maximum at 1200 metres (4000 feet) appeared to come from outside the galaxy. Most important, RAE 1 found that the free electrons floating round in space (one every 10 or so cubic centimetres/ $\frac{3}{5}$ cubic inch), absorbed very long-wave radiation. This set an unavoidable limit to further radio exploration and ended the radio astronomers' determined push to ever-longer wavelengths.

As the use of satellites spread to include ultraviolet and X-ray astronomy, the subject came to be dominated by American or American-launched systems. There were several reasons for this: in particular the increased technical sophistication required was more easily found in the United States than in the Soviet Union. This fact was reflected in the respective national priorities, and only a handful of the thousand Kosmos satellites placed in orbit between March 1962 and March 1978 were for astronomical use. Satellite-borne astronomy rapidly divided into three sections. Most experiments studied the Sun, and in the study of cosmic radiation sources ultraviolet and X-ray equipment was sharply differentiated.

The studies of the Sun required only the simplest telescopes of a few centimetres aperture. These were coupled to spectrographs to give high resolution in wavelength and increasingly good spatial resolution across the Sun's disc. The sixth Orbiting Solar Observatory, OSO 6, for instance, operated in two modes. It could be directed at one position on the Sun and scan the spectrum from 0·03 to 0·14 micrometres in 10,000 steps or be set at one particular wavelength and map the brightness changes across the Sun's disc and inner corona at roughly 4000 separate points. Instruments like these were flown on a series of solar astronomy satellites and were highly successful, but the research groups concerned nearly drowned in the flood of data that came back.

The astronomical satellites necessarily needed more effort on telescope design than the solar astronomy satellites, and they achieved their success in spite of bad luck: two of the first four failed to reach orbit. The first success, Orbiting Astronomical Observatory 2 (OAO 2), was launched in December 1968 and carried eleven telescopes. Four of these were 20 centimetres (8 inches), four were 30 centimetres (12 inches) and three were 40 centimetres (16 inches) in aperture. The large number was not to produce a satellite that could look in all directions at once but to avoid, as far as possible, the use of moving parts. Things that move are a thorough nuisance in satellite design, especially in the earlier systems, where every gramme of unnecessary weight was ruthlessly pared away. Motors were heavy and used scarce power, and bearings were liable to seize up in the vacuum of space. In addition, Newton's third law insists that it is impossible to move a component on a satellite one way without the rest of the satellite moving the other way. Compensation for this reaction uses up the limited ability of the satellite to remain stable and accurately pointed.

The second successful orbiting astronomical observatory, OAO 3, was able to exploit bigger rockets and was heavier and more complex. The satellite carried an 80-centimetre (31-inch) telescope with specially coated mirrors to give enhanced performance in reflecting ultraviolet radiation. It was launched on 21 August, 1972, and renamed to celebrate the fifth centenary of Copernicus's birth. (The Russian Interkosmos 9 was christened Kopernik to commemorate the same event: it studied the long-wavelength radio signals from solar flares.) Copernicus was followed by two further telescope-bearing satellites, the Dutch ANS and the British-designed, European-sponsored and then American-launched "International Ultraviolet Explorer". The former was launched in 1975 and the Explorer in 1978. They carried a 22- and a 45-centimetre (9- and 18-inch) aperture telescope respectively.

The data returned from the "ultraviolet" satellites has necessarily been less exciting than their X-ray counterparts have achieved. As their results are in a region of the spectrum adjacent to the visible region so intensely studied from the ground, the possibility of discovering new and strange objects is much reduced. The satellites have, in general, extended our understanding of ordinary stars, especially the hotter types that radiate strongly in the ultraviolet. The most interesting measurements, made by Copernicus, have been on the interstellar medium. The sensitivity and resolution of Copernicus's combined telescope and spectrograph allowed the detection of the rare isotope of hydrogen, called deuterium. This rare atom is very easily destroyed in the nuclear furnaces inside the stars, so the deuterium atoms in the interstellar medium must be a relic of the origin of the universe in the Big Bang. The precise concentration of deuterium atoms depends on how "loud" the Big Bang was. If the explosion was slow, as a very heavy universe blew apart, then there would have been enough time for deuterium to be destroyed in the heat of the explosion, and consequently there would be very little deuterium about today. A less massive universe, exploding more rapidly, would have moved too fast to destroy the fragile deuterium. The Copernicus measurement, that there are fourteen deuterium

The launch of the Uhuru X-ray satellite (right), *the 42nd satellite of the Explorer series. This particular rocket was launched by a joint US–Italian team from the coast of Kenya.*

(Below) *The final touches being put on NASA's first X-ray Explorer satellite, Uhuru, which was launched in 1972. Circling the Earth at an altitude of about 500 kilometres (320 miles), it discovered a large number of X-ray sources both within and beyond the Milky Way, a virtual revolution in the subject of X-ray astronomy.*

(Right) *Final preparation of the second High-Energy Astronomy Observatory (HEAO 2), the Einstein X-ray Observatory. Its telescope, the first large-aperture X-ray telescope to give high-quality images, has provided information concerning distant galaxies, neutron stars and black holes as well as other highly energetic types of stellar behaviour.*

atoms for every million ordinary hydrogen atoms in the interstellar medium, therefore gave a figure for the mass of the universe. The average density of the universe came out at one atom to every 2 or 3 cubic metres (70 or 110 cubic feet) of space. The total mass, for the 10^{71} cubic metres of the universe, is twenty billion times the mass of our galaxy. This is rather a small universe, too light for its gravity to overcome the expansion resulting from the Big Bang, so that it will expand for ever.

In December 1970, two years after the launch of OAO 2, the first X-ray astronomical satellite was placed in orbit. "Small Astronomical Satellite 1" or SAS-1, was renamed Uhuru, a Kenyan name in accord with its launch from a platform off the Kenyan coast. Uhuru was designed to survey the sky for new X-ray sources and to improve the accuracy of X-ray star position measurement. The satellite carried two X-ray detectors, each 20 centimetres (8 inches) square. One was masked to see a patch of sky five degrees square, the other more tightly limited to a strip five degrees long and only half a degree wide. Uhuru raised the number of X-ray sources known by roughly a factor of ten. Both the Andromeda galaxy, M31, and the more distant giant radio galaxy Centaurus A (NGC 5128) were found to emit X-rays. More exciting was a group of pulsating sources within our galaxy, among them Cygnus X-1, Hercules X-1 and Centaurus X-3. These objects were very massive binary stars, and in the latter two the companion star was a neutron star. The most interesting of these stars was that in Cygnus. This had all the characteristics of a pair of stars rotating about one another. Observations in the visible part of the spectrum showed one star in the Cygnus X-1 system to be a supergiant star, but there was no sign of its companion. Analysis of the orbit showed the invisible partner to be at least eight times the mass of the Sun. If it were an ordinary star it would be easily seen; white dwarfs and neutron stars cannot exist with masses more than three times the mass of the Sun. The inference was that the missing star has to be the first known black hole. The X-rays are emitted as matter is torn off the giant star and whirls around the nearby hole before it falls inexorably in.

The other major discovery of the Uhuru satellite was emission of X-rays from clusters of galaxies. Galaxies are not uniformly distributed in space, but tend to come in clumps, the largest known containing 2500 galaxies. The vast majority of the X-ray sources outside our galaxy proved to be such clusters. Closer examination showed that it was the intergalactic gas, a thousandth the density of the interstellar gas and at a temperature of some 100 million degrees, that was the source.

The coming of age of another "new astronomy" – X-ray astronomy – followed the launch of Uhuru and led to a series of X-ray satellites. The difficulty set by the masked counter style of "telescope" is that sensitivity comes only with increased size. By the launch of HEAO 1 in 1977, the X-ray satellite had reached a limit: HEAO 1 was 2·4 metres ($7\frac{4}{5}$ feet) in diameter and weighed 2·7 tonnes. The solution was to build true X-ray telescopes, using the fact that X-rays would reflect off polished surfaces if the incident beams were very nearly parallel to the surface in the first place. The idea was tested with small telescopes in rockets and then with a 22-centimetre (9-inch) diameter solar X-ray telescope on Skylab. Finally, a giant X-ray telescope was built for the HEAO 2, now called the Einstein X-ray Observatory. Launched in November 1978, this satellite carries four nested X-ray telescopes, the outermost 56 centimetres (22 inches) in diameter. It has proved as productive as its builders hoped. The ability to image X-rays has given a resolution of a few arc seconds: where Uhuru simply detected radiation from the Andromeda galaxy, "Einstein" shows sixty-nine separate sources in the galaxy. It has shown that all quasars emit X-rays, and nearby stars have X-ray-emitting outer envelopes, similar to the solar corona.

The speed of development of X-ray astronomy has been staggering. It took 340 years to get from Galileo's telescope to Palomar, a gain of about three million in sensitivity. The same gain in radio astronomy, from Jansky to the Very Large Array, took fifty years. X-ray astronomy went from initial discovery to the Einstein satellite in only sixteen years.

The present state of satellite astronomy is relatively quiet: hopefully it is the quiet before another storm of discoveries. Orbiting observatories have short lives and no major instrument is at present operating. The next satellites scheduled for launch are the European X-ray satellite Exosat, and the first major infra-red satellite IRAS. Following these is the Space Telescope, a 2·4-metre (96-inch) aperture optical and ultraviolet telescope to be sent up in the controversial Space Shuttle reusable spacecraft. This telescope should see farther and more clearly than any optical or ultraviolet telescope yet built. Virtually all major instruments since 1610 have achieved breakthroughs in discovery and understanding, so it is safe to predict that the Space Telescope will achieve the unpredictable. What this will be we do not know, but perhaps there is a "new astronomy" of the visible region just around the corner.

THE SPACE TELESCOPE

The Space Telescope, which is due to be placed in orbit round the Earth by the Space Shuttle in 1983, will be able to observe, record and measure astronomical phenomena in the absence of seeing effects and the faint light of the airglow. It will have ten times better resolving power than ground-based telescopes and will be able to see objects that are fifty times fainter than those observable with any existing telescopes on Earth. In addition it will be able to observe in wavelength regions that are not detectable from the ground, particularly in the ultraviolet.

The Space Telescope consists of four parts: the two dominant parts are an optical telescope – a 2·4-metre (94-inch) Ritchey-Chrétien reflector – and an instruments package containing two cameras, two spectrometers and a photometer. These are serviced by a support systems module which will supply the means of converting the telescope images into electronic signals for transmission back to the ground as well as providing the satellite guidance and control. The fourth component, the solar panels, will generate the power needed to operate the satellite. It is expected that, given regular servicing, either in space or back on Earth, the Space Telescope will have a lifespan of fifteen years.

The Space Shuttle being inched slowly out of the Vehicle Assembly Building at the Kennedy Space Center, Florida. The Shuttle is a reusable manned space vehicle, whose importance is that it should enormously reduce the cost of putting a satellite into orbit.

The Space Telescope's 2·4-metre (94-inch) diameter primary mirror being ground. The mirror blank is about 30 centimetres (12 inches) thick, with a centre hole of some 60 centimetres (24 inches) aperture, and weighs more than 900 kilograms (2000 pounds).

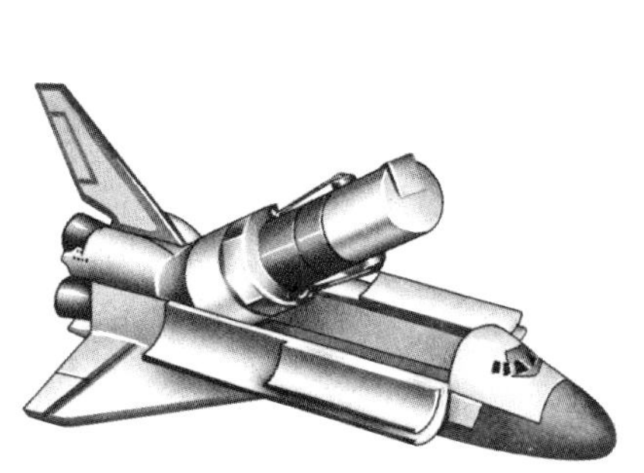

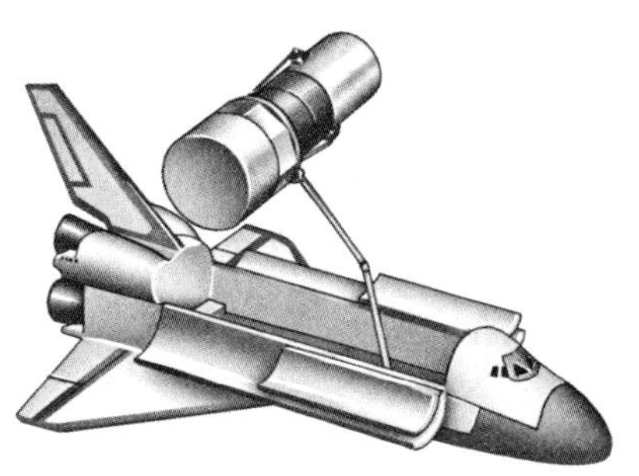

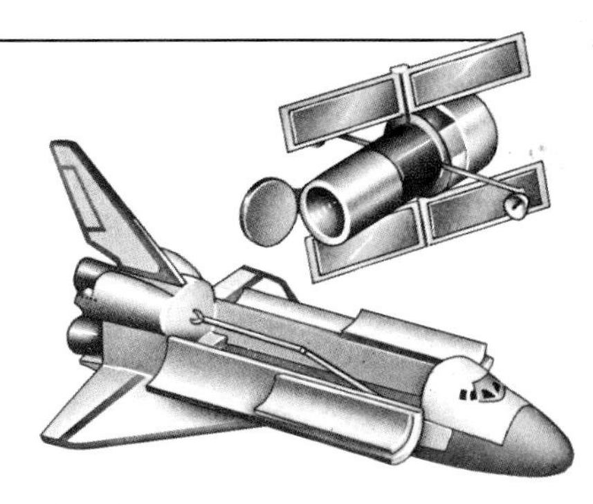

The 10,000-kilogram (22,000-pound), 13-metre (43-foot) long Space Telescope will be transported into space within the Shuttle Orbiter vehicle. By means of a remote manipulator arm, the telescope will be lifted clear of the Orbiter and positioned in the required orbit. The telescope will then disengage, and its solar panels and radio antennae will deploy as it commences to orbit the Earth. The data collected will be converted into radio signals, which will be transmitted to the ground via a number of tracking and data relay satellites orbiting the Earth.

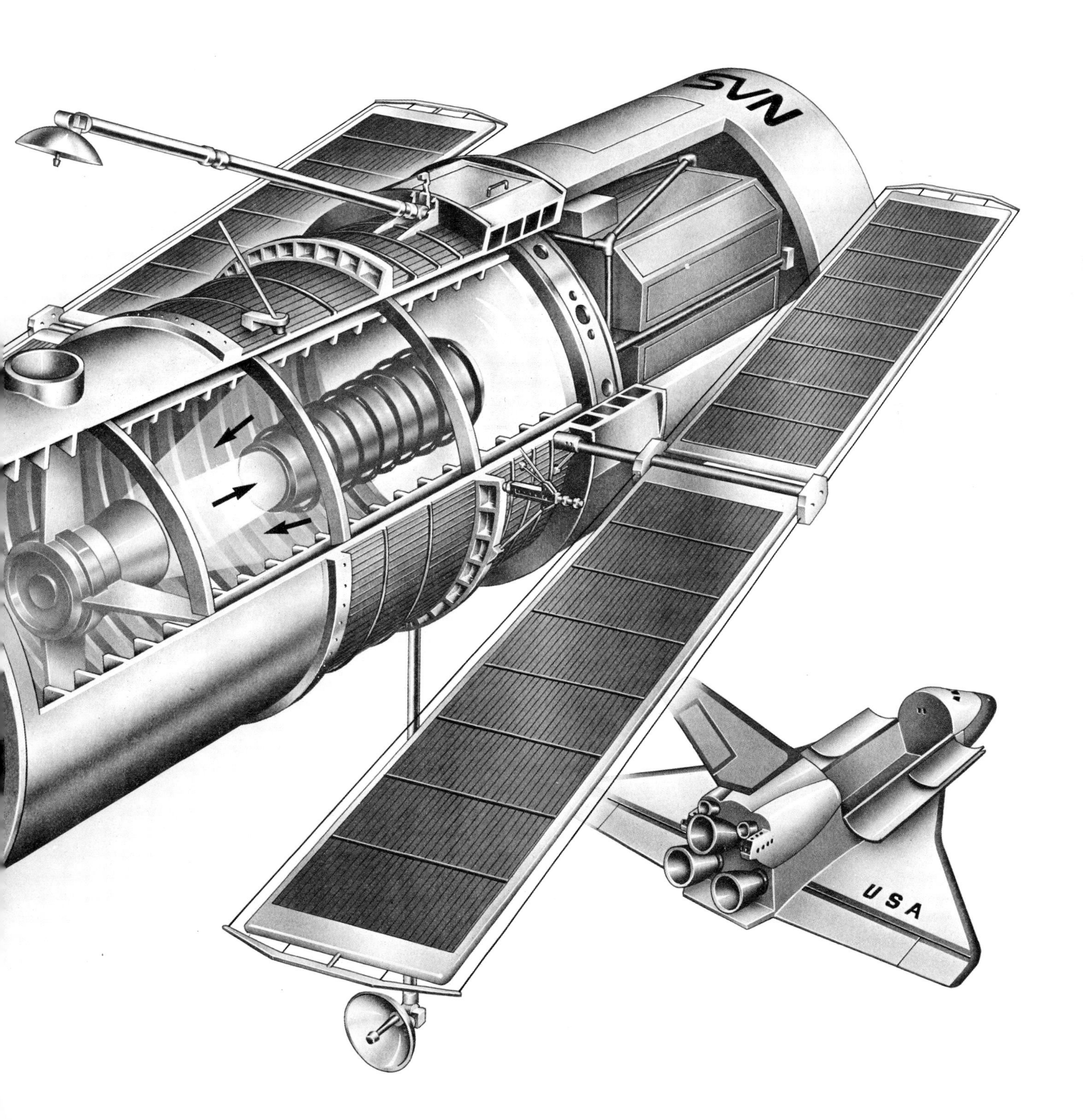

CHAPTER FIFTEEN

DIVERSITY

Solar and infra-red telescopes with unusual features: balloon-borne astronomy, 1950 to the present-day

"Research into the nature of the solar interior is a subject that a few years ago seemed well tidied up. . . . The evolution of the Sun since it arrived on the main sequence some 5 billion years ago was well described, and its ultimate fate when it will evolve through the red giant stage in another 5 billion years was completely predictable. . . . The models are theoretically self-consistent, agree with all known aspects of stellar evolution as shown by stars of other types and ages, and agree with all known parameters observed on the Sun.
"Except one. That is the now famous case of the missing neutrinos, and as of this writing it is threatening to topple the entire structure of stellar interior theory."

Robert W. Noyes in
Frontiers of Astrophysics, *1976.*

All astronomical understanding depends ultimately on the precise measurement of the positions of stellar objects. This branch of astronomy is called astrometry. At the simplest level it is necessary to establish whether an object discovered by radio or X-ray astronomers is the same as one known in the visible region of the spectrum. If this is the case then the powerful techniques and telescopes of optical astronomy can provide vital data. Thus the identification of the radio-emitting quasars with visible "stars" led rapidly to two discoveries: quasars must be very small, as their brightness varies on a short time scale; and they are not nearby objects because their positions are fixed and show no sign of the "proper motions" that are the hallmark of the Sun's near neighbour stars. Later came the identification of the quasar red shift from the analysis of the visible spectrum, a measurement that could not have been made on the original radio data, as there are no single wavelength lines in the radio emissions from quasars.

In a similar vein, the vast energy output of the quasars is determined by first measuring the minute fraction of that energy that reaches the Earth and then allowing for dilution over the enormous distance from the quasars to us. All astronomical distance calculations depend on a series of steps, the first of which is the measurement of parallax of nearby stars.

The products of decades of classical observation were summarized in 1963 and 1964 with the publication of two "astrometric" catalogues, one of star positions and the other of star parallaxes. The *Fundamental Catalogue Number 4*, or *FK4*, contained the positions of 1535 reference stars, determined to an accuracy of around a fifth of an arc second. This accuracy has slowly faded away, as uncertainties in our knowledge of the precession of the equinox (and therefore of the catalogue's co-ordinates to the real sky) and the proper motions of the stars have accumulated. The partner to *FK4*, the 1964 supplement to the *Catalogue of Stellar Parallaxes*, brought the number of "known" parallaxes to 6422. To measure parallax one must determine the small changes in star positions and as a result the accuracy is better, around 0·016 arc second. Thirty per cent of the stars included, however, have parallaxes less than 0·018 arc second, and for these the listed figures are of very little significance.

The *Fundamental Catalogue* was based on measurements using the traditional instrument, the transit telescope. Since 1963 several improvements have been made to this instrument, starting with its modification for photographic use, principally by S. Laustsen in Denmark and

M. S. Zverev of Pulkovo Observatory (but working in Chile). L. Sukharev, also of Pulkovo, tackled the problem of flexure of the telescope tube by turning to a fixed horizontal telescope which is fed with light by a pivoted flat mirror. The latter is the only moving part and for this the bending problems are much less important. The principal step, however, was taken in Denmark, where the photoelectric transit telescope was developed, a few kilometres from the site of Romer's original transit instrument of 1684. This new system, in which the star position is determined without requiring the astronomer to make any direct settings, is the basic tool of the new *Fundamental Catalogue Number 5* (*FK5*), which is expected to appear in 1984. The new list will both contain more stars (about 5000) and work to higher accuracy than its predecessor, largely because the proper motions of the stars are now better determined.

There has been a revolution in parallax measurement, too, after forty years of rigorous adherence to the methods pioneered by Frank Schlesinger. It came with the commissioning of an entirely new telescope. K. A. Strand at the United States Naval Observatory broke with the tradition of long-focus refractors and developed a 1·5-metre (60-inch) reflector specially for precision measurement. The telescope has been highly successful and has shown that, provided sufficient care is taken, the usual faults of the reflector – lack of rigidity and sensitivity to temperature – can both be overcome. The gain in precision over the older methods is about a factor of four: accuracy to 0·004 arc second is now a matter of careful routine. This improvement brings more than fifty times as many stars within reach of reliable parallax determinations, and direct measurement of interstellar distances now reaches out towards 250 parsecs of fifty million times the distance from the Earth to the Sun. The fundamental need for measurements of this type is perhaps best illustrated by the fact that the first telescope to be installed at the international observatory in the Canary Islands will be a new astrometric reflector.

Another stage in the divorce of measurement and observer has been the development of automatic plate-measuring machines. These measure the position of star images on photographic plates with great precision and speed. Peter Fellgett of Reading University pioneered such machines with "Galaxy", completed in 1968, and a series of ever more sophisticated devices have followed. The latest is the computer-driven Advanced Plate Measuring (APM) facility, built by Ed Kibblewhite at Cambridge University. The APM scans photographic plates with a narrow laser beam: it is much faster and more accurate than – and almost as intelligent as – an average astronomer. When presented with a plate to measure it can distinguish galaxies from stars and recognize defects such as meteor trails or dust specks on the plate. The machine can identify single well-resolved stars and will find the centre of these star images to a precision of one micrometre on the plate. It can then repeat the process for thousands of stars an hour. The effect has been to make routine star plates, taken with orthodox telescopes, the raw material for precision measurement. With the aid of this machine Andrew Murray at the Royal Greenwich Observatory has been able to search for stars of significant proper motions or parallax using the plates from the 1·2-metre (48-inch) UK Schmidt telescope in Australia.

The radio astronomers have also raised the precision of their measurements of position, in this case by further developments of the two-telescope interferometer. The necessary steps were pioneered in Canada. In previous long-baseline work, for instance between Jodrell Bank and Malvern, radio links were used to connect the telescopes. This technique is not practical, however, when the baseline runs to thousands of kilometres. The crucial step was the development of extremely stable "atomic" clocks that could be transported between radio-astronomical observatories. With the aid of these clocks and high-speed tape recorders, synchronized observations can be made by the two telescopes while they are observing the same source. Using a computer, the tapes are compared to produce the result that would have occurred if the two telescopes had been linked during the observation. The new technique can be used regardless of the distance between the radio telescopes. The first observations used a baseline that stretched right across Canada from Penticton in British Columbia to Algonquin, near Toronto. Soon the links were extended to intercontinental dimensions – Australia, California, West Virginia, Massachusetts, Jodrell Bank, Sweden and Simeis in the Crimea were all interconnected. Telescope separations only a little less than the diameter of the Earth could be boosted even further by reductions in wavelength: 75 centimetres (30 inches) in 1967, 6 centimetres ($2\frac{2}{5}$ inches) in 1969 and 3·5 centimetres ($1\frac{2}{5}$ inches) for the US–USSR link in 1971.

Position measurement using this network rapidly matched that of the optical astronomers: 0·02 arc second precision was achieved for several "point" radio sources. Differences in position, which are easier to determine, soon reached a precision beyond that of optical measurement.

The archetypal quasar, 3C 273, had long been known to be a pair of radio sources – the quasar itself and its protruding jet of radiating gas. The new techniques showed that the quasar itself was composed of four sources – the closest pair only 0·0015 arc second apart and each less than 0·0005 arc second across. (As a comparison, the latter represents the thickness of this page seen from a distance of 2 kilometres/1¼ miles.) Astronomers repeated Sir Arthur Eddington's test of Albert Einstein's general relativity – the gravitational deflection of radiation by the Sun – using interferometers. Their result was within one per cent of the theoretical prediction, a measurement about fifty times more precise than the 1919 result.

The chief problem at present is to bridge the gap between radio and visible position measurement. The stars used in the *Fundamental Catalogues* are all relatively bright – typically about seventh magnitude. The radio measurements are chiefly on quasars, which are not among the brighter visible stars. Apart from 3C 273, which is of thirteenth magnitude, the brighter quasars are sixteenth and seventeenth magnitude – fainter by a factor of 10,000. This makes comparison very difficult: in the visible range it is the difference between a one-second and a three-hour exposure with the same telescope. It is virtually impossible to take a plate showing both *Fundamental Catalogue* stars and quasars side by side. Much of present-day astrometric research is directed to the establishment of a series of carefully tested stages of intermediate brightness, a step that is vital if the two techniques are to be linked together.

The combination of two telescopes as an interferometer has also been exploited in the visible region to measure the sizes of stars. The first such work was, in fact, a direct extension by Richard Hanbury Brown of his work with radio telescopes. The resulting device, known as an intensity interferometer, relies on the comparison of the fluctuations in the signals received by a pair of optical telescopes. If the telescopes are close together and the star is an almost point-like source of light, then there is a very slight correlation between the two sets of fluctuations. If the star is larger, or the telescopes farther apart, even this minimal correlation is lost. The advantage of the method is that it works in spite of the effects of seeing. The disadvantage is that it takes hours of observation to establish the existence of any correlation between signals, even for the brightest stars. After a brief controversy over the method, which some critics thought to be incompatible with the fundamental principles of quantum mechanics, Hanbury Brown built a pair of telescopes at Narrabri in Australia.

The intensity interferometer is searching for a very small effect that depends on the separation between two optical telescopes. To achieve adequate sensitivity,

The two 6·5-metre (250-inch) reflectors of Professor Hanbury Brown's intensity interferometer at Narrabri, New South Wales, Australia. The distance between the two main mirrors can be varied by moving them along a railway track. The interferometer is not sensitive to the quality of the star images, so does not use domes to prevent wind-induced vibration or high towers to avoid poor seeing near the ground.

Hanbury Brown had to build two 6·5-metre (250-inch) instruments and mount them on railway tracks. Although the telescopes were, by a considerable margin, the largest optical telescopes in the world, they did not need to be of very good quality. All that was needed was to get the light from a bright star on to the photomultiplier at the prime focus of the telescope. This proved possible with a primary mirror built up as a mosaic of 270 small hexagonal glass elements, at the combined focus of which the star image was a smudge about 2·5 centimetres (1 inch) across. The railway was a circular track 188 metres (615 feet) in diameter with a telescope shed at one point. There were no domes, the telescopes being garaged during the day or in bad weather. The telescope pair found the predicted signal correlation, measured it, and so achieved phenomenal angular resolution. At their extreme separation of 188 metres, the telescope-to-telescope distance was 420 million wavelengths, and this gave tremendous resolving power to the measurement of the diameters of stars. Prior to Hanbury Brown's work the sizes of only half a dozen stars were known and these were all giants, very different from the great majority of stars, which are those on the main sequence. The smallest angular diameter measured with the new telescopes was 0·00047 arc second (for Delta Scorpii) and the highest precision was an error of twenty-two millionths of an arc second (for Beta Crucis). Hanbury Brown found that Vega (Alpha Lyrae) is 0·00308 arc second in angular diameter, a long way below Galileo's first estimate of five arc seconds. Galileo's measurement was a skilful estimate of the seeing in Renaissance Pisa, not a measurement of the size of the star.

The Narrabri interferometer has now been dismantled, as by 1972 it had studied all those stars bright enough to yield results in reasonable observation times. In its place a new method of combining the light of two telescopes has emerged. Developed by Antoine Labeyrie in France, the new technique, called speckle interferometry, is both more difficult and very much more sensitive than Hanbury Brown's intensity interferometry. In this form of observation the effects of seeing are overcome by taking extremely short exposures, so short that the turbulent motion of the atmosphere is frozen: the time allowed is about a hundredth of a second. The method finds wide application because of its ability to add the effects of many of these brief exposures to give a useful sensitivity. The name "speckle" interferometry comes from the speckled appearance of each brief picture of the star, and this detailed structure within the one- to two-arc-second seeing blur reveals information on scales less than the size of the blur. The first applications were directed to improving the angular resolution of single telescopes and were of particular importance to the study of double stars. If there were no atmospheric effects, a 4-metre (155-inch) telescope should be able to resolve two stars only 0·025 arc seconds apart. The application of speckle interferometry by H. McAlister using the Kitt Peak 4-metre has yielded data on stars as close as 0·037 arc second. A study of Betelgeuse, using the same telescope, found the first hints of bright and dark patches on the surface of a star other than the Sun.

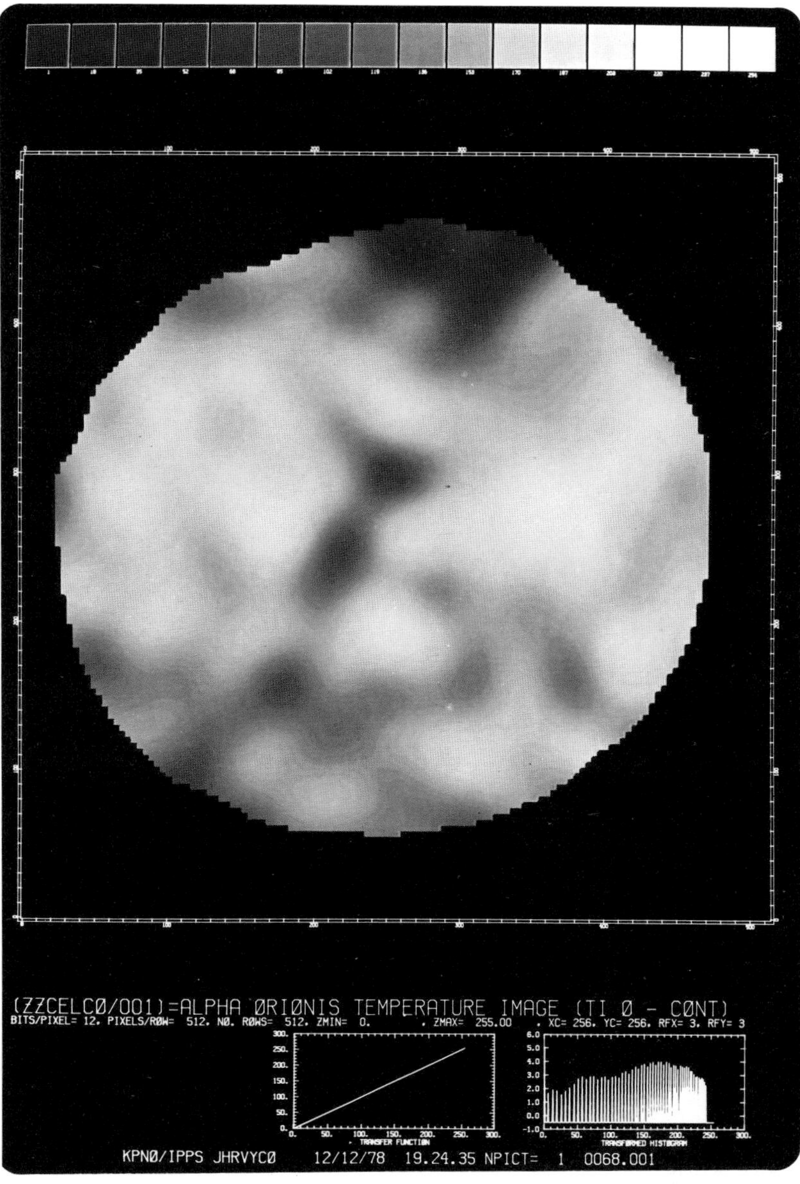

The giant star Betelgeuse, in the constellation of Orion. This image of the star's surface was produced by the technique of speckle interferometry: it is generated by computer analysis of a set of short-exposure photographs. The light and dark patches suggest that the star is not uniformly bright, though the structure is at the resolution limit of the 4-metre (160-inch) telescope used.

Labeyrie at Nice Observatory made the pioneering attempts to use speckle methods to combine the light from two telescopes and so reach a resolution corresponding to the distance between the two. The technical challenge that has to be met is that the light path from the star via one telescope to the point of observation has to be kept equal in length to the path from

THE McMATH SOLAR TELESCOPE

The solar telescope at Kitt Peak National Observatory is the largest and most powerful instrument of its kind. The unorthodox design is the result of two separate decisions: first, like most solar telescopes, it has a very great focal length (90 metres or 300 feet) and must therefore be fixed in position. Second, unlike most solar telescopes, it uses a simple single mirror heliostat to guide sunlight to the instrument rather than the two-mirror system used for the classical vertical telescopes. This requires the telescope axis to be pointed in the direction of the north celestial pole.

The 2-metre (80-inch) diameter plane mirror of the heliostat is driven (at one revolution in twenty-four hours) to follow the motion of the Sun across the sky. The outer skin of the telescope tube is not only a windshield; it has an elaborate cooling system comprising a network of pipes containing cooling fluid, water and anti-freeze, which has been installed in order to minimize the effects of solar heating on the telescope performance. More than a million watts of solar power are incident on the structure under normal conditions, yet the temperature within the telescope is equal to that of the outside air.

(Left) *The observation room of the Solar Telescope. The converging beam of light from the Sun is reflected vertically downwards by a third mirror (at the top of the picture) so that the 85-centimetre ($33\frac{1}{2}$-inch) diameter image of the Sun at the final focus of the telescope is correctly positioned for analysis.*

(Above right) *A drawing of the heliostat that is on top of the 30-metre (100-foot) high telescope tower. The 2-metre (80-inch) plane mirror that rotates with the Sun is mounted so that the light beam passes through the hole in the centre of the bearing. The two adjoining 90-centimetre (36-inch) mirrors are used to give additional solar images.*

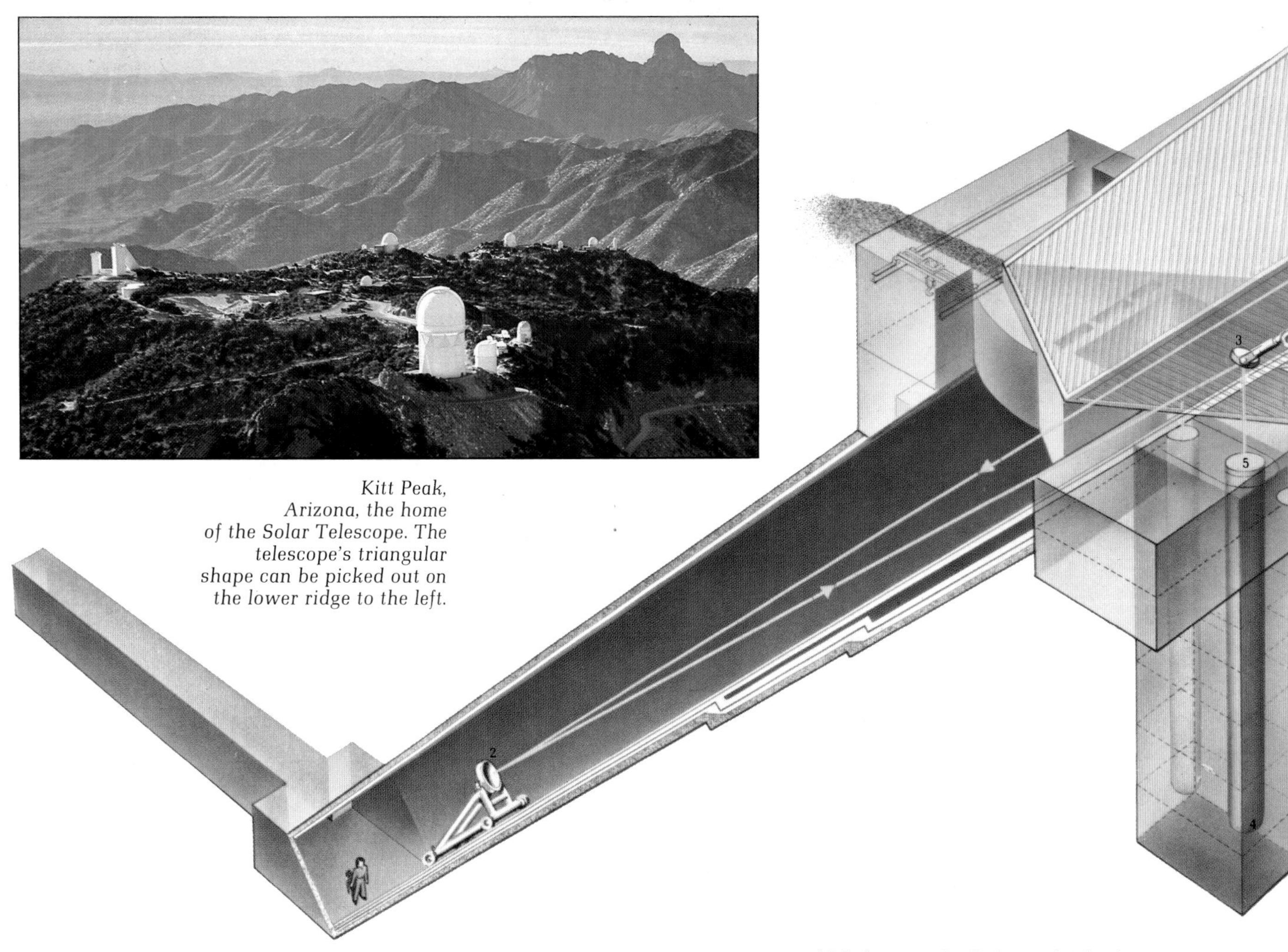

Kitt Peak, Arizona, the home of the Solar Telescope. The telescope's triangular shape can be picked out on the lower ridge to the left.

The optical arrangement of the telescope is such that sunlight falls on the flat heliostat mirror (1). *This reflects the light rays down the long sloping tube to a concave mirror* (2), *which focuses the light. A third (plane) mirror* (3) *reflects the image towards the spectrograph* (4). *The telescope forms an image of the Sun on a table at the top of the spectrograph tank* (5). *From this image the astronomer chooses features for detailed study.*

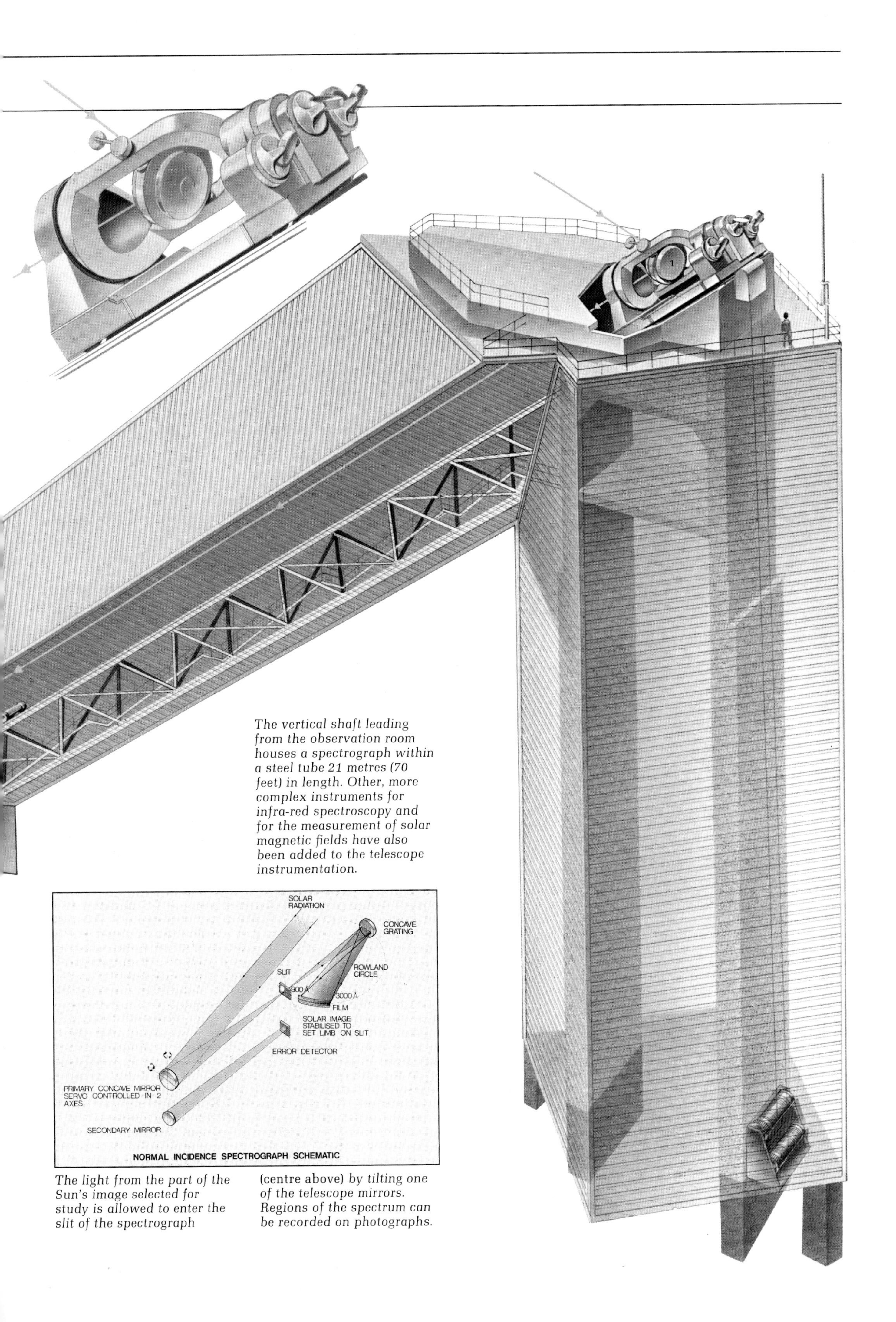

The vertical shaft leading from the observation room houses a spectrograph within a steel tube 21 metres (70 feet) in length. Other, more complex instruments for infra-red spectroscopy and for the measurement of solar magnetic fields have also been added to the telescope instrumentation.

The light from the part of the Sun's image selected for study is allowed to enter the slit of the spectrograph (centre above) by tilting one of the telescope mirrors. Regions of the spectrum can be recorded on photographs.

the star via the other telescope to the same point. As the star is moving across the sky either one of the telescopes or the apparatus in which the two beams are combined also have to move. The precision required is a few micrometres, and is most easily achieved if the two telescopes are aligned in a north–south direction.

The two telescopes used by Labeyrie had an aperture of 25-centimetres (10 inches). They have since been moved from Nice to a high-altitude site at Calern Observatory in southern central France, where they are mounted on a precision railway 40 metres (130 feet) long. The interferometer gave diameters for bright northern stars, including Deneb and Castor, and its success encouraged the construction of a bigger interferometer. The new device is planned to reach an angular resolution of 0·0003 arc second for seventeenth magnitude stars. The two telescopes, each of 1·5 metres (60 inches) aperture, are of unorthodox design and material. They are flask-shaped reinforced concrete structures supported in giant ball-and-socket joints instead of the normal equatorially mounted steelwork.

The two-telescope speckle interferometer at Calern in France. The two small telescopes move on the railway in the foreground and the light beams are combined in the central hut. In the background is a flask-shaped 1·5-metre (60-inch) telescope, which is to be used in a larger interferometer.

Robert McMath at the observing desk of the solar tower telescope of the McMath-Hulbert observatory.

Speckle interferometry is not of immediate application to the Sun. First, the method only works for a selected narrow wavelength region, usually less than ten per cent of the visible range. Second, the method of analysis which removes the effects of seeing only works if the image is small. The Sun is more than half a degree across and well outside the range attainable. The main impetus of solar studies has therefore been dominated by other attempts to improve the observation of fine detail on the solar surface. The first of these, by Martin Schwartzshild of Princeton University in the 1950s, overcame the seeing effect with a 30-centimetre (12-inch) telescope carried above the worst of the atmosphere by a giant high-altitude balloon. The project, called Stratoscope, was a modest success, though the telescope's performance was severely degraded by solar heating as the telescope was suspended in a near-vacuum 24 kilometres (80,000 feet) above the ground.

The next attack was much more in the tradition of the Hale–Mount Wilson solar telescopes. A group under the chairmanship of Robert McMath recommended to the United States National Science Foundation that two large solar telescopes, with apertures of 91 and 206 centimetres (36 and 81 inches), should be built at the new National Observatory on Kitt Peak. The telescopes were to be matched to high-resolution spectrographs which would reveal turbulence and magnetic fields in the solar atmosphere. In the event, three telescopes were built, one of 1·5 metres (60 inches) aperture and two of 90 centimetres (35 inches), all mounted on the same structure. As the Sun is always to the south of the observatory and always lies within a limited range of altitudes, it was not necessary to build a telescope that would follow the Sun across the sky. Instead, a heliostat system comprising three flat mirrors directs sunlight into the three telescopes, which are fixed. The main telescope is on a grand scale: in order to feed light efficiently into the largest practicable spectrograph, the 1·5-metre telescope must have a focal length of 90 metres (300 feet). To avoid the bad seeing caused by sun-heated hot air near the ground, the main heliostat mirror is on a tower 30 metres (100 feet) high. The telescopes share a common tube which slopes down from the tower at the same angle as the latitude of Kitt Peak. The lower two-thirds is buried underground, and the upper third is surrounded by a 30-tonne metal windshield, the temperature of which can be controlled to reduce air turbulence in the light path.

The McMath telescope is still the largest solar telescope in the world, and is unrivalled for some aspects of solar study, not least because of the sophistication of the instruments that act to analyse the sunlight collected. One major programme has been concerned with sunspots and another with the distribution and changes in the solar magnetic field. The Kitt Peak solar astronomers have also studied the abundance of rare and unusual elements such as lithium and an odd isotope of helium, both of which are sensitive indicators of the reactions occurring in the solar interior. One aspect of the big telescope was a disappointment. The builders hoped that it would show very fine detail on the Sun's surface, but the combination of the site, the long air paths in the telescope and the problems of solar heating of the optics have meant that the seeing is only

rarely better than about one and a half arc seconds. (See also pages 178–179.)

The next step in the search for better seeing was achieved by Richard Dunn at the solar observatory on Sacramento Peak, in the mountains overlooking Alamagordo in New Mexico. His new telescope incorporated three improvements. First, the telescope was on a higher mountain, which was already known to offer good daytime seeing and a very clear sky. Second, the top end of the telescope was raised even farther in the air – to 41 metres (135 feet) above the ground. Third, and most important, the deleterious effects of bad seeing within the telescope were avoided by the simple but expensive decision to remove all the air within the instrument. The result is an evacuated tower telescope with an aperture of 76 centimetres (30 inches), which, although smaller than the Kitt Peak instrument, is able to see much finer detail. At the top is a two-mirror heliostat in an evacuated turret. The telescope mirror is set 90 metres (300 feet) vertically below a 10-centimetre (4-inch) thick entrance window, at the bottom of a 250-tonne steel pipe that forms the vacuum tank. Like the 2·5-metre (100-inch) Hooker telescope on Mount Wilson, this telescope (and its 20-metre/65-foot long spectrograph) floats on mercury. In this case the float allows the whole tube to rotate slowly in order to compensate for rotation of the solar image during the day.

This telescope has shown an enormous amount of detail in the complex structures at various levels in the surface of the Sun and has allowed astronomers to study the individual convection cells of the solar surface rather than a blurred-out average over a larger area. Its success is attested by the construction of several other evacuated telescopes for solar observation.

While the evacuated telescope has led to a better understanding of the Sun's surface, a very different "telescope" has given observations that are extremely difficult to reconcile with the theory of the solar interior. This upsetting instrument is a neutrino telescope. Unlike optical telescopes, it is buried 1500 metres (1 mile) beneath the Earth's surface in the Homestake mine in Lead, South Dakota. Neutrinos are elusive sub-atomic particles emitted in the nuclear reactions that occur deep inside the Sun. The light emitted in these reactions is successively rapidly reabsorbed and emitted by the hot gases such that radiation takes a million years to reach the solar surface. The neutrinos, in contrast, pass straight

The balloon-borne telescope Stratoscope I, shown being prepared for launch in 1957. The balloon carried the telescope and telemetry equipment to an altitude of 25,000 metres (83,000 feet) in order to take photographs of the surface of the Sun, showing fine detail normally obscured by the blurring effects of the lower atmosphere.

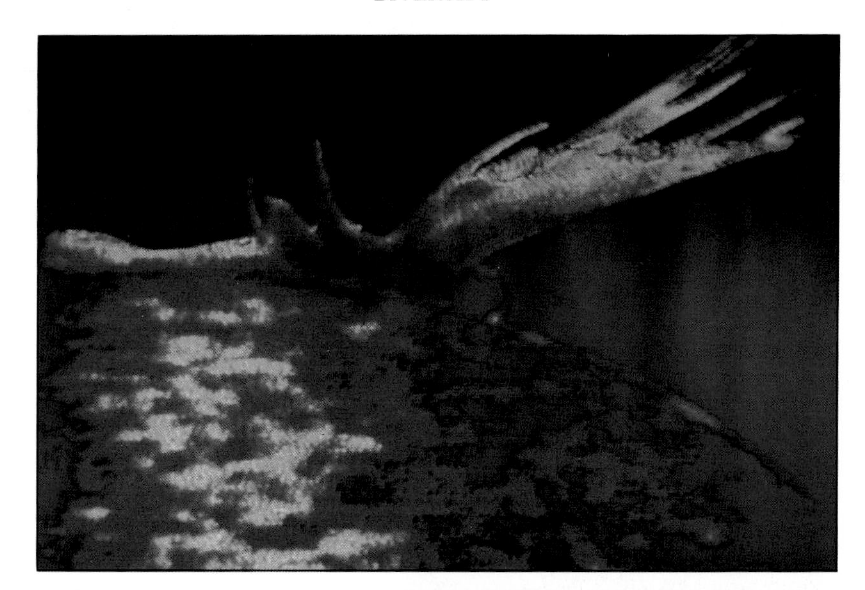

through the entire mass of the Sun and immediately escape into space. The number of neutrinos emitted is enormous: every second about 100 million pass through an area the size of a full stop on this page. The number is much the same day or night because the particles pass through the entire Earth (and the reader) without any significant losses. It is extremely difficult to detect particles that scarcely interact with the entire universe. A method has been found, however, and it has been exploited by Raymond Davis and his collaborators at the Brookhaven National Laboratory. There is a rare nuclear reaction in which a chlorine atom absorbs a neutrino and is transmuted into an argon atom. Given enough neutrinos, enough chlorine atoms and a certainty

(Above) *The Sun photographed by the solar telescope of Skylab, the first manned astronomical observatory in space, which was launched in 1973. The flow of hot gas surging out from the solar surface (top right) is strongly influenced by the local magnetic field.*

The underground neutrino "telescope" (right). *The detector is buried deep underground to screen it from other subatomic particles that would reduce the instrument's sensitivity. As the results obtained disagree strongly with the accepted theoretical model of the interior of the Sun, an alternative system using a different detector, pure gallium metal, is being built.*

that no other nuclear reaction is upsetting the results, it is possible to detect neutrinos. For Davis, enough chlorine atoms means a tank containing 500,000 litres (11,000 gallons) of cleaning fluid (ethylene chloride). This tank is deep underground to screen out all particles from space except the penetrating neutrinos.

The results have been a major embarrassment. Theorists were confident that they understood the processes that generate the Sun's energy, and the theory predicted that Davis's 600 tonnes of liquid should collect five or six neutrinos a day. Ever more careful measurements have shown that the neutrino flux is no more than two neutrinos a day, less than a third of the expected value. This was a disastrous discrepancy – a similar difference between the theoretical and actual solar heat and light output would plunge the Earth into a fatal ice age, when not only the oceans but the carbon dioxide in the atmosphere would freeze solid. The difficulty in reconciling experiment and theory has now led to a suggestion that it may be the neutrino that is not understood rather than the Sun. The neutrino may in fact be a chameleon-like particle that continually changes its character so that it is a standard solar neutrino for only a third of the time and something else for the other two-thirds. The difficulty of this desperate hypothesis is that to perform this trick the neutrino cannot have zero mass, as physicists have always assumed. It must weigh something, and a fifty-thousandth of the mass of the electron, the lightest particle, would be enough. Unfortunately, the Sun and stars have been pouring out neutrinos since the beginning of time and there must be so many around that even neutrinos with such a tiny mass would together weigh almost as much as the rest of the universe.

The neutrino may be elusive, but there is another predicted influence from space which has yet to be detected at all. This is the influence of gravitational waves – fluctuations in the local gravitational field on Earth due to the motion of heavy objects at interstellar distances. The slow, steady orbits of planets round the Sun or of the Sun round the galaxy are not efficient sources of such radiation. Only massive binary stars in close and rapid orbit or the catastrophic explosion of a supernova are likely to be detectable. The effect of such radiation is twofold: to make two independent masses move together or to make a single mass change in length. The first search for gravitational radiation was started in 1966 by an American physicist, Joseph Weber. Weber looked for the second of these effects and set out to monitor the size of carefully isolated 1·5-tonne blocks of aluminium (his "telescopes"). Each passing gravitational wave should have produced a pulse in the telescope's output. This pioneering device also produced a large number of spurious pulses and there was a great deal of argument about the computer techniques that should be used to separate real events from accidental artifacts. Since then several other groups – IBM, Bell Telephone Laboratories and academic groups in Rochester, New York, Glasgow, Munich and lastly Moscow – have joined the hunt for gravitational waves. Detectors are now a million times more sensitive than Weber's original equipment and are able to detect even more minute changes in the length of large blocks of metal. At present a change that is less than a hundredth the size of an atomic nucleus – itself a hundred thousandth of the size of one atom – is measurable. However, the detectors have not confirmed any of Weber's events and their negative results have led to the conclusion that his pulses were spurious. At the same time, the theory has improved and it too is rather discouraging. The detectors now appear to be within a factor of ten of the sensitivity needed to detect gravitational waves from a supernova in our own galaxy, but unfortunately such events are rare, occurring at a rate of one every ten or twenty years. On the positive side, there are groups at Stanford, Louisiana and Rome Universities developing the next generation of detectors. In these, all the measurements will be made in equipment cooled close to absolute zero (−273°C), which should give a major gain in sensitivity. More encouraging has been the identification of a binary star in which two neutron stars, each more massive than the Sun, orbit one another in just under eight hours. This pair of stars is gradually losing energy, and at the rate expected for loss by gravitational radiation. Although this radiation has not been detected directly, there is now a probable source to look at.

The remaining study for which special-purpose telescopes have been built, infra-red astronomy, has not been faced with the low signal strengths of neutrino and gravitational wave astronomy. Instead, like the first X-ray and radio sources, the early infra-red discoveries were sources much stronger than expected. The infra-red region covers a wide range of wavelengths between the radio and visible regions – from roughly one micrometre to one millimetre – and is of vital importance in astronomy because cool objects (those cooler than "red-hot") radiate most of their energy as this form of radiation. After preliminary skirmishes with insensitive detectors, starting with the use of Rosse's 1·8-metre reflector in 1869, the subject came of age in the mid-1960s.

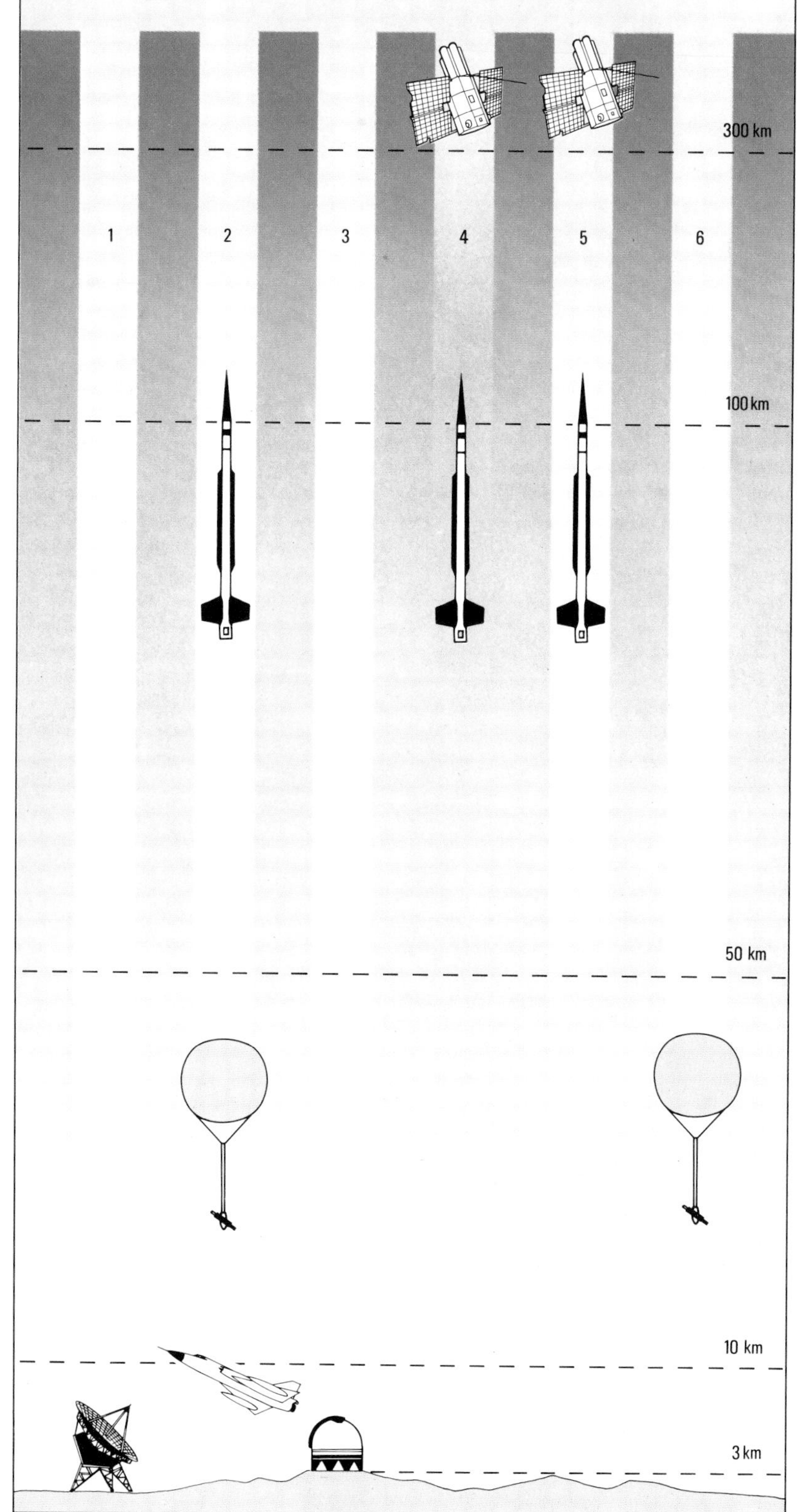

PENETRATION OF ELECTROMAGNETIC WAVES

Long-wavelength radio astronomy must be done using satellites, as the ionosphere will not pass any signals, while radio telescopes working between 10 metres (33 feet) and 1 millimetre (0·04 inch) can be ground based (1). In the far infra-red, balloons are needed (2), and at shorter infra-red wavelengths aircraft or high-altitude mountain sites are used. In the visible region mountain sites well away from city lights are essential (3). Near ultraviolet studies again require balloons that attain heights above the absorbing ozone. At shorter wavelengths in the ultraviolet the atmosphere is exceedingly opaque, and rocket- or satellite-borne telescopes are necessary (4). X-rays (5) and gamma rays (6) are more penetrating, but satellites and balloons are needed because the radiation still does not reach the ground.

Apart from the problem of the low sensitivity of infra-red detectors, two problems faced the astronomers who started to explore this region of the spectrum. First, all objects radiate in the infra-red, not just stars but telescopes and domes and people as well. The second difficulty was that there are only a handful of narrow "windows" in the infra-red, wavelengths at which the atmosphere is transparent. The main sources of atmospheric absorption are water vapour and carbon dioxide, both of which are largely to be found in the lower atmosphere. In this case infra-red astronomy was possible from telescopes carried by aircraft or balloons. For all systems at all altitudes the problem of infra-red radiation from the telescope itself remained. The worst offender was the steelwork supporting the secondary mirror, which radiated at a wavelength of ten micrometres with a brightness several thousands of times that of the stars.

The crucial events in establishing infra-red astronomy were the "Two Micron Survey" by Gerry Neugebauer and Robert Leighton of the California Institute of Technology and the development of a very sensitive spectrometer by Pierre Connes in France. The survey was carried out using a purpose-built telescope and worked at a wavelength of two microns (micrometres) to exploit a narrow window of atmospheric transparency. The survey found 5500 stars of significant infra-red brightness, and only a few of these were bright at the shorter visible wavelengths. The number of stars detected was relatively low, so that individual stars were usually more than a degree apart, and the detector was not nearly sensitive enough to register the weak radiation from the night sky between. It was thus not necessary to use a high-grade telescope, and a cheap and simple instrument was specially fabricated. The main component of the telescope, the primary mirror, was formed by spinning liquid epoxy resin at a steady speed for three days until the plastic hardened and set. The final image size was 100 arc seconds in diameter and the sensitivity of the infra-red detector about one thousandth that of detectors for visible light.

The survey highlighted a large number of curious objects demanding further study, among the most interesting of which were cool objects immersed in thick, warm dust. These appeared to be stars in the process of formation – the opposite end of the stellar evolutionary sequence that leads eventually to white dwarfs and neutron stars.

The other step forward in the 1960s was a breakthrough in spectroscopic techniques for the infra-red. This had been pioneered in the 1950s by British physicist Peter Fellgett, who both developed and demonstrated the new technique. He then suffered ten years of others telling him that it would not, could not and should not work. The successful application to astronomy was eventually made by Pierre Connes in France, who produced a series of spectra of dramatically high quality. The detection of such unexpected and exotic species as hydrofluoric acid in the atmosphere of Venus added momentum to the cry for cheap and simple telescopes for infra-red observation.

A series of low-cost telescopes was built, usually on lines similar to Hanbury Brown's Narrabri instruments. Mosaic primary mirrors made from cheap materials such as plate glass formed the main radiation collectors of what were known as "flux buckets". The largest of these was a 4·9-metre (190-inch) French instrument built to collect infra-red radiation for Conne's spectrometer.

These telescopes were overtaken in only a few years by the development of detectors. The night sky radiates a great deal of infra-red – as much as the full Moon – and hot engines in tanks and aircraft radiate even more. In consequence, the military developed infra-red detectors for night use, and they spent vast (almost astronomical) sums of money. When the new detectors emerged from their secret lairs, the flux-buckets were no longer useful. The detectors were now sensitive enough to respond to radiation from a very small slice of the night sky, and they were swamped by the quantity that the poor optics of the early telescopes scrambled in whenever a star was observed. The arrival of the new detectors caused a rapid revolution: the last of the mosaic mirror flux buckets, built in Meudon in France, was obsolete as soon as it was finished and the telescope was never used.

At this point, in about 1970, infra-red telescope design divided into three strands. The first approach was to build moderately good special-purpose instruments on mountain sites above the worst of the atmosphere. The second consisted of modifying existing telescopes by the addition of alternative secondary mirrors and mirror cells, while the third made its chief priority the avoidance of atmospheric absorption and invested in aircraft-, balloon- and rocket-borne telescopes.

The high-altitude work started with rocket-borne 16·5-centimetre ($6\frac{1}{2}$-inch) aperture telescopes launched by groups at NRL and Cornell University to study the cosmic background radiation. These instruments were designed to work at wavelengths around a tenth of a millimetre. The problem of radiation from the telescope itself was avoided by cooling the entire system, mirror structure and all. In a later development of the same theme a sky survey used rocket-borne telescopes of similar aperture, this time cooled to only two degrees above absolute zero. The absence of unwanted radiation from the telescopes made the total observing time, under an hour, as effective as a year of work with a ground-based telescope ten times the size but at room temperature.

The rocket-borne telescopes were severely restricted in aperture and this is a particularly critical limitation in the longer wavelengths of the far infra-red. Once again, as in the radio region, the long wavelength makes it impossible to obtain highly detailed images with small apertures. A 16-centimetre (6-inch) telescope observing at a wavelength of a tenth of a millimetre cannot resolve detail finer than twenty arc minutes across. The next step in size, developed by an American astronomer, Frank Low, was a 30-centimetre (12-inch) aperture telescope carried to an altitude of 12 kilometres (40,000 feet) in a Lear Jet aircraft. Its success led to the Kuiper Airborne Observatory – a Lockheed C141A (the "star lifter") carrying a 91-centimetre (36-inch) aperture telescope. This aircraft is able to fly for three and a half hours at 14·5 kilometres (48,000 feet) altitude.

Greater altitudes and longer observing time came from the balloon-borne telescopes, a dozen of which appeared in the mid-1970s. Four of these instruments, two American, one British and one German, have apertures around 1 metre (40 inches), with the Imperial College, London, system marginally the largest. The balloons, which hold enormous quantities of helium, can carry the telescope at an altitude of about 30 kilometres (75,000 feet) for ten or twelve hours.

The special infra-red telescopes sitting on the ground and working at shorter wavelengths started with medium-size instruments like the 1·3-metre (50-inch) at Kitt Peak and the 1·5-metre (60-inch) British telescope sited on Tenerife. These telescopes used lightweight structures and thin mirrors and gave images a few arc seconds across. Radiation from the telescope was minimized by using small secondary mirrors behind which the warm support structure was hidden, and by the careful use of cooled baffles near the detector. The small secondary mirrors were also incorporated in "phase sensitive" systems which could distinguish between radiation from the sky and that from the star being observed. In these systems, a small secondary mirror (a "wobbly secondary") could be tipped to and fro many times a second to allow the detector to alternately sample "star and sky" and "sky alone".

Balloons containing vast quantities of helium can be used to lift several tonnes of apparatus to altitudes of 35 to 40 kilometres (20 to 25 miles). This is the basis of infra-red studies as the equipment is carried above the absorbing components of the atmosphere.

This style of telescope rapidly increased in size and became situated on higher-quality sites. The present leaders are the 3-metre (120-inch) NASA Hawaii telescope and the 3·8-metre (150-inch) United Kingdom Infra-Red Telescope, UKIRT. Both these telescopes have image quality almost as good as orthodox optical telescopes, while the 3-metre has a tracking and guidance system of very considerable sophistication. The telescopes are both sited on top of Mauna Kea, a volcanic peak in Hawaii. This site is very high and very dry, a combination that gives very good telescope performance.

The range of objects studied by the new telescopes includes not only cool stars but also the dust surrounding many of them. This has added a whole new aspect of stellar structure based on low temperatures and molecules to the classical study of hot stars and atomic processes. Although the cool stars are by far the most numerous star type in the galaxy, they are not the most fascinating of the bright infra-red sources. That is, without a doubt, the centre of our galaxy.

Infra-red radiation has the ability, as a consequence of its long wavelength, to penetrate the dust that obscures the centre of the galaxy in visible light. The infra-red astronomer can "see" radiation from ordinary stars and detect spectrum lines from gas which allow him to make velocity measurements. The picture of the galactic centre that has emerged is one of a fantastic region crowded with stars and gas. The innermost parsec of the galaxy holds a million stars packed together a hundred times more closely than those in our neighbourhood. Observation of an infra-red spectrum line at 12·8 micrometres shows this assembly to be swirling round once in 10,000 years. (The Sun, 10,000 parsecs from the centre, takes 200 million years to complete an orbit.) Much of the fascination comes from the calculation that this observed period requires the centre of the galaxy to contain eight million times as much matter as the Sun, while the observed stars account for only a quarter of this. Many of the objects seen near the galactic centre can be identified as cool stars or warm dust clouds. One object, however, that at the position of the strong and point-like radio source Sagittarius A, fails to fit any standard pattern. The combination of a strange compact object and a discrepancy of six million solar masses starts ringing "black hole" alarm bells and the prospect of such an odd entity so near the Earth is indeed intriguing. As S. Faber of California University said when arguing for a 10-metre (400-inch) aperture telescope – "Infra-red observations of the galactic centre are among the most exciting to be made."

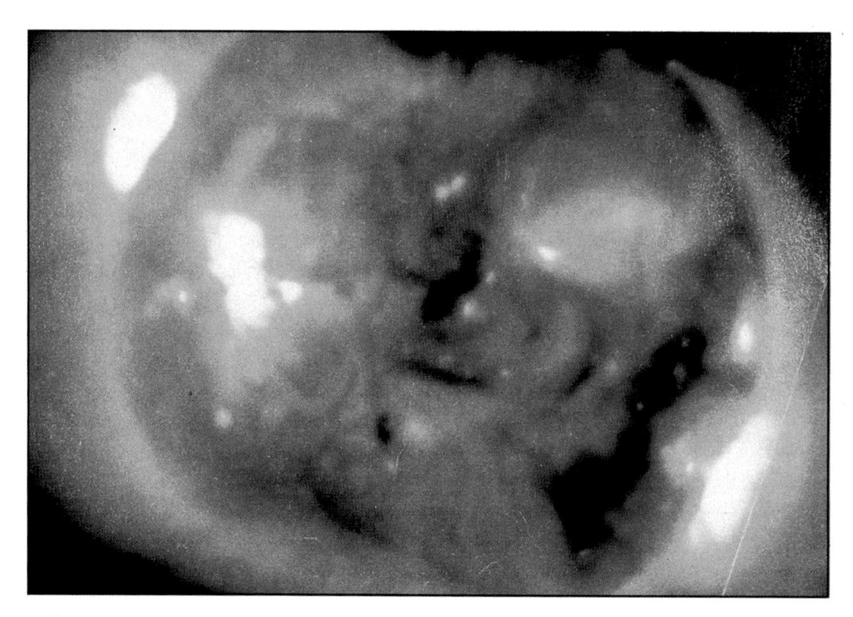

(Above) *A photograph of the Sun taken using an X-ray telescope carried on a satellite. The bright patches seen on the solar disc are "active" areas from which X-rays of very short wavelength are emitted.*

Several of the world's large infra-red telescopes are sited on the peak of Mauna Kea in Hawaii in the Pacific.

CHAPTER SIXTEEN

NOT EVERYONE HAS TEN MILLION DOLLARS

Modern amateur astronomy and the uses of small telescopes

"Mr Roy Panther in the garden of his home at Walgrave, Northamptonshire, beside the telescope with which he discovered a comet on Christmas night."

Caption in The Times,
London, 8 January, 1981.

"Mr. David Branchett, a 26-year-old production worker at the Pirelli cable factory at Eastleigh, Southampton, spotted the exploding star which is henceforth known as Nova Branchett 1981 through a pair of binoculars from his bedroom window."

The Guardian,
London, 22 January, 1981.

Ever since the fifteenth century the history of astronomy in general, and of the telescope in particular, has been woven of four strands. Three of the strands have been the concerns of professional men who have made their living from astronomy. The fourth has concerned the amateurs, pursuing their particular spare-time activity because they enjoyed the subject. These amateurs have probably had more influence on astronomy and its technology than have amateurs in any other discipline.

The three types of professional are, first, the permanent worker at an observatory, usually funded by the state (Tycho Brahe, Flamsteed or Bessell, for example); second, the academic, who both teaches and pursues astronomy as his research interest (Galileo, George Hale or Sir Martin Ryle); and, third, the instrument designer and builder (Ramsden, Fraunhofer or Ritchey). In the past all three groups have gathered recruits from among "amateur" astronomers, though that phase has now virtually ended with the spread of accessible higher education.

In the seventeenth century British astronomy recruited an enthusiastic young amateur, John Flamsteed, to be the first Astronomer Royal. In the next century a musician, William Herschel, so outstripped his contemporaries as to justify a special professional appointment – Royal Astronomer.

Many years later Friedrich Bessell, a harbour clerk who studied astronomy in the hope that it would lead to a career as a navigator and so give him a chance to see the world, became Director of Königsberg Observatory. Norman Lockyer was appointed a lecturer at Imperial College, London, in October 1881, thirteen years after he had discovered helium in the solar spectrum. At the time of his major discovery he had been a clerk in the War Office. In the more recent past Milton Humason, one of the most skilled and respected observers at Mount Wilson, started out as a mule driver working on the construction of the observatory.

The influence of the amateur has, however, been at its most effective in the development of the telescope, and this has been brought to bear in many different ways. The least involved, though not the least useful, are the wealthy men whose money provides telescopes for others to use. The best known of these are Lick and Yerkes, who funded the world's two largest refractors. Some, such as Newall in England and Bischoffsheim in France, have compromised, buying world-class telescopes for their own use and allowing others to use them. The wealthy amateur with a more wholehearted commitment to observation has been an important figure ever since William IV of Hesse Cassell, the influential friend of Tycho Brahe. Soon after the invention of the telescope, another such amateur, Johannes Hevelius, pioneered the very long-focal-length refractor, and converted the roof of his house in Danzig into the finest observatory in Europe.

Other amateurs, such as Samuel Molyneux, who found the first evidence for the aberration of light, aided the development of the telescope because he had no need for the obsessive commercial secrecy of professional instrument-makers like his contemporary James Short. With John Hadley, Molyneux published all the information available on mirror-making, including his own work. It was from these instructions that William Herschel first learned how to cast and polish speculum metal mirrors. Molyneux's most influential telescope, the first high-quality zenith sector, was inside his house rather than on top of it, and required holes to be cut in the roof and intermediate floors to accommodate the 7·5-metre (25-foot) vertical tube. Not all amateurs published their work, though. One of the more notable exceptions was the barrister Chester Moor Hall, who invented the achromatic lens before its first commercial manufacturer John Dolland, but whose reticence lost him the credit for the invention.

The amateur, unlike the professional entrusted with state funds, does not have to err on the side of caution, and can thus experiment with new ideas. One such experiment was Jesse Ramsden's first and only large equatorial, which was completed for Sir George Shuckburgh after Ramsden's death in 1800, but was not a success. On the other hand, twenty years later, Tulley's 15-centimetre (6-inch) refractor, built for Sir James South but sold to W. H. Smyth, used a cross axis rather than Ramsden's yoke as mounting and successfully pioneered the use of a clock drive on the polar axis to keep the telescope tracking the stars automatically.

In the Victorian era, the great age of the private fortune, many amateurs were attracted to astronomy and applied the new technology of their times. Lord Rosse, James Nasmyth and William Lassell – landowner, engineer and brewer respectively – all developed the reflecting telescope at a time when the professionals were convinced that only achromatic refractors were relevant to astronomy. In the same era other amateurs, in particular Henry Draper in Boston and William Huggins in London, led the professionals in the introduction of photography and spectroscopy.

Since those days, the distribution of wealth has become more even. The modern Nasmyth may no longer make so great a fortune that he can retire at forty-eight to pursue his interest in astronomy. Equally his Manchester factory will not be close to the hellish city described by Engels in 1845 in that devastating classic *The Condition of the Working Class in England*. Big telescopes are now beyond the reach of even the most wealthy. In the United States, the giant Carnegie Trust, long sympathetic to astronomy, has recently limited its activities to the support of better components rather than entire telescope systems and it has in particular funded the development of more sensitive light detectors. The last of the wealthy amateurs was perhaps Robert McMath, a civil engineer whose enduring influence may be seen at the McMath-Hulbert Observatory in Michigan and in the McMath solar telescope at Kitt Peak. This last instrument owes its status as the largest solar telescope in the world in part to the enthusiasm of a man who knew that the best bridges were the biggest ones and applied the same ethos to telescopes.

An interest in astronomy, however, is not restricted to the very wealthy. A great deal of satisfaction can be obtained from an investment comparable with many other hobbies. An adequate pair of binoculars costs little more than a tennis

Milton Humason examining a spectrogram. He began life as an amateur astronomer, earning his living working on the construction of the Mount Wilson Observatory. Later he became the janitor at the Observatory and then assistant observer, rising to become one of the most skilled and respected members of the staff.

racket, and a worthwhile telescope not much more than a good set of golf clubs. The first and most important point is that one may gain a great deal of pleasure without any special equipment at all. The possibilities for a satisfying hobby include almost all stages of astronomy, from those first reached in 1500 to those attainable in the present century. At the simplest level the aim is to find one's way about the night sky. This requires little more than a map, a torch and some clear sky.

Much quiet pleasure can be derived from just knowing the general layout and characteristics of the night sky. Roughly two-thirds of the entire sky is visible from the middle northern latitudes at one time or another, and this contains sixty per cent of the constellations. Some of these are well known – Orion, for example – though even this constellation spreads beyond the usually recognized corner stars, Betelgeuse, Bellatrix, Rigel and Saiph. Other constellations, usually modern ones which were fitted into the gaps between the classical pictures, are hard to find, especially if city lights are near to brighten the sky. These faint groups include some animals: Vulpecula, the little fox; Equuleus, the little horse; Lacerta, the lizard; Leo Minor, the little lion; and Canes Venatici, the hunting dogs.

With the constellations as signposts, two main circles can be found in the sky – the Milky Way and the Zodiac. The Milky Way is dramatically obvious from a good site as a glowing band of light, but it is invisible in competition with city lights. The Zodiac is an imaginary line marking the band in which planets may be found. Notes on which planets are visible, the phases of the Moon and other matters of interest are set out once a month in some newspapers (often in a map too small for convenient use) and in magazines such as *Sky and Telescope.*

At this point, or a little earlier if the equipment is available, the naked eye deserves some assistance – initially almost certainly not a telescope, since a good telescope is too costly and a poor one is useless. A well-made pair of binoculars is much easier to use than a telescope, and far better value at the same price: the field of view is broader, making it easier to find what you are looking for. Too high a magnification is not a help: any unsteadiness in holding the binoculars is magnified just as much as the view. With a good pair of 8 × 50 binoculars the observer is

A small part of the mixture of dust and stars making up the Milky Way, which is seen on clear nights as a band of faint light stretching across the constellations. The dark areas are clouds of gas and dust that obscure the stars at the centre of our galaxy.

approximately level with Galileo, and he will be able to see the moons of Jupiter, the stars in the Milky Way and many of the members of star groups such as the Pleiades, as well as craters and other details on the Moon. The Moon is best observed when it is not full and there are strong shadows at the boundary between "day" and "night" halves. When the Moon's surface is illuminated from directly above, as at full Moon, there are no shadows and surface detail is lacking.

The most obvious object in the sky – the Sun – should NEVER be looked at through binoculars. Looking directly at the Sun with only the naked eye for any sustained period can cause permanent damage as its heat is concentrated on the retina. A 50-millimetre (2-inch) objective lens collects almost 1000 times as much light as the 2-millimetre ($\frac{1}{12}$-inch) pupil of the eye in daylight and therefore raises the light and heat level far above the danger point.

The added light grasp of the binoculars makes the colours of stars more obvious: Betelgeuse and Rigel are clearly different. Binoculars will not show planetary detail such as the rings of Saturn or dark markings on Mars, and they will reveal just a slightly fuzzy outline for the Orion nebula. Only the Andromeda galaxy, M31, is large enough to show any appreciable size in binoculars, and appears as a diffuse elliptical glow. It appears larger than the diameter of the full Moon, justifying the seventeenth-century description "like a candle seen through horn".

The next step upward in amateur observation does require a telescope. The simplest equipment is based on a telescope which is small enough to carry and can be mounted on a portable but rigid tripod. There is an enormous number of manufacturers of such equipment and an equally wide spread in the quality of their products. Both types of telescope are of relevance to the amateur: achromatic refractors in the range 7·5 to 10 centimetres (3 to 4 inches) aperture, and Newtonian reflectors of 15 centimetres (6 inches) aperture. To these classical telescopes modern technology has added more complex refractor-reflector systems related to telephoto lenses: such Schmidt-Cassegrain or Maksutov-Cassegrain telescopes are usually between 10 and 20 centimetres (4 and 8 inches) in aperture.

The choice from among this diversity of telescopes is difficult as there is no single simple test of the quality of a complete instrument. Any telescope that offers a worthwhile gain over binoculars is expensive. It is also a precision device which will only perform well if its components are both well made and properly aligned. If a particular telescope fails to satisfy these conditions its faults will become much more evident when one compares it with another similar instrument. Hence, quite apart from the pleasure of associating with like-minded enthusiasts, it is invaluable to exchange experience and advice on available instruments with fellow observers. National astronomical associations can usually direct a first-time telescope buyer to a local organization which can give a great deal of helpful advice. As with many other devices, second-hand telescopes are cheaper than new ones and home-made telescopes are cheaper than either. None of the three options, however, is worth the money if it cannot give a high-quality image.

The size of the telescope needed to see significantly better than a pair of binoculars is fairly closely defined by two conflicting requirements: the instrument should be of a maximum possible aperture but be only of moderate length. Small refractors are more efficient than small reflectors, but are longer for a given aperture. A refracting telescope with an object glass less than 7·5 or 8 centimetres (about 3 inches) in diameter will do little more than a pair of binoculars. On the other hand, if the aperture is much more than 10 centimetres (4 inches), the telescope is

Features on the surface of the Moon show up particularly well during the first and last quarter. The Moon is shown with north at the top although, when seen through an astronomical telescope, it appears upside down. The dark Mare Imbrium can be clearly seen and the plains are scarred with mountains and craters.

clumsy and no longer portable. The reflector as a first telescope is more tightly constrained. A Newtonian telescope 15 centimetres (6 inches) in aperture is the traditional compromise between power and portability, and there is little scope for change. The third class of telescope, the Schmidt- or Maksutov-Cassegrain, is much more compact for a given aperture: some care is needed as it is a design that is sensitive to the quality and alignment of the increased number of components.

A refractor or reflector able to show considerably more detail than binoculars will necessarily have a much smaller field of view and its higher magnification means it is much more susceptible to shake. The telescope is useless unless it is set up on a solid support, for example a heavy tripod or a fixed pillar in the garden. A good small telescope will resolve two stars less than a couple of arc seconds apart: if one end of its 1·5-metre (5-foot) long tube is shaken to and fro by only 25 micrometres (0·001 inch) the angular extent of the vibration is all of seven arc seconds. This does not prevent observation, but it certainly precludes the making of any measurements. There is no doubt that the ability of a telescope to move smoothly, freedom from vibration and the firm mounting of components are all incompatible with lightweight design. It follows that good small telescopes tend to be heavy. With an instrument of modest size, firmly mounted, it is possible to see detail on the brighter planets. Venus shows only the phases and is usually otherwise blank, but it can be seen in daytime as well as at night. Jupiter shows bands and the red spot and Saturn reveals its rings and larger satellites. Mercury and Mars give very little detail, but as the observer's equipment is comparable in performance to Herschel's, Uranus can just be seen as a disc. Looking beyond the solar system, a large fraction of Messier's list of nebulous objects acquires some structure. As a guide, over half the Messier objects are brighter than Neptune (magnitude 7·85).

The most important attribute for an amateur astronomer is patience. In the first place the natural perversity of inanimate objects means that it will be cloudy on spare evenings and clear on busy ones. Not everyone is prepared to allow astronomy as high a priority as Steven Groombridge, who would leave the dinner table to make observations with his transit circle and return when the measurement was complete. Another reason for patient observation is that cosmic objects move slowly. It is well known that the Moon always keeps the same face turned towards the Earth. However, a little more care in observation shows that from week to week it appears to rock slightly from side to side and up and down. Observations lasting longer than an evening's hurried glances will eventually reveal almost sixty per cent of the Moon's surface, as well as giving views of the main features in many different angles of illumination. There are other transient appearances in the sky, for example dust storms on Mars and comets and novae.

A very large number of amateur astronomers, for whom funds are short and the time and space are available, have found it entirely practicable to make a telescope. This can either be done by buying the critical components – lenses or mirrors – and building a mounting or by making all the components oneself. A completely home-made telescope is almost necessarily a 15-centimetre (6-inch) Newtonian reflector. It is entirely possible to grind and polish a near-perfect 15-centimetre mirror at home, while it is almost certainly not possible to make a 10-centimetre (4-inch) achromatic objective for a reflector. The difference is not that the lens has two surfaces but lies in the difficulty of producing a high-quality spherical surface whose radius is specified in advance. If a mirror gives a good

An early 20th-century advertisement for a 10-centimetre (4-inch) refracting telescope with an equatorial mounting. This type of instrument has changed very little in the last 100 years.

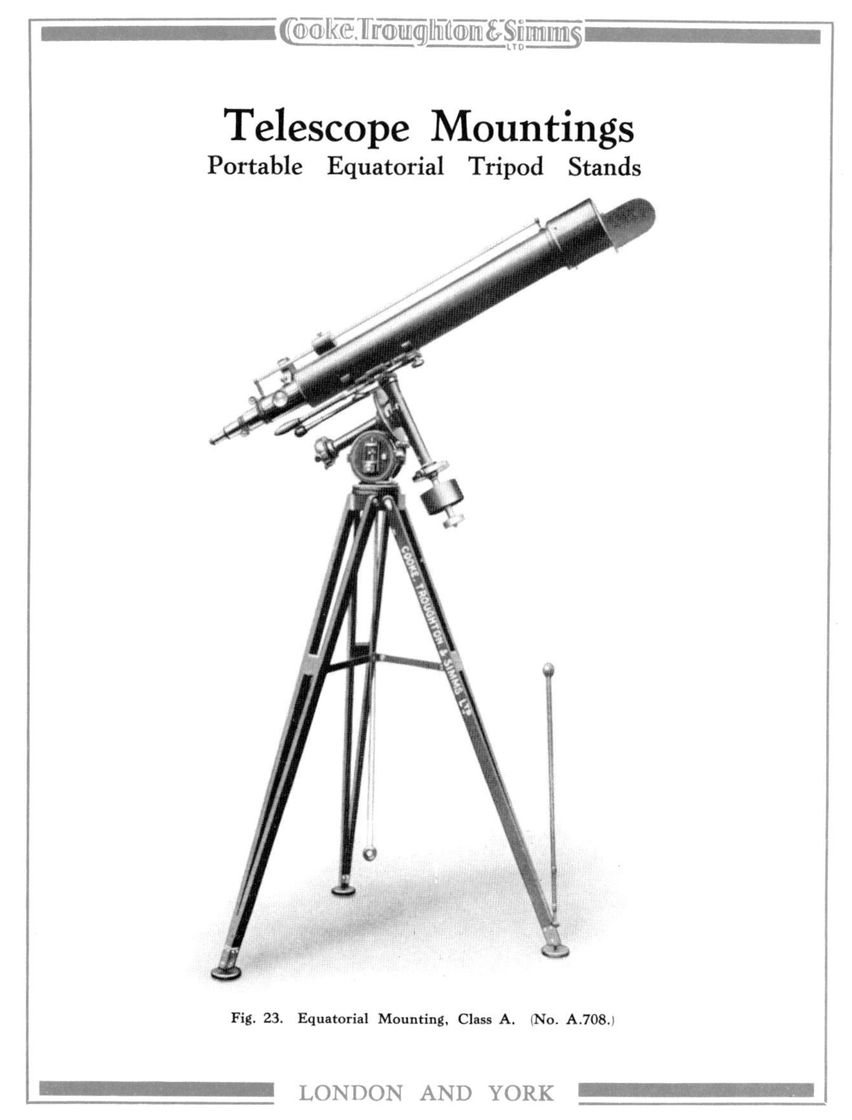

image but the focal length is 10 centimetres too long or too short, then the telescope length can be adjusted to fit. If half of an achromatic doublet lens has the wrong focal length, the two components cannot be assembled to give a high-grade lens. For a first attempt at telescope-building, ambition should be held in check. While it is possible to make a good 40-centimetre (16-inch) aperture mirror at home, it cannot easily be achieved.

The techniques of mirror-making and of simple telescope manufacture have been described in many books. The salient points are that both are possible without either exceptional dexterity or an elaborate workshop. With mirror-making the two essentials are cleanliness and patience. One needs cleanliness because even a single grain of grinding abrasive picked up by the final polishing tool will produce a serious scratch on the polished mirror and the whole process must start again. Patience is needed because the polishing process is slow: since it is necessary to work the mirror surface to a fraction of a micrometre, and each stroke of the polishing tool removes considerably less than this, a great deal of polishing is required. There is no point in making some components: eyepieces, for instance, are much better bought. The principal point to bear in mind here is that eyepieces need to be easily interchanged in the dark whenever one requires to change the magnification of the telescope. A frantic three-dimensional jigsaw puzzle to find the adaptor that fits one eyepiece to a mount designed for another is guaranteed to take thirty seconds longer than the interval of good seeing that justified the change.

There is a wide range of observation open to the amateur, but it needs to be related to the equipment available. Any amateur who decides to work on a programme of serious observation has one great advantage over his professional colleague. The professional is not a full-time observer, and he has difficulties in gaining access to large telescopes. First he must submit a research programme to a telescope allocation committee – there is no point in an astronomer asking for telescope time just because he feels that it would be nice to do some observing. The programme has to be worthwhile astronomy and of a type for which the telescope concerned is well suited. Two particularly important points are the number of nights needed and whether it matters if the Moon is above the horizon and hence lighting up the sky. Even if all aspects of the programme are satisfactory, and this is agreed by independent reviewers of the proposal, all is not yet plain sailing. A typical large telescope is oversubscribed by a factor of four, so no programme is certain.

J. Hooke, a British amateur astronomer, with his 10-centimetre (4-inch) refracting telescope (with the smaller guide telescope) for photography of the Moon. The smooth drive needed for photography is supplied by a standard gramophone motor.

In contrast, the amateur gets 100 per cent of the time on his own telescope for the programme he himself has chosen. This gives him the freedom to do speculative or monitoring work. This includes the patient search for rare events – the appearance of comets and novae, for example – but the range of other studies is considerable. Amateurs with moderate telescopes can monitor changes on planetary surfaces, including dust storms on Mars, bright spots on Saturn and variations in the rate of rotation of the red spot on Jupiter. All these require regular and careful observation, but are not observations made by professionals. Amateurs also undertake useful programmes involving the stars, including the measurement of the brightness variation of several types of variable stars (especially important in the case of the irregular variables), determination of binary star orbits and, particularly fascinating, lunar occultations. Some of these programmes require telescopes at the large end of the amateur range: double star measurement or regular observation of Mercury require instruments with apertures of 30 centimetres (12 inches) or more.

The main problem with a telescope that is too large to carry is that it has to have a permanent mounting. The ideal place is as far as possible from obstructions that block the horizon, so the first choice of site would be right in the middle of the lawn or on the roof. It is also necessary to keep off the weather with some sort of observatory: the simplest and most practical type is a shed on wheels which is rolled away when the telescope is used. Telescopes need to be kept dry; many buildings with hinged roofs and ninety-nine per cent of domes do not keep out the rain. (This applies even to some domes on multi-million dollar telescopes!) The alternative of putting the shed in one corner of the garden and moving the telescope out to observe has not been much used. It should be practicable if the telescope trolley is low, heavy and fitted with screw jacks to anchor the mounting when in use (as on the dishes of the Very Large Array radio telescope).

"Large" amateur telescopes – that is, instruments of aperture from 20 to 50 centimetres (8 to 20 inches) – are invariably reflectors. A 25-centimetre (10-inch) aperture achromat lens would need a telescope tube nearly 4 metres (13 feet) long and the cost of housing such an instrument is very high. Personal experience as the owner of a 25-centimetre aperture lens that is still in a desk drawer, not in a telescope, makes this problem very well understood.

When a choice has to be made of the appropriate style of reflector required, the amateur has a freedom denied the professional. In a professional observatory a small telescope, usually a 40- or 50-centimetre (16- or 20-inch) Cassegrain reflector, has to satisfy a closely defined specification. It must have accurate and well-calibrated setting circles, a high-quality drive which operates both smoothly and at the correct rate, and it must be usable without needing a long list of special instructions. The amateur may build or commission a telescope to suit his own observation programme, his own views on design or any other idiosyncrasy. It follows that large amateur telescopes are as varied as their owners.

Examples of every form of mounting known to the telescope engineer (and some others) can be found. The only rule is that the larger the aperture the more vital it is to use an equatorial mounting: the telescope then needs to be moved about only one axis while observing. The modern professional trend to alt-azimuth mountings is based on the use of small computers to run variable-speed drive motors on two axes simultaneously. Now that a home computer costs less than a colour television there is no reason why this should not be a practical solution for amateurs too. There is a limit, however, to the number of skills that can be mastered at once. A new year resolution to learn mirror-making, computer-programming, the layout of the Ephemeris and telescope-building while also reshaping the back garden to house a 40-centimetre telescope is unlikely to be kept.

Among the amateur programmes that require a larger telescope, that with the most enduring usefulness is probably the determination of double star orbits. A remarkable fact that has emerged from the theory of stellar structure is that the mass of a star is its single most important statistic. Give a theorist the mass of a star and he can calculate its entire life cycle, from birth in a collapsing dust cloud via the main sequence to collapsed remnant. Ironically, the mass of a star is very hard to determine, the only method being the analysis of the motions of two stars in orbit round one another. These obey Kepler's laws and therefore, given enough data, the observations can yield the masses of the two stars. The basic building bricks in the process are the regular and precise measurement of the separation and orientation of double stars as they pursue their orbits. This type of work, begun with Struve's Fraunhofer refractor at Dorpat, has now been taken over by the well-equipped amateur. Some binary stars, including Eta Cassiopeiae, Omicron Eridari, Epsilon Bootes, have not yet completed one circuit of their orbits since Struve began measurements. It follows that the subject was not by any means "finished" during the latter half of the nineteenth century when such measurements were one of the dominant forms of observation in professional observatories.

Amateur photography through the telescope, to yield a permanent record of observation, is a disappointing pursuit, at least at first try. It is discouraging to find how small the planets are when photographed. If a film is mounted directly at the focus of a typical 15-centimetre (6-inch) reflector, of focal length 1.5 metres (5 feet), the Moon will give an image 15 millimetres ($\frac{1}{2}$ inch) across. This fits quite nicely the frame of a 35-millimetre reflex camera, but the result will not show the detail visible to the observer's eye at the telescope. In such prime focus photography, Jupiter, the largest planet, will be a dot a third of a millimetre across and Saturn and Mars only specks on the emulsion. The battle to do better is a conflict between magnification and exposure time. Enlarging the image with additional lenses makes it fainter, but a long exposure makes accurate guiding of the telescope essential. Few published pictures taken by amateurs have used instruments less than 30 centimetres (12 inches) in aperture.

The photographic film is not the only alternative detector to the eye. The simplicity of modern electronics and the sensitivity of the photomultiplier are now well within the reach of the modern amateur. This means he can now measure with precision the brightness changes of variable stars, for instance. A more ambitious programme, which would need to exploit the developments in the use of cassette tape recorders for data logging, is to study lunar occultations with a photomultiplier. It is worth remembering that astronomy is not necessarily a solitary pursuit. Just as someone with an interest in soccer will find others to make up a team, so it is possible to form an astronomical team. A hi-fi enthusiast whose latest amplifiers have reached an unimaginable perfection may well be the man to join a team interested in astronomical electronics. This has other advantages. Almost every observatory has a record player in the dome and while eyes and hands are busy at the telescope, ears are free to listen to Mozart. Earphones not only give high-quality reproduction, they also keep your ears warm on cold, frosty nights.

The classical programmes (in astronomy, that is, not music) for the amateur with modest equipment and a reluctance to involve himself with the eccentricities of high technology, are regular sky searches. The goal of every amateur is to discover either a new comet arriving from the outer limits of the solar system, or a novae, the sudden explosion of a faint star into temporary brilliance. Both these searches need an intimate familiarity with the background stars to ensure that the interloper is quickly spotted.

The interest in novae comes chiefly because they are not well understood. A better knowledge of their properties needs observations with large professional telescopes as soon as possible after the initial explosion. If a new nova appears, there is an established astronomical telegram system to notify professional observatories.

The number of discoveries is only modest: in a typical year about two novae and four comets will be discovered. About half the comets are found by professional astronomers on survey plates taken for some other reason, but the remainder – an object every three months or so – are found by amateurs. The preferred instrument for searches of this type are binoculars or telescopes with a wide field of view. For his sister Caroline, William Herschel built a special "comet seeker", a telescope notable for being pivoted at the eyepiece and thus particularly convenient to use. The cousins of the comets, meteors (shooting stars), can be observed with no equipment at all. Here data from a wide network of amateur observers is of considerable use, as meteors are so low in the sky – typically at an altitude of about 100 kilometres (60 miles). Two observers more than a couple of hundred kilometres apart will, therefore, make quite different observations.

The other field in which large equipment is an embarrassment is the study of the Sun. A telescope more than 10 centimetres (4 inches) in aperture has to be masked down to avoid quite literally "cooking" the results. Small refracting

Roy Panther, an amateur astronomer, who discovered a comet on Christmas night 1980 with his 20-centimetre (8-inch) telescope. The Comet Panther, seen in the constellation of Lyra, was the first comet to be found by a British astronomer for fifteen years.

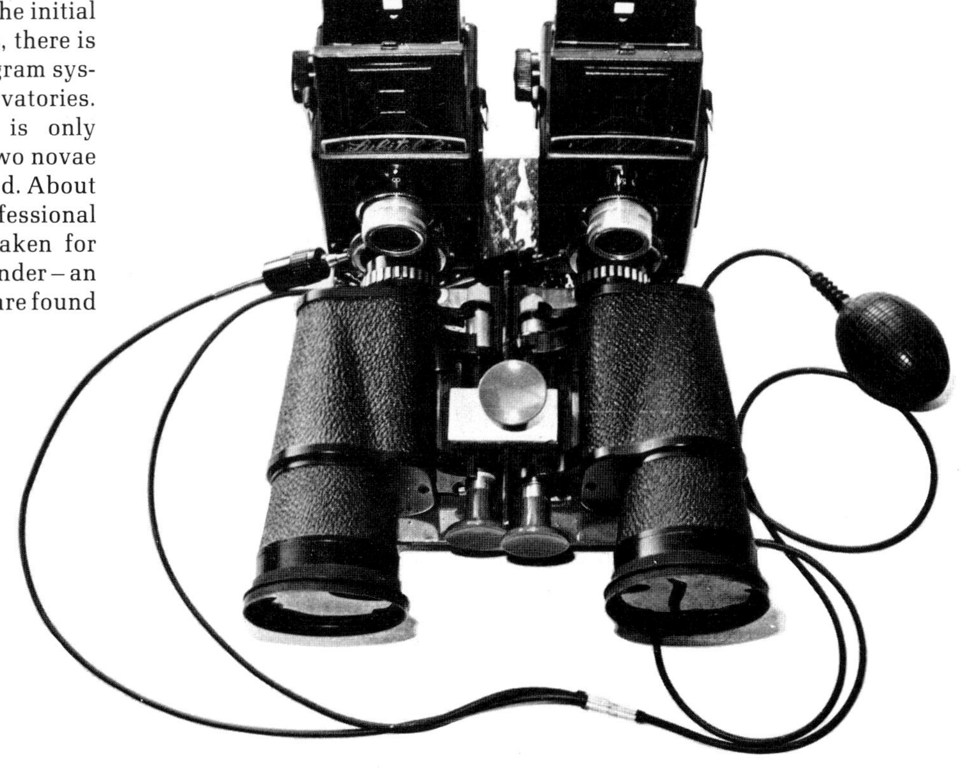

Binoculars can be utilized for much interesting astronomical work at reasonably low cost. Here a pair of ordinary binoculars has been modified to photograph an eclipse of the Sun. The magnified image is directed from the binoculars into the camera. The reflex camera viewfinders should only be used at low light levels such as at an eclipse.

telescopes are used in the study of sunspots and solar rotation. The telescope is slightly modified to project an image on to a screen in order to avoid the danger of direct viewing. This is the only safe method. A dense filter over the eyepiece is nothing more than an alternative method for damaging the eye since the heat of the Sun can fracture the glass, spitting out sharp splinters straight into the eyeball. An image of the Sun 15 centimetres (6 inches) across can be used for survey work, while higher magnification will show the smallest sunspots – little dots called pores. Many amateurs have modified their telescopes to show the solar spectrum and to study solar prominences. The clarity and intensity of colour in a high-quality solar spectrum is quite beyond any colour photograph or printed picture. In contrast to the planets, solar photography is straightforward as there is a large amount of light available and exposures are so short that guiding problems and telescope shake are of little concern. The main technical problem is that if the camera shutter is near the focus of the telescope it will become extremely hot. When pictures are not being taken, the telescope objective should be covered.

Amateur astronomers cannot easily study regions of the electromagnetic spectrum other than light. The remarkable sensitivity, negligible cost and general tolerance of the eye put observation in the visible into a quite different class from those in the ultraviolet or infra-red. The exception is the long-wavelength end of the radio astronomy spectrum. Dipole aerials a few metres in size will detect radio waves from the active Sun and from Jupiter. Aerial arrays, typically about the size (and aesthetic quality) of an iron bedstead, will give greater sensitivity. With arrays that do not swamp a modest back garden it is possible to pick up radio waves from the two strongest sources outside the solar system, Cassiopea A and Cygnus A. It is also easy to build two-aerial interferometers that can observe the Sun or Jupiter in transit. The tolerances are large as the wavelengths are so long and there is no need for accurate guiding of the aerial systems. The practicability of amateur radio astronomy is perhaps best illustrated by the fact that at one time – in the latter half of the 1930s – the whole of radio astronomy was an amateur pursuit, in the sole hands of an American, Grote Reber. Reber's great efforts in the development of

Viewing an eclipse from the Greek island of Santorini in 1976. An old gunsight was used to project the image on to a white screen – the simple and safe method of viewing the Sun.

radio astronomy have been at least as influential as the amateur discoveries of earlier times. Among these are the spiral structure of some galaxies and ultraviolet stellar spectra in the nineteenth century, aberration and the planet Uranus in the eighteenth and observations of the acceleration of the Moon's motion and the transit of Venus in the seventeenth century. For the present generation, there is the hope that some star on this side of the galaxy will explode into a supernova. This cataclysmic eruption would be 10,000 times brighter than an ordinary nova (and an "ordinary nova" is 100,000 times brighter than the Sun). Supernovae occur in our galaxy at a rate of about half a dozen a century. Like the novae, only a small percentage are on our side of the galaxy and the great majority are unseen because their light is absorbed by the dust of the interstellar medium. The last two were seen by Tycho Brahe in 1572 and Johannes Kepler in 1604, and no supernova has been visible in our galaxy since the invention of the telescope.

What is virtually certain is that when the next supernova does occur, it will be seen first not by a professional but by an amateur astronomer.

A unique sequence of observations of the eruption of Nova Cygni during the period of its brightening, taken by amateur astronomer Ben Mayer in 1975. No star was visible at the location of the arrow in the top picture. The next night (centre) *the nova became visible and brightened steadily. After two nights the nova reached second magnitude, the brightness of Deneb, which can be seen on the right* (bottom).

An amateur radio astronomer with his telescope (below), *built in his back garden. The "bowl" which picks up the signals is 5 metres (15 feet) wide and is made of wire netting stretched across a metal-and-wood frame.*

CHAPTER SEVENTEEN

THE FUTURE

The Large Space Telescope,
US and European plans for very big mirrors,
the Next Generation Telescope 25-metre

"The Committee recommends to the Director of Kitt Peak National Observatory that he,
Continue work directed toward the ultimate construction of a large optical-infra-red telescope;
Set as a design goal an aperture in the 25–30 metre range;
Continue to pursue a number of design concepts until the most cost-effective means of achieving the desired collecting area is clear."

KPNO Scientific Advisory Committee, January 1980.

The enormous wavelength range covered by modern astronomy splits the subject into three complementary areas: radio, space and optical. The three types of astronomy are in strongly contrasting conditions at the present time. The radio astronomers have recently achieved major improvements in their equipment, with the completion of the American Very Large Array and the construction of two high-precision dishes for millimetre-wavelength astronomy. The expectation is that the subject will now move into an exciting period of new observation. Until it is clear what puzzles emerge and which of them require new kinds of radio telescopes, it will be difficult to accurately define what type of instruments will be needed.

The subject of space astronomy is, as ever, a high-risk business. There is always the chance that a rocket will fail on launch and the hundreds of man-years of effort put into the satellite will be lost in seconds. This risk is slight, however, compared with the uncertainties on the political side of the subject. Space astronomy is "big science" – so big that the budgets are matters of governmental or international review. At a time of recession the axe is wielded with vigour: the number of telescopes likely to get into orbit in the 1980s is less than half the number launched in the 1970s. There is still a reasonable sequence of telescopes, mostly for special rather than general application, on the timetable, and the space astronomer is sitting with his fingers crossed hoping that telescopes that are already designed will survive and be built and launched.

Both radio and space astronomers have recently completed the design of the equipment they need in the near future. This is not, however, the case in optical astronomy. The optical telescope today stands at a crisis point similar to that reached at the end of the last century, when the construction of a series of steadily larger, but basically similar, refracting telescopes culminated in 1897 and 1898 with the building of the Lick and Yerkes refractors. With their completion it became clear that the design had reached a limit and larger lenses were not a practicable proposition. At the present time the same evolutionary process has led to a host of new 3- to 4-metre (120- to 160-inch) aperture reflecting telescopes all based on the same design concepts: monolithic primary mirrors and equatorial mountings. Now, as in 1900, the astronomer has a host of problems which require him to collect more light, and the simple recipe "make one like the last lot, but bigger" will not suffice.

There are several reasons for this. The most important is that many of the problems crucial to our understanding of the universe require a considerable increase in telescope aperture, a jump even larger than that from the 2.5-metre (100-inch) Hooker to the 5-metre (200-inch) Hale telescope. The astronomy that underlies this need for a major jump in telescope size is a result of the uneven distribution of matter in the universe. To take an example, a great deal is known about the ordinary main sequence stars in our galaxy, many of which lie within a few hundred parsecs of the Sun, but to apply the knowledge to the nearest similar system, the Andromeda galaxy, it is necessary to collect enough light to offset the much increased distance. Since the Andromeda galaxy is almost 700,000 parsecs away, a modest increase in telescope size would be of very little help. There is also a tendency for interesting objects to be too far away to yield their secrets to existing telescopes. At present astronomers studying globular clusters can only analyse the composition of the bright giant stars, while it is the fainter main sequence stars that should hold the key to the cluster's origin and evolution. Clues to the origin of the universe itself are necessarily to be sought in the study of the most distant galaxies and quasars, and the most distant are necessarily the faintest. A report in 1980 listed a couple of dozen topics for which a much larger telescope was critically necessary and even this report scarcely touched any problems other than those immediately concerned with the structure and evolution of our universe.

The struggle for greater sensitivity cannot follow any further the route taken for the last quarter of a century. Existing instruments in the 3- to 5-metre class have achieved steadily more impressive performance, not because the telescopes themselves have changed very significantly but because the light detectors used on those telescopes have improved enormously in sensitivity. This process is now approaching a fundamental limit as the detectors get within reach of 100 per cent efficiency. As this point is approached it is only possible to achieve a major gain in performance by collecting more light.

The reason that it is necessary to break with tradition and seek an alternative technology is that the final engineering and optical tolerances on a telescope are not a proportion of its size, but are fixed in terms of the wavelength of light: while the telescopes get bigger the tolerances do not. This increasing difficulty is most easily expressed in terms of money. There is an approximate relation between the costs of traditional equatorial reflecting telescopes of different aperture and this shows that doubling the aperture increases costs sixfold – a 2-metre telescope costs six times as much as a 1-metre, and a 4-metre costs six times as much again.

M82 (above) is one of the nearest active galaxies. To collect as much light from M82 as we now can from the nearest galaxy, Andromeda, requires a telescope mirror sixteen times the area of existing ones.

Part of the Virgo cluster (below), a concentration of 2500 galaxies about 20 million parsecs away. The search for larger scale order, to see if distant clusters form superclusters, requires instruments of very great aperture.

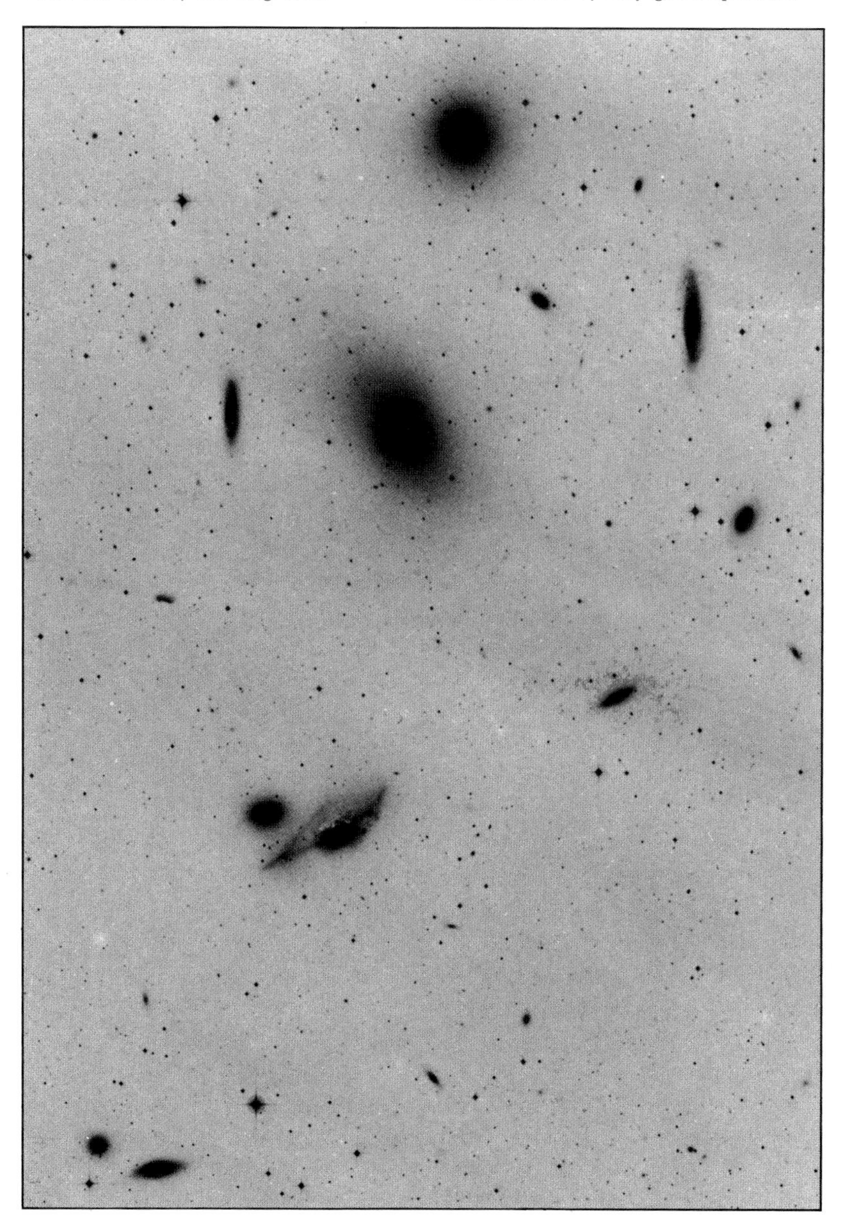

Extrapolation of this multiplying effect to a 25-metre aperture, the ideal telescope for the 1990s, gives a cost 120 times that of a 4-metre telescope, or just under $2500 million. This figure is not at all accurate, as the method of calculation is crude and unreliable, but none the less it demonstrates that a giant telescope cannot just be a bigger machine in the traditional mould. No one – neither governments nor even a syndicate of governments – is going to pay that much for a colossal equatorial telescope. If a giant telescope is going to be built, then it must be done for less than a tenth of this extrapolated cost.

A giant ground-based optical telescope could not be built in less than ten years. Orbiting telescopes take considerably less time to build, and although only a small number of space telescopes are due to be launched in the 1980s, this will not necessarily be the case towards the end of the century. On the other hand, a giant ground-based telescope would not at present cost as much as a satellite-borne instrument of only 1-metre (40-inch) aperture. Nevertheless, one question that has not faced previous generations of telescope builders is "can it be done better in space?" The answer briefly is that there is a division of potential achievement, with ground-based telescopes being admirably suited to general-purpose research, while the near future of space astronomy will be dominated by special-purpose instruments: the Space Shuttle is scheduled to launch ultraviolet, infra-red, solar and astrometric telescopes. Only one of the satellite-borne instruments – the 2.4-metre (96-inch) Space Telescope – is a general-purpose machine. This telescope will have one very important advantage: it will not have to look through the atmosphere and so will avoid "seeing" problems, atmospheric absorption and the light of the airglow. It will therefore be able to pick out a faint star from a crowded field of stars and study it over a wider wavelength range than is possible from the ground. On the debit side, a telescope in orbit is an instrument of limited stamina: a ground-based telescope will work for fifty years, but no satellite can achieve that, even if Shuttle crews service it at regular intervals. The Space Telescope is also inflexible: the number of auxiliary instruments is limited and cannot easily be changed.

The critical argument, however, which stresses that space and ground-based telescopes must be in alliance, not competition, is that the queue to use space telescopes is of almost infinite length. If space telescopes were used for experiments that can be done from the ground for very much less money, it would limit the time the satellite can spend on problems that it is uniquely able to attack. This is particularly true for infra-red observations. In the near infra-red region the balance between a space telescope and a giant ground-based telescope in terms of resolution tilts slowly in favour of the latter as the wavelength increases. In the near infra-red the atmospheric seeing gets better, and at the longer wavelengths the Space Telescope is able to resolve less detail because of its smaller mirror. If the wavelength is 10 micrometres the ground-based seeing limit has improved to half an arc second, while the wavelength-limited resolution of a 2-metre (80-inch) space telescope has risen to one arc second. The image quality is no longer better from orbit provided that the competing ground-based observation is made using any one of a dozen large existing telescopes.

The special infra-red telescopes planned for space work in the 1980s will operate in a different mode. They will be cooled to a very low temperature to gain sensitivity, but will not achieve high resolution as the apertures will be small. The first system, Infra-Red Astronomical Satellite, IRAS, is a telescope of 60 centimetres (24 inches) aperture which will simply survey the sky for new infra-red sources. The follow up, German Infra-Red Laboratory, GIRL, is only 40 centimetres (16 inches) aperture and is designed to do studies of spectra of individual objects.

The two optical telescopes – apart from the Space Telescope – that will probably be flown are special-purpose instruments and mark the return to instrument styles that have been obsolete on the ground for centuries. The solar telescope is to be a Gregorian reflector, utilizing a concave secondary mirror. The problems of solar heating in a spacecraft are tremendous and the energy concentrated at the intermediate focus of a Gregorian can be reflected harmlessly away, something that is not possible with a conventional Cassegrain telescope where there is no intermediate image. Hipparcos, the last on the list of probable satellites, is an astrometric instrument, sponsored by the European Space Agency. It will be a modern version of the astronomical sextant, measuring its way round the heavens by looking at stars in two directions at once.

The diverse capabilities of these instruments will augment, but not replace, giant ground-based telescopes. The enormous light grasp attainable on the ground more than offsets the difficulties due to seeing and the airglow. This is particularly true in detailed study of spectra. Spectral lines can reveal a star's velocity, and it is very difficult to measure the velocity shift of a faint star from a satellite whose own velocity changes by 50,000 kilometres (30,000 miles) an hour during the measurement. New techniques, such as speckle interferometry, can allow a ground-based telescope to partially overcome the blurring effect of seeing—a 25-metre (1000-inch) aperture telescope could resolve 0·004 arc second. The ground-based system also has great advantage in its use of auxiliary instruments. A prototype of a new detector, for instance, has only to survive the rigours of a car journey, and not a space launch, to be tried on the telescope.

Large ground-based telescopes will undoubtedly have a continuing role for at least the next quarter of a century, regardless of the level of space activity. To produce new, larger instruments, the astronomical world needs a revolution in telescope technology. The 3- and 4-metre telescopes built in the last few years will be the last of their breed, just as the refractors of the 1890s marked the end of their branch of the evolutionary tree.

One direction the revolution will take is already clear. A major part of the cost of any telescope stems from the mechanical sophistication needed to guide the instrument to follow accurately a celestial object as the Earth turns. The advantage of equatorial mounting is that it gives a simple uniform drive rate about one axis, the polar axis, and once set needs no motion at all about the other. The disadvantage is that the whole telescope swings round an axle inclined at 30 to 45° (depending on the observatory's latitude), and this motion twists and flexes the telescope structure and the mirrors. If instead of the polar axis the telescope is mounted on vertical and horizontal axes (alt-azimuth mounting), then the flexure problem is vastly simplified. This simplification is at the heart of the 6-metre (240-inch) telescope at Zelenchuk in the Caucasus, at present the largest optical telescope in the world. The Russian designers explored the use of both equatorial and alt-azimuth mountings and found the structure needed to hold an equatorial mount rigid was unacceptably heavy. The alt-azimuth design was less than half the cost, most of the saving being from dispensing with the enormous tonnage of steelwork required for the polar axis. One of the prices paid for this simplification is that both axes must be driven, and at different and non-uniform rates. But non-uniform drive rates can now be handled with a computer. The other problems are nuisances rather than disasters. The field of view in an alt-azimuth telescope rotates as the telescope tracks an object across the sky, and it is impossible to look straight upwards to the overhead point as the drive rate becomes impossibly large. Since any star that crosses the zenith at a particular time on a given date will not do so a few weeks earlier or later, this is not a serious limitation and simply requires more care in planning a programme of observations.

The move to alt-azimuth designs has been rapid and complete. All the design studies for a 25-metre aperture telescope use the simpler system. A 4·2-metre (165-inch) British telescope has now been designed in the new style for the Roque de los Muchachos Observatory on La Palma in the Canary Islands. So have designs at present under review in places as far apart as China and Texas. The near-compulsory virtue of alt-azimuth design for large instruments has now come to influence smaller telescopes. A recently published study on a 2-metre telescope for Kitt Peak Observatory showed that the rejection of equatorial mounting led to a major saving in the cost, even for such a small aperture.

While the choice of the mounting to be used for a future telescope is generally agreed, the decision on how to build the optical system is not. It appears at present that any group of n telescope designers will produce at least $n+2$ possible telescopes, all with some attractive features and all needing closer study. The problem

The improvement that results from good seeing and a dark sky is so large that the 2·5-metre (100-inch) Isaac Newton telescope is being moved from southern England to a mid-oceanic mountain site in the Canary Islands, where the new observatory in under construction.

that lies at the heart of all these alternatives is how to achieve the effect of a giant primary mirror, given that a single monolithic slab of glass or quartz is impossible to fabricate, polish or support in a telescope without distortion. The answer must be to subdivide the mirror into a set of sub-units. It is at this point that the diversity appears. The primary mirror can be segmented to give a mosaic; or the aperture can be made up of several independent telescopes on the same set of bearings; or it is possible to combine the light from a set of separately mounted, independent telescopes.

The first serious attempt to design a 25-metre (1000-inch) aperture telescope came in the summer of 1976. The initial difficulty was one of sheer disbelief. The 5-metre (200-inch) Hale telescope had needed a year of modifications before it was working, the 6-metre (240-inch) Zelenchuk instrument had an even harder time, including the breaking of the mirror blank. A 25-metre aperture is roughly the size of two tennis courts and the shape of this enormous surface has to be controlled so that a one-arc-second star image can be achieved at the focus. Outside the design group there was a general feeling that anyone who tried to find solutions could not really understand the problems involved. Several astronomers, when asked what research would be possible with such an instrument, replied in a general tone that they had better things to do than answer silly questions.

The first design was, in consequence, as conservative as possible. The primary mirror was to be a mosaic of small mirrors shaped as part of a sphere (rather than a paraboloid), and it was supported so that gravitational loads on the system were constant regardless of the direction in which the telescope was pointed. A primary mirror support system was chosen that avoided any highly stressed components or convoluted design. The result, the Rotating Shoe, was massive – 6500 tonnes, mostly of reinforced concrete, supporting a mirror made of 1800 almost-hexagonal elements. A secondary mirror system collected the light from a 25-metre-diameter patch of the oversized primary and reflected the image to laboratories placed on either side of the mirror shoe.

The design is a far cry from conventional optical telescopes and is just one

The Steerable Dish (below) is one possible design concept for the 25-metre (1000-inch) aperture optical telescope. The primary mirror is a mosaic of smaller mirrors, held in alignment by servo control, while the main structure is developed from radio telescope experience.

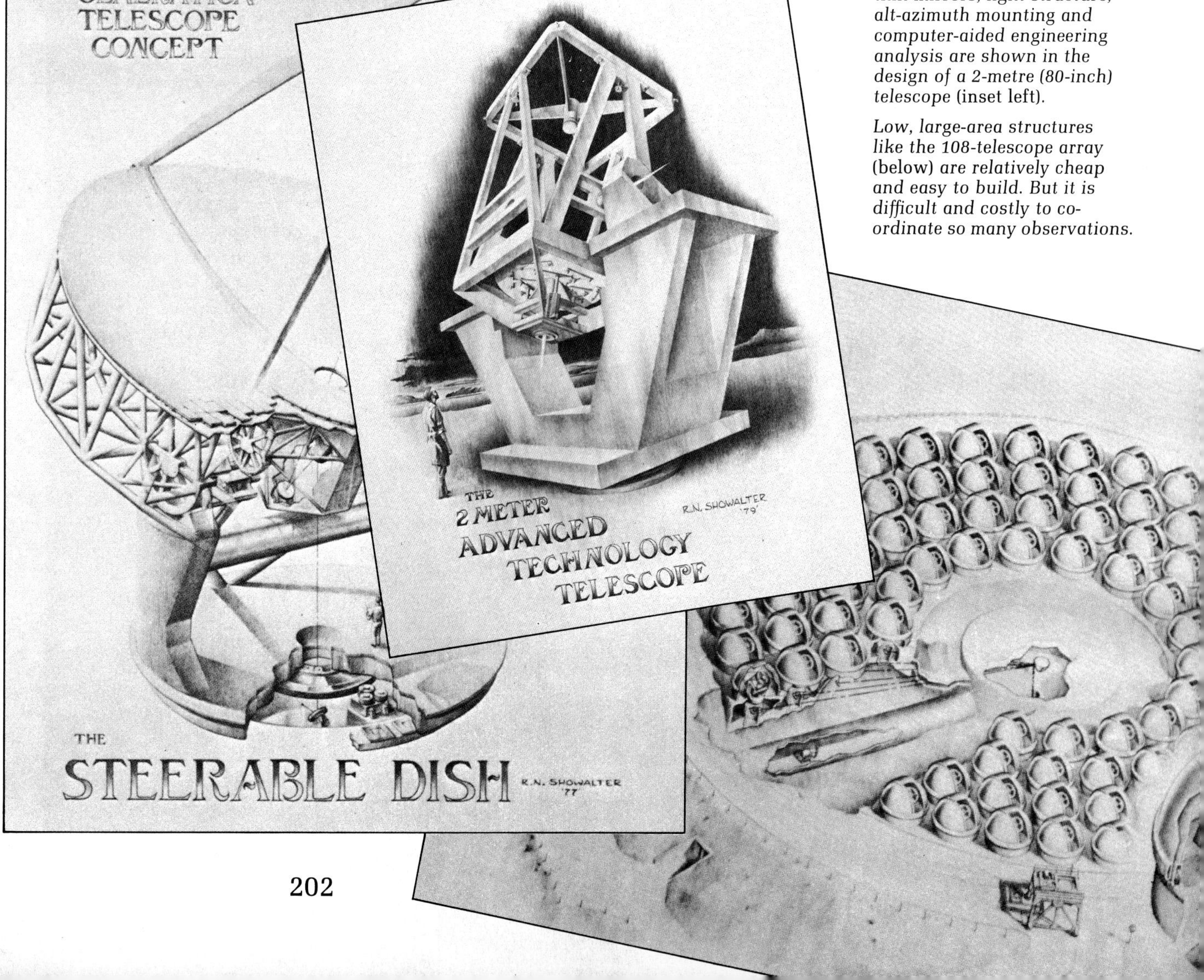

New design techniques for large telescopes have influenced smaller instruments. The effects of thin mirrors, light structure, alt-azimuth mounting and computer-aided engineering analysis are shown in the design of a 2-metre (80-inch) telescope (inset left).

Low, large-area structures like the 108-telescope array (below) are relatively cheap and easy to build. But it is difficult and costly to co-ordinate so many observations.

step away from the fixed bowl of the radio telescope at Arecibo. A complete bowl of mirrors is not a practical approach to a giant optical telescope because the secondary mirror has to be pivoted to swing about the centre of the bowl. It would not be possible to build across the 100-metre (330-foot) diameter of the bowl a bridging structure to hold the secondary with sufficient rigidity and control to give final star images only one arc second in diameter during a long exposure. Slicing away most of the bowl reduces this problem, though it is still the most difficult aspect of the shoe design. The primary mirror is part of a sphere and so suffers from spherical aberration, which would spread the star "image" at the primary focus to a smudge 25 centimetres (10 inches) in diameter. The secondary mirrors correct this, and so are part of the design.

The shoe design did achieve one major victory. No one could immediately demolish the concept as needing either impossible optics or engineering technique beyond the state of the art. This gain in credibility led the group to consider a handful of alternative designs. These were less pessimistic in approach than the shoe, and one of the schemes went so far as to examine the most difficult solution. Once again, the model was a radio telescope, but this time similar to the instrument at Jodrell Bank rather than Arecibo. The primary mirror would again be a mosaic, but its shape could be a paraboloid, similar to the primary of a classical Cassegrain telescope. The concept analysis for this Steerable Dish engendered a healthy respect for the difficulties of the design, but it did not identify any aspect that was impossible. The Steerable Dish has, in fact, many astronomical advantages, but its difficulties are reflected in the fact that it would probably be the most expensive of the options considered.

The other two telescope designs differed a great deal from both the Steerable Dish and the Rotating Shoe. One was the Multiple Mirror Telescope, in which several telescopes were mounted on a single frame and their light was combined at a collective focus. An important point in its favour was that the Multiple Mirror Telescope was no mere figment of the imagination. A prototype was being built on Mount Hopkins in southern Arizona. The other possible system, "The Array", also had a long history as several earlier studies, particularly in Britain and Canada, had looked at the concept.

The Rotating Shoe (below) was an early solution to the problem of a giant telescope design. Massive in scale, the main difficulty is the support of the secondary mirror—itself a 3·2-metre (130-inch) telescope on the end of a 25-metre (1000-inch) swinging arm.

The Array design involved a set of individually mounted telescopes feeding light to a single laboratory. It has obvious advantages. The individual telescopes can be much the same as existing instruments, so no new problems will arise in their construction. In addition, no colossal engineering problems will need to be solved at, or on the way to, the site, which is certain to be at the top of a mountain a long way from anywhere.

In a previous study of an array of telescopes, Mike Disney of University College, Cardiff, had suggested that an assembly of individual instruments in the 2- to 2·5-metre (80- to 100-inch) class would give high performance at minimum cost. Infrared astronomers, however, had a strong preference for an array based on a smaller number of larger telescopes. Three array designs were eventually considered, two in Arizona and one in Canada. These were an array of 108 telescopes of aperture 2·4 metres (95 inches); a more compact design, using sixteen 6·25-metre (250-inch) telescopes; and the Canadian system, consisting of seventy-two 3-metre (120-inch) telescopes, which was based on earlier work by Harvey Richardson.

The 108-telescope array turned out not to be as good an idea as preliminary inspection suggested. The system can only be used if each telescope is separately instrumented and the output from all the telescopes added in a computer in the central laboratory. There are severe problems in reflecting all the images so that they coincide and add together. At the final focus in the central laboratory the maximum field of view is very tightly restricted and can scarcely exceed the size of Jupiter (less than one arc minute). More important, extremely tough engineering tolerances must be met in order to superimpose 108 images with the one-arc second accuracy demanded. These tolerances require the individual telescope mirrors and bearings to be far superior to existing telescopes and this severely erodes the whole basis of the array concept, namely the use of ordinary telescopes. Roger Angel of the Steward Observatory, Arizona, had already demonstrated that optical fibres could be used to raise the efficiency of ordinary telescopes, and he explored the use of much longer fibres to link together the telescopes of an array. This approach has many advantages, but loses heavily in wavelength coverage, as the fibres are not transparent for much of the infra-red.

The alternative approach is to use photoelectric light detectors at each telescope in the array and to add these signals in the computer. This runs into heavy financial trouble. As auxiliary instruments become more sophisticated they rise in cost, and there are already auxiliary instruments that cost as much as 2-metre telescopes. To supply 108 such instruments would annihilate all the economies of small-scale construction that give the array a possible cost advantage. It is clear, however, that the faults of the array concept diminish as the size of the individual telescopes increases.

The results of all these studies emerged at an international conference in Geneva in December 1977. The most noticeable effect was that the conference discussed 10-, 16- and 25-metre (400-, 630- and 1000-inch) aperture telescopes as reasonable engineering structures. The indifference or incredulity that met the first attack on the "1000-inch" problem only sixteen months earlier had evaporated. The conference accepted the need for alt-azimuth mountings and multi-element mirrors and agreed to two further points. These were that low cost meant the use of lightweight structures and that servo-controls were needed to adjust the mirror surfaces and the position of the focus. Lightweight telescopes had to have thin mirrors, so particular attention was paid to telescopes like the UK Infra-red Telescope (UKIRT), which has a primary mirror much thinner and much less massive than the "traditional" telescopes of the early 1970s.

Two years later, in January 1980, there was another conference, this time in Tucson, Arizona, at which a whole range of advanced telescope concepts, covering apertures from 2 to 25 metres, were presented. The most important news was that the Mount Hopkins 4·5-metre (180-inch) Multiple Mirror Telescope (MMT) was assembled and operating. The new telescope, effectively the third largest in the world, has its primary mirror area divided into six individual mirrors, and uses an alt-azimuth mount. The significance of the MMT is that it has been built for a budget very considerably below that needed for a traditional 4·5-metre.

The MMT consists of six 1·85-metre (72-inch) aperture telescopes mounted in an elaborate steel crate. The telescopes are not held in alignment by the steelwork, but are all supported on servo-driven actuators. The alignment is continuously checked with a complex laser system. If the six images are not coincident at the final focus, the individual mirrors are automatically adjusted. The chief difficulty of such a system is to achieve sufficiently reliable operation: if any part of this elaborate servo system develops a fault, the overall performance suffers. The most encouraging aspects of the MMT, however, are not concerned with the sophistication of the laser controls. One of the matters of particular interest does not even concern the telescope. This is the

The Multiple Mirror Telescope on Mount Hopkins in Arizona. The telescope, which combines six 1·85-metre (72-inch) telescopes in a single complex frame, achieves an effective aperture of 4·5 metres (180-inches): it is the third largest telescope in the world. The new structure, its servo controls and a revolutionary dome combine to make the telescope a working test-bed for many concepts relevant to future developments.

use of an inexpensive box-shaped building in place of a dome. The MMT is not enclosed in a spacious, costly, hemispherical structure inside which the instrument can pivot. Instead it has a close-fitting building which itself rotates with the telescope. One of the biggest problems with seeing is the turbulent mixing of layers of air near the ground, and there had been much concern that this new shape – with all the aerodynamic subtlety of a brick – would prove particularly troublesome. In the event the MMT building, poised on the topmost point of Mount Hopkins, has proved entirely satisfactory. A classical hemispherical dome costs about as much as the telescope housed within, and the cheap, easily built MMT structure is a major move towards the cost reduction necessary if a giant telescope is to become a reality.

The Tucson Conference demonstrated a surge forward in telescope design, with scarcely a mention of the heavyweight equatorial designs that had dominated the scene only five years earlier. Designs were presented for 7- and 10-metre (275- and 400-inch) telescopes, one for the University of Texas, the other for the University of California. The two designs offered a strong contrast. The Texan design explored the limit which one could reach without doing anything dramatic. The starting point was to insist that the primary mirror be a single element as large as possible, the size being limited only by the capacity of the casting and annealing furnaces. The second constraint was that the telescope be made to work without complex servo controls – the shape of the mirror and the exact point of the focus would remain constant by design, not by electronics. Apart from this, full advantage was to be taken of modern materials and of computer-assisted structural analysis of the components.

These conditions produced a design for a 7-metre telescope, which pushes almost all the design parameters beyond the limits set by previous instruments. The large aperture is achieved with a short focal length and a thin mirror. The latter choice was particularly important, as it turned out that the removal of 1 kilogramme ($2\frac{1}{4}$ pounds) from the mirror reduced the total weight of the telescope by 8 kilogrammes ($17\frac{1}{2}$ pounds). The short focal length, only twice the primary mirror aperture (a focal ratio of 2), also pushes the telescope farther along the road taken by almost every telescope designer since Herschel began building big reflectors with a focal length ten times the aperture.

The Texas telescope design is a halfway house on the road towards giant telescopes. It uses an MMT-style "dome", an alt-azimuth mounting and a monolithic mirror with entirely mechanical support.

NEXT GENERATION TELESCOPE CONCEPT

Six telescopes of aperture 10 metres (400 inches) together have the same light-gathering power as one 25-metre (1000-inch) instrument and an individual telescope of such an array is much easier to design and build than a single super-giant instrument. The light from the six telescopes can be combined either by mounting auxiliary instruments on all six telescopes and forming the resultant image from the separate signals by addition in a computer, or the starlight can be combined at a central coudé focus shared by all six telescopes. The need for a common focus for all six telescopes requires that the array should be as compact as possible. In the design shown here the telescope-to-telescope distance is 45 metres (150 feet).

As with the competing design for a 25-metre (1000-inch) telescope, the instrument uses altitude (A) and azimuth (B) axes. It also employs a lightweight segmented primary mirror (C). As the mirrors for the coudé focus (D) have to be behind the primary mirror and on the telescope axes, the altitude axis is also behind the mirror. The bearings are enlarged into giant D-shaped structures (E), as this provides a counterweight to balance the telescope as well as being an inherently stiff lightweight structure. The telescopes can be used independently and auxiliary equipment would be carried on the Nasmyth focus platform (F, G). There is no prime focus as this is incompatible with both lightweight design and a compact dome. In addition to the versatility that comes from independent use of the components of the array, it is also possible to combine pairs of telescopes to form a two-telescope interferometer.

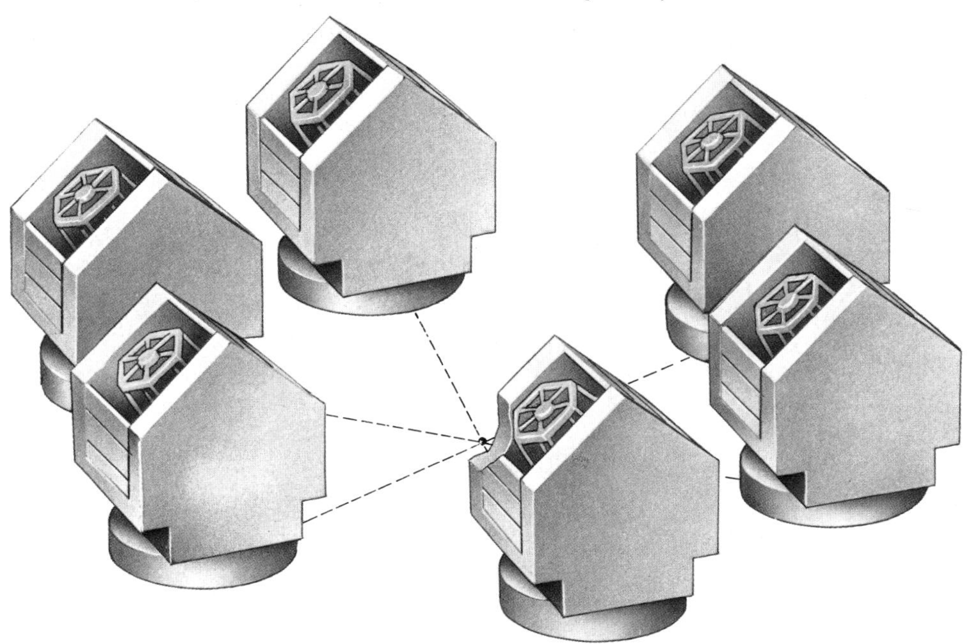

The "domes" of the array of telescopes are not the traditional hemispherical buildings used for classical equatorially mounted telescopes. Lightweight, close-fitting rectangular buildings that rotate with the telescopes are used instead, as these are less expensive, and the prototype, the dome of the Multiple Mirror Telescope, has proved very successful.

The array uses a total of six mirrors to guide starlight to the central focus. Mirrors 1 and 2 form an image at the Cassegrain focus close behind the primary mirror. This image is then relayed by three further mirrors, 3, 4 and 5 near the telescope to the central laboratory, where a mirror deflects the beam to the final focus at the centre of the array (shown above).

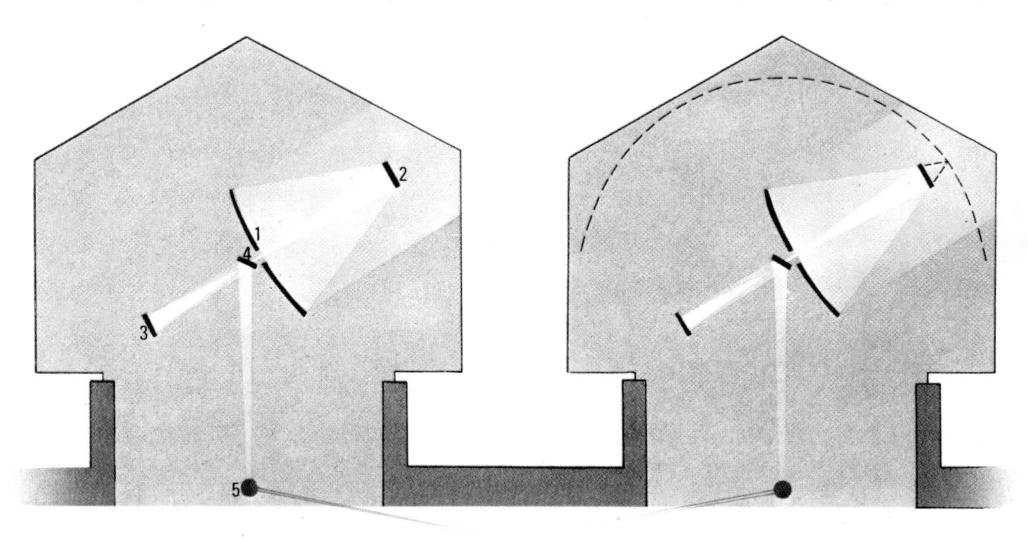

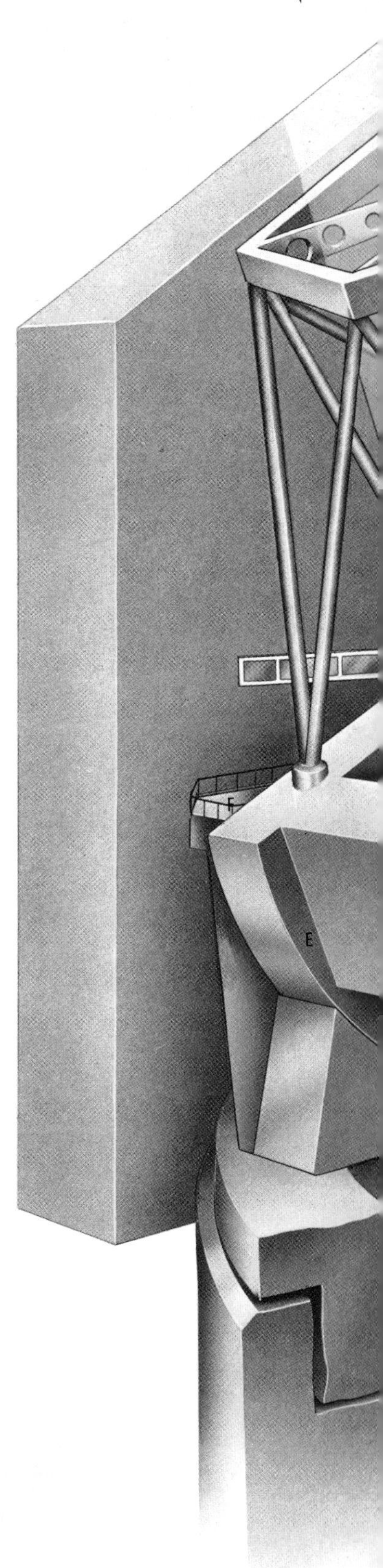

The array telescopes design uses a primary mirror built up from a central disc 5 or 6 metres (200 or 240 inches) across and one or two rings of "petals".

To renew the reflecting coating, the array telescope mirror is removed from beneath and lowered on to the basement, where one coating system serves all six telescopes.

The California design aims for a larger aperture, 10 metres (400 inches), and takes two further steps to achieve the increased size. These are to use a segmented primary mirror and to servo-control the alignment of the segments. Going beyond the Texas design, the mirror has an even smaller focal ratio (1·75) and is thinner, with the mirror segments only 10 centimetres (4 inches) thick. A large part of the design development has been concerned with two topics. First has been the building and testing of a trial set of servo-controls and actuators for the sixty movable segments which form the primary mirror. The second topic has been the difficulty of making the segments of the paraboloidal mirror so that they assemble to give a simple high-quality image. This has been achieved by bending each mirror blank, polishing it under stress and then letting it spring back to give the desired portion of a parabolic surface.

Beyond these two programmes, both of which have a good chance of being funded, several other developments towards an even larger aperture have been explored. The European Southern Observatory has examined a 16-metre (630-inch) telescope design which is similar to the California 10-metre concept. The MMT team at the University of Arizona and the Smithsonian Institution have extrapolated beyond the existing Multiple Mirror Telescope to look at a related structure that supports eight separate 5-metre (200-inch) telescopes in a single frame.

A model of the 10-metre (400-inch) telescope designed by the University of California. Servo controls and alt-azimuth mounting have done away with "traditional" Serrurier trusses and polar axes to give a structure that is clearly not a development of the instruments built in the thirty years following the 5-metre (200-inch) Hale telescope.

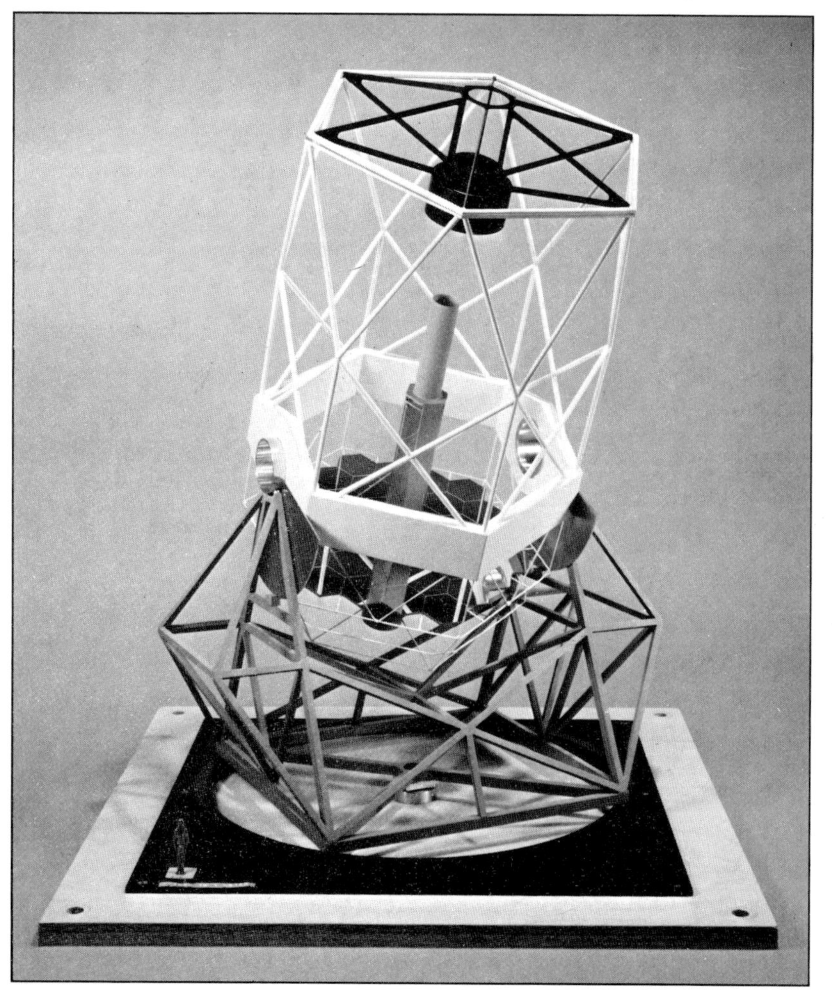

The push towards a 25-metre aperture has benefited from, and been of benefit to, these other programmes. Three different telescope concepts are still possible contenders for the most cost-effective solution. The first of these is the Steerable Dish – the radio telescope style of telescope with a mosaic primary mirror and a simple secondary mirror. The second is a further enlarged multiple mirror telescope carrying either six 10-metre (400-inch) or eight 8·8-metre (350-inch) telescopes on a single frame. The third is an array of six 10-metre telescopes feeding a central laboratory. All three designs have a lot in common – all use lightweight optical components, alt-azimuth mountings and segmented primary mirrors.

At present the choice between the three concepts is impossible: each has advantages and disadvantages. The six-telescope array is the most versatile but the most expensive in terms of auxiliary instruments. The enlarged MMT has the benefit of a successful prototype, but involves a very large moving structure. The Steerable Dish is compact but has to satisfy tighter tolerances to make the single primary mirror achieve a resolution of an arc second. Each of the three designs has its supporters. The next stage is to quantify the potential of each concept for astronomical research and to make a careful assessment of their relative costs.

My own personal hope is that the array of six independent telescopes will prove the most effective system. An array of half a dozen world-class telescopes, combined into a single system, offers unrivalled versatility. In addition to using the six instruments in combination, it is also possible to operate them independently. This would allow different simultaneous observations, either of the same object at different wavelengths or of different objects. The latter capability means that astronomers could survey a set of similar objects with exceptional rapidity. It should also be possible to combine the telescopes in pairs to do speckle interferometry, which gives the six-telescope system a particular advantage over the other 25-metre designs.

The future of astronomy lies with a new generation of giant telescopes, but not just with the hope that they will succeed with the programmes that are obviously needed and will help justify their funding. If the history of the telescope over the past 373 years is a guide, we may hope that they will achieve the unexpected. The reason for building instruments with capabilities beyond anything available is that we cannot predict the outcome.

GLOSSARY

aberration The very small annual displacement of a star from its true position in the sky. It is caused by the Earth's orbital velocity making the light "drag" and enter the telescope at a small angle from the normal.

absolute zero The temperature at which an atom would have no energy of motion. It corresponds to $-273°$ on the Centigrade scale, and to $0°$ on the Kelvin or Absolute.

absorption spectrum When light from an incandescent body such as the Sun passes through a cooler gas, atoms in the gas absorb radiation of certain colour or *wavelength*. If the light is spread out to form a spectrum, a series of dark absorption lines is seen, each representing an atom in the gas.

achromatic lens (achromat) Light passing through a single lens is dispersed into its component colours, a fault known as *chromatic aberration*. By combining two lenses made from different types of glass, the aberration of one lens can be neutralized by that of the other. Such a lens is achromatic.

aerial telescope The early single-lens telescopes were often many metres long and the *objective lens* high above the ground (aerial) had to be supported with an arrangement of poles and rigging.

airglow A very faint illumination of the night sky caused principally by the emission of light energy from oxygen atoms recombining into molecules after having been separated during the daytime by the Sun's energy.

alt-azimuth mounting A telescope mounting whose axes allow the tube to be moved in a vertical and horizontal direction. To follow a celestial object across the sky, the telescope must be adjusted simultaneously on both axes.

angular resolution The angular distance between the closest stars or points that a telescope can separate. With large telescopes, the unsteadiness of the atmosphere frequently prevents the instrument from achieving its potential resolution.

aperture The working diameter of the main lens or mirror in a telescope.

apparent magnitude A measure of the brightness of a star as it appears from the Earth. A brightness ratio of 100 corresponds to exactly five magnitude divisions (one magnitude division corresponds to a brightness ratio of about $2\frac{1}{2}$, the fifth root of 100). The faintest naked-eye stars are of about magnitude 6; the largest telescopes can detect stars as faint as about magnitude 25.

arc minute A measure of angular distance, equal to 1/60th of a degree.

arc second A measure of angular distance, equal to 1/60th of an arc minute or 1/3600th of a degree.

asteroid A word meaning "star-like": the term first given to the small planetary bodies the majority of which orbit the Sun at distances between those of Mars and Jupiter.

astrolabe Portable sighting instrument brought to perfection by the early Arab astronomers with which the local time, or the observer's latitude, can be found by observing one of a number of standard bright stars.

astronomical unit The mean distance of the Earth from the Sun, equal to 149,597,870 kilometres (92,955,807 miles).

azimuth Angular distance around the horizon, usually from north (0°) round by east (90°), south (180°), and west (270°).

band spectrum When light from an incandescent body such as the Sun passes through a very cool gas (such as the Earth's atmosphere) in which atoms have combined into molecules, some sections of the radiation will be absorbed. If the light is spread out to form a coloured band, or spectrum, this absorbed radiation will be represented by dark bands.

big bang theory Theory of the explosive origin of the universe, based originally on the discovery that all the galaxies are flying away from each other and would have been close together some 20,000 million years ago.

binary star Two stars revolving around each other under the bond of gravity. It is believed that about a quarter of all the stars in our galaxy are members of such systems.

black hole The term given to the theoretical result of a massive body collapsing into solid nuclear matter. Radiation is unable to escape, so that the body, although still exerting a gravitational effect, is invisible.

blue shift The apparent compression of the light-waves reaching an observer when the source emitting them is approaching at a high velocity. Since blue light has the shortest *wavelength* of visible light, colours are in theory made bluer. However, only when the speed of approach attains a large fraction of the velocity of light can any colour change be noticeable; normally, the effect is detected by measuring the wavelength of particular features in the light emission.

Cassegrain telescope A design of reflecting telescope announced in 1672 in which the light focused by a large concave mirror is refocused by a much smaller convex mirror, usually passing back through a hole cut in the centre of the large mirror. Most large, modern instruments employ variations of the Cassegrain system because it is so compact.

celestial equator The projection of the plane of the Earth's equator on to the sky.

celestial poles The projection on to the sky of the Earth's north-south axis. All the celestial objects appear to rotate about these poles once a day. The bright Pole Star lies very near the north celestial pole.

celestial sphere The imaginary sphere which carries the stars and other celestial objects and is marked with lines of celestial latitude and longitude. It appears to rotate around the Earth once a day.

Cepheid variables An important family of very bright pulsating stars which change their luminosity in a regular period. This period is linked to the average luminosity. By determining the period of brightness variation (typically a few days), the luminosity of the star can be calculated and its distance from Earth deduced by measuring how bright it appears in the sky.

chromatic aberration The colour of light determines the amount of bending, or *refraction*, it undergoes when passing through a lens or prism, blue light being bent more than red. Therefore, when a single lens is used to focus light, a coloured image results, a defect known as chromatic aberration. It can be minimized by using a combination of two or three lenses (an *achromat*).

collector plate The last of a series of plates in a *photomultiplier* tube that release successively increasing numbers of electrons in response to a small light input. The electric circuit generated in the collector plate is measured.

comet A diffuse body of very low mass (equivalent to a solid body only a few kilometres across) consisting mainly of volatile icy and dusty material, orbiting the Sun. Most known comets have a very eccentric orbit, and at their closest approach to the Sun they emit clouds of gas or dust or both, often in the form of a tail. Relatively few comets ever attain naked-eye brightness.

constellation One of the eighty-eight fanciful groups of stars into which the sky has been divided.

continuous spectrum The pure rainbow-like appearance produced when the light from an incandescent body is spread out into a band of different colours.

corona The Sun's outer atmosphere, which becomes visible with the eye only when the Moon covers the solar disc at the time of a total solar eclipse. It consists of atoms of oxygen, nitrogen, iron and many other elements at a very high temperature, much higher than that of the Sun's surface. Its form varies with the sunspot cycle.

coronagraph A device developed and perfected by Bernard Lyot at the Pic du Midi Observatory in the Pyrenees in the 1930s. By reducing stray light within the telescope and observing from a high altitude above the dustier layers of the atmosphere, it proved possible to record the brighter inner corona of the Sun at times other than a total eclipse.

cosmic dust Fine grains of material, mostly carbon, silicon, or iron, that pervade interplanetary space and slowly descend through the Earth's atmosphere.

coudé telescope A telescope employing one or more flat mirrors to bring the image to a convenient observing position, usually one that is fixed regardless of the orientation of the instrument itself.

declination The celestial equivalent of terrestrial latitude: the position of an object in the sky in degrees north or south of the *celestial equator*.

Doppler effect The apparent change of frequency or wavelength of the radiation from an object that is either approaching or receding from the observer. It can apply to sound or light waves. If the object is approaching, the frequency is increased; if the source is receding, the frequency falls.

double star Two stars lying very close together in the sky, either through a line-of-sight effect or because they are physically connected (a *binary star*). There is no strict criterion for classification; a naked-eye double star will have to be several arc minutes apart if they are to be distinguished, whereas the separation of a telescopic double may be less than one arc second.

eclipse Strictly speaking, this term refers to the passage of one body through the shadow cast by another (such as an eclipse of the Moon in the Earth's shadow); but it is also applied to the total or partial passage of the Moon across the face of the Sun.

ecliptic The projection on the sky of the plane of the Earth's orbit; in other words, it is the apparent path followed by the Sun around the sky in the course of the year. Since the orbits of most of the major planets are inclined to that of the Earth by only a few degrees, they are always to be found within a few degrees of the ecliptic, in the region known as the *zodiac*.

electromagnetic radiation The propagation of energy through space in waves of tiny "packets" known as photons. Such energy is classified according to the wavelength or the distance between successive intensity maxima.

electron Sub-atomic particle of very small mass and negative charge, which orbits at the periphery of an atom.

element One of the ninety-two naturally occurring types of atom from which all known matter is composed. By far the most common element in the universe is hydrogen.

ellipse The geometrical shape obtained by running a pencil around inside a loop of string dropped over two pins. Every planet moves in an elliptical orbit. The less like a circle, the greater the eccentricity of an ellipse.

emission spectrum If a rarefied gas is heated until it glows, it emits radiation at discrete *wavelengths* only. If this radiation is then spread out to form a coloured band, it will appear as a series of bright lines indicating the wavelength of the various emissions.

equatorial mounting A telescope mounting in which one axis (the polar axis) is set parallel with that of the Earth. Since the Earth spins on its axis once every twenty-four hours, rotation of the telescope around its polar axis in the opposite direction but at the same speed will neutralize the Earth's spin and keep the telescope pointing the same direction.

equinox The instant at which the plane of the Earth's equator passes through the Sun. This happens twice a year, on or around 21 March (the vernal equinox) and 23 September (the autumnal equinox). From the vernal until the autumnal equinox the Earth's north pole is inclined towards the Sun and it is spring and summer in the N. Hemisphere.

field of view The circle of sky that can be seen at one view through a telescope or covered by a camera lens.

fixed stars An outmoded term, once used to distinguish between the planets (which have their own motion across the sky) and the stars, which are virtually stationary with respect to one another.

flash spectrum The phenomenon seen at the beginning and end of a total eclipse of the Sun, when the Sun's inner atmosphere, or chromosphere, shines at the edge of the Moon although the solar surface is obscured. Since the chromosphere emits light only at certain discrete *wavelengths*, a spectrum consisting of a number of bright lines of light is seen.

focal length The distance between the lens or mirror and the image it forms of an infinitely distant object.

focal point The place where the rays of light focused by a lens or mirror converge to form an image of the distant object observed.

Fraunhofer lines If the radiation from the Sun is spread out into a coloured band, or spectrum, it is seen to be crossed by a great number of dark lines. Each line represents the absorption of a certain light-emission by the atoms of elements in the Sun's atmosphere.

galaxy A self-contained aggregation of stars and star-producing material (gas, dust), which probably condensed soon after the universe came into existence. The Milky Way galaxy to which the Sun belongs, with a diameter of about 25,000 parsecs and a stellar population of the order of 100,000 million stars, is typical of the spiral type of galaxy.

Galilean telescope The form of telescope used by Galileo for his astronomical observations commenced in 1609. The image formed by the main convex lens is examined by a small concave lens acting as the eyepiece. It gives an erect image, but has an extremely restricted field of view, showing only a small area of sky at a time.

geocentric Literally "Earth-centred": the ancient idea of the way the universe was constructed. Used now to indicate space distances measured from the Earth's centre.

globular cluster A cluster of hundreds of thousands of stars, usually a member of a group of similar clusters forming a halo around the central mass of a spiral galaxy. The clusters contain Population II stars, which are old objects similar to those found near the nucleus of a spiral galaxy. In this they are completely different from the "open" or "galactic" clusters found in spiral and irregular galaxies, which contain many recently formed stars.

Gregorian telescope A telescope design suggested by James Gregory in 1663, in which the light focused by a large concave mirror is reflected by a much smaller concave mirror, usually passing back through a hole cut in the centre of the large mirror. The Gregorian gives an erect image and used to be popular as a terrestrial telescope. Astronomically it is obsolete because of its extra tube length over the *Cassegrain* type.

heliocentric Literally "Sun-centred": the theory, first widely publicized by Copernicus, that the Earth and the other planets orbit the Sun. Used currently to indicate a space distance measured from the centre of the Sun.

heliometer A device used in the eighteenth and nineteenth centuries for measuring accurately the angular distance between two fairly close objects in the sky. The telescope lens consists of two half-lenses, each of which produces a separate pair of images. One half-lens is moved sideways until the image of one object produced by the first lens is superimposed on the image of the other object produced by the second lens. The amount by which the lens has to be moved indicates the angular distance being measured.

heliostat A single flat mirror which is motor-driven so as to reflect the same point in the sky into a fixed telescope. It was originally used for solar work, particularly at the time of an eclipse.

Hertsprung-Russell diagram A chart showing the real luminosity of a group of stars plotted against their surface temperature or colour. If the sample is a large one, the points representing the stars will fall into one of a number of separate areas on the diagram. Most will lie on or near a diagonal line running from the very luminous, hot stars to the very dim, cool ones (the *main sequence*), but there will also be some very luminous, cool stars (*red giants*) and some very hot, dim stars (*white dwarfs*), in addition to other sub-groups.

image intensifier An electronic device for increasing the brightness of an optical image produced at the focus of a telescope. The image is projected on to a metallic surface from which electrons are released in accordance with the intensity of light falling on it; the electrons are then beamed to form a new, brighter image on a fluorescent screen.

inferior planets The name given to the planets Mercury and Venus, whose orbits are smaller than that of the Earth.

intensity interferometer A device for measuring the very small angular diameters of some of the largest nearby stars. The fluctuations of brightness of the images of the star formed by two separate but nearby telescopes are compared; the more nearly they agree, the smaller the source must be.

interferometer A device for determining the distance between close radiation sources or the size of a single small source. The radiation from the object is received by two or more widely spaced collectors and then mingled; the pattern obtained enables the separation or width of the source to be calculated. The farther apart the collectors are, the closer the sources that can be separated.

interstellar medium The thin residue of material that is distributed between the stars. It consists mainly of hydrogen atoms, but organic compounds of hydrogen and carbon are also present.

infra-red radiation The name given to the band of invisible radiation beyond the red end of the spectrum, whose *wavelength* lies in the range 0·001–1 millimetre. Infra-red waves are emitted mainly by the cooler bodies in the universe, such as young stars that are still contracting from clouds of material.

ionization In its normal state, an atom contains equal numbers of electrons (negatively charged particles) and protons (particles carrying an equal but positive charge). Excitation by heat energy, or interaction with other atoms, can remove or add electrons, producing a condition of ionization in which the electric charges within the atom are no longer neutralized.

ionosphere The layer in the Earth's atmosphere at a height of about 75–150 kilometres (50–90 miles) in which, due to solar energy, atoms have become positively charged. This layer acts as a reflector for short-wave radio signals, and can be destroyed by violent solar activity.

Keplerian telescope Instead of using a concave lens for an eyepiece, Kepler (in 1611) proposed using a convex, or positive, lens. Although the view is inverted instead of erect, much more of the sky can be seen at one time.

latitude The distance in degrees of a point on a planet's surface, measured from the equator towards the north or south pole.

light year A largely obsolete measure of distance, corresponding to the distance a light-wave would travel in one year at its space velocity of 299,792 kilometres per second (186,282 miles per second). It is equal to about 9·5 million million kilometres (6 million million miles), or 0·31 parsec.

line spectrum The pattern of lines seen when light is either emitted or interrupted by a rarefied gas and then spread out to form a coloured band or spectrum.

longitude The distance in degrees measured around a planet's surface in a plane parallel to that of the equator of a given point on its surface from an arbitrary line or meridian joining both poles. On the Earth, this meridian passes through Greenwich, England.

lunar eclipse The partial or total passage of the Moon through the shadow cast by the Earth. This can occur only at "full", when the Moon is on the opposite side of the Earth from the Sun. Not all full Moons are eclipsed.

magnitude The brightness rather than the size of a celestial object. The magnitude scale is graded so that a ratio of five magnitudes corresponds to a brightness ratio of exactly 100 times; one magnitude step is equivalent approximately to a brightness ratio of $2\frac{1}{2}$ (the fifth root of 100).

main sequence The name given to the class of stable stars that form the majority of stellar objects to be found in the galaxy. Within the main sequence, luminosity, temperature and mass are all consistently related. The sequence extends from red dwarfs of less than a thousandth the Sun's luminosity, to brilliant white stars that are hundreds of times brighter than the Sun.

major planets The name given to the nine principal bodies circling the Sun.

meridian On a planet, an arbitrary line joining its poles from which a system of *longitude* is reckoned. It also refers to the line joining the observer's northern and southern horizon and passing directly overhead. A meridian circle is a telescope mounted so that it can be pointed only along the meridian.

meteor The brief flash of incandescent air seen when a meteoroid (an interplanetary body usually only a few millimetres across) enters the Earth's atmosphere at a speed of up to 50 kilometres (30 miles) per second and is vaporized.

meteorite The remains of a large meteroid that survives its passage through the atmosphere and reaches the Earth's surface. Its composition can be metallic, stony or a combination of the two.

Milky Way The name given to the faint, irregular band of light that appears to encircle the night sky. It is caused by the combined light of innumerable stars in the nearby arms of the galaxy. Its patchy appearance is due to intervening masses of dark material.

minor planets The small planetary bodies the majority of which orbit the Sun at distances between those of Mars and Jupiter.

mural arc A device used for determining stellar positions. A telescope, attached to a graduated arc which could be anything from a quarter to a complete circle, was mounted on a wall aligned accurately north-south, so that the instant at which a star crossed the *meridian*, as well as its altitude at that moment, could be determined.

nebula An interstellar cloud consisting principally of hydrogen, but also containing small quantities of other elements. Cool nebulae may appear dark if seen projected against a starry background, or bright if they are reflecting light from nearby stars. Emission nebulae are made self-luminous through the excitation of their atoms by very hot stars; this is the case with planetary nebulae, which are shells of gas expelled from massive stars in a supernova explosion.

neutron A sub-atomic particle carrying no electric charge, which together with a *proton* forms almost all the mass of an atom.

neutron star The collapsed remains of a massive star that has been destroyed in a supernova explosion. Its material falls inwards by gravitational force so violently that its solid nuclear matter is crushed together, attaining a density of about 200,000 tonnes per cubic millimetre. The star at the centre of the famous Crab nebula is believed to be a neutron star, the remains of the supernova explosion observed in the year 1054.

Newtonian telescope The form of the first practical telescope using mirrors, constructed by Isaac Newton in 1668. The light reflected by a large concave mirror is reflected through a hole cut in the side of the tube by a small flat mirror. The Newtonian system is still used by amateurs but it has no particular advantages for professional astronomers.

nova A violent stellar outburst, believed to be caused by material from one relatively cool member of a close pair of stars impinging on the surface of its much hotter companion. This influx of nuclear fuel results in the outer shell of the star being blown away, and the total brightness, as seen from the Earth, may rise by 10,000 times in one or two days. Maximum brightness is usually sustained only for a very few days, after which the nova fades, taking many years to sink back to its original brightness.

nucleus The core of an atom in which the neutrons and protons (equally massive particles with no electric charge and a positive charge respectively) are located.

nutation A small oscillation of the Earth's axis, with a period of eighteen years, induced by the Moon's regular monthly travel north and south of the plane of the Earth's orbit.

objective A term applied to the main lens in a refracting telescope. It is also known as an object-glass.

occultation The passage of one celestial body in front of another. An eclipse of the Sun is, therefore, strictly speaking an occultation of the Sun by the Moon.

orbit The imaginary path traced out by an object moving through space under the gravitational control of another body.

parallax The apparent shift of position of a nearby object relative to the distant background when the observer's location is changed. The first measurements of stellar distance were made in 1838 by observing the position of a star against the background stars (assumed to be infinitely remote) at six-monthly intervals, thereby using the diameter of the Earth's orbit as a baseline.

parsec A standard astronomical unit of distance, being the distance from which the radius of the Earth's orbit (the Astronomical Unit) would appear only one arc second (1/3600°) across. It corresponds to 30,857,000,000,000 kilometres (19,174,000,000,000 miles), or 3·26 light-years.

perihelion The point on the orbit of a planetary body that is closest to the Sun.

phase This usually means the apparent change of shape of a planetary body due to the presentation of its sunlit hemisphere.

photocathode A metallic plate housed within an evacuated tube, from which electrons are released when it is affected by light.

photo-electric effect The ejection of electrons from certain metals when they are affected by light. Since an electric current is caused by a transfer of electrons, the intensity of the light falling on the metal can be calculated by measuring the current.

photoheliograph A telescope designed exclusively for photographing the Sun.

photometer A device which can be visual, photographic or electronic, for measuring light intensity.

photomultiplier A highly sensitive light detector and amplifier combined. Electrons released from the metallic plate on which the light falls then free other electrons from other plates in a cascade effect. Amplification of the order of one million can occur.

photon The basic packet of energy which constitutes light and other similar forms of radiation.

planet A non-luminous body of relatively low mass, orbiting a star.

planetary nebula A cloud of gas probably ejected from a massive star at the time of its final explosion as a supernova. The name derives from its appearance as a faint, circular planet when seen through a small telescope.

precession The very slow oscillation of the Earth's axis with a period of about 26,000 years. It is caused by the unequal gravitational attraction of the Sun and Moon upon the parts of the equatorial bulge that lie north and south of the orbit plane.

primary mirror In a reflecting telescope, the large concave mirror which reflects the light from an object and forms the primary image (which may be refocused by a secondary mirror).

proper motion The term given to the drift against the "fixed" sky background of some particularly near and fast-moving stars. Although all the visible stars have space velocities of many kilometres per second, they are so distant that only a few thousand have had reliable proper motions determined. The star of largest proper motion, a faint object in the constellation of Ophiuchus known as Barnard's Star, moves across the sky at a rate of 10·3 arc seconds per year – equivalent to traversing the Moon's angular diameter in about 180 years.

proton A sub-atomic particle forming part of the nucleus of an atom, carrying a positive electric charge.

pulsar A stellar object releasing pulses of light and radio emission at regular intervals of not more than a few seconds. It is believed to be a rapidly spinning, collapsed stellar core known as a neutron star, in which the constituent atoms have been crushed to their nuclei and the whole body is only a few kilometres in diameter.

quadrant A family of early instruments, some of them pre-telescopic, which were used to determine the positions of stars and planets. The essential feature was a quarter-circle divided into fractions of degrees, either mounted permanently in an observatory or in portable form for use by navigators.

quasar A "quasi-star" or "quasi stellar object" (QSO). An extra-galactic energy source of enormous power and relatively small size. Evidence for their enormous distance comes from their apparent speed of recession, which suggests that some quasars are near the boundary of the observable universe.

radio astronomy The study of objects at *wavelengths* of from a few centimetres to hundreds of metres. One of the most important wavelengths used by radio astronomers is 21 centimetres ($8\frac{1}{4}$ inches), the emission due to cool hydrogen. Many important objects, faint or undetectable in visible light, are powerful radio emitters.

red giant The phase through which a star passes once it has transformed its usable supply of hydrogen into helium by nuclear reactions. Its outer shell expands, attaining a diameter of tens or hundreds of millions of kilometres. As a result, the material becomes extremely rarefied and the temperature falls, with a resultant reddening of the star's colour. This stage lasts until the new helium reactions in the core emit so much energy that the shell is destroyed in the so-called "helium flash".

red shift The apparent stretching of the light waves reaching an observer when the source emitting them is receding at a great speed. Since red light has the longest *wavelength* of visible light, the colour is technically reddened. However, only when the velocity of recession reaches a large fraction of the velocity of light can any colour change be noticeable; normally, the effect reveals itself only when the wavelength of individual emissions from the object are carefully measured. The red shift of light from remote galaxies gave the first indication that the universe is expanding.

reflection The rebounding of a wave motion at a sharp interface. Electromagnetic radiation – emission of energy packets or photons from atoms – is reflected best by metallic surfaces.

reflector A general term describing any telescope that forms an image by using a mirror. In the classical reflector (*Newtonian*) the image is focused by a single concave mirror. Large modern reflectors usually use two curved mirrors, one concave and the other convex, arranged in the *Cassegrain* form. Supplementary plane mirrors may be added to reflect the beam to a suitable observing position. The advantages of the reflector over the refractor (which forms its image using lenses) include compactness and the formation of an image free from all colour defects.

reflex zenith tube A telescope specially designed to observe stars near the overhead point, or zenith, where atmospheric effects upon the image are at a minimum. In the reflex design, the star's image is reflected by a perfectly horizontal pool of mercury.

refraction The change of direction of a light ray when it passes through adjacent media of different density. Refraction of starlight by the Earth's atmosphere has the effect of making the star appear to be slightly higher in the sky than it really is. A glass lens uses refraction to focus light from an object to form its image.

refractor A telescope that forms its primary image by refraction through its main lens, or object-glass. In astronomy, refractors have been in use continuously since their invention in 1608, although practically all modern large instruments have been of the reflecting type.

relativity The theory proposed by Albert Einstein in his two famous papers of 1907 and 1915 that Newton's concept of the universe as containing "absolute" motion must be replaced by the concept of "relative" motion, so that no fixed frame of reference exists.

resolution (resolving power) The ability of a telescope to resolve, or separate, two adjacent objects. In visible light, the closest angular separation that can be achieved is equal to 11·8 arc seconds divided by the diameter of the telescope's lens or mirror in centimetres. The resolving power of a radio telescope is much less acute, but can be improved by using an *interferometer*.

retrograde motion Orbital motion in a direction opposite to the sense in which all the major planets move. This would mean that, as viewed from a point in space to the north of the general plane of the solar system, an object in retrograde motion would be moving anti-clockwise. The term also applies to the apparent back-tracking of a superior planet against the star background when the Earth is carried past by its greater orbital velocity.

right ascension The celestial equivalent of terrestrial longitude, measured in hours around the celestial equator, which is the projection on the sky of the Earth's orbital plane. It is measured in hours because the framework appears to rotate around the Earth once a day. The line marking 0 hours passes through the constellation Pisces, although it is referred to as the First Point of Aries, a gradual drift having taken place since the constellations were first drawn up by Arab astronomers in the early Middle Ages.

Ritchey-Chrétien telescope A modification of the original *Cassegrain* design of reflector, with a large concave primary mirror and a much smaller convex secondary mirror. The Ritchey-Chrétien design uses special hyperbolic curves on both mirrors, allowing the telescope to cover a wider area of sky with good definition.

satellite A natural or artificial body orbiting a planet.

Schmidt telescope More correctly termed a camera, this is the standard astronomical instrument for photographing large areas of sky (diameter up to 20° or so) on a single plate. The photograph is taken directly at the focus of a spherically curved concave mirror whose defective definition is corrected by a thin optical lens situated at the upper end of the tube. The largest Schmidt telescope in the world is the 2-metre (79-inch) aperture instrument at Tautenberg, East Germany.

scintillation The flickering of starlight due to the passage of irregular currents across the light path. Naked-eye scintillation is caused by streams of warm and cold air intermingling at heights of several kilometres in the Earth's atmosphere. Another form, interplanetary scintillation, noticeable at radio wavelengths, is due to matter ejected from the Sun passing between the object and the observer.

secondary mirror The smaller of the two curved mirrors used in reflecting telescopes such as the *Cassegrain*. Its function is not to collect light (this is done by the larger primary mirror) but to refocus it in an appropriate way and to improve the quality of the image.

seeing A general term used to describe the steadiness of the atmosphere at the time of observation.

sextant A graduated arc of 60°, originally used by astronomers such as Tycho and Flamsteed to determine the position of one star relative to another. In highly modified and compact form, it is now a navigational device used for taking the altitude of celestial objects.

solar eclipse The passage of the Moon across the Sun's disc, which happens at new Moon on those occasions when the Moon is in or very near the plane of the Earth's orbit. Partial eclipses can be seen from a wide area of the Earth's surface, but to see a total solar eclipse the observer must be on the path of the Moon's shadow, which is usually less than 200 kilometres (125 miles) wide. The average duration of totality is about three minutes, and a total eclipse occurs somewhere on the Earth's surface almost every other year.

solar flare Intense emission from a small area of the Sun's surface, associated with the development of a sunspot group, and usually lasting for only a few minutes.

solar prominence An uprush of glowing hydrogen from the Sun, usually associated with sunspot activity. Some prominences last for days with little change, while others are ejected at velocities of many hundreds of kilometres per second and may escape altogether from the Sun's gravitational pull.

solar system The Sun and its planetary family. Once thought to be unusual or even unique, there is increasing evidence that planetary systems around stars may be a common phenomenon, the star and the planets originally condensing simultaneously inside the same cloud of material.

solstice The instant when the Earth's north or south pole is inclined at the maximum possible angle towards the Sun (on or about 21 June and 21 December respectively). In the Northern Hemisphere, these dates correspond to the summer and winter solstice.

space probe A general term applying to any space vehicle carrying instruments to transmit data from other regions of the solar system.

speckle interferometry A technique in which the true image of a star, rather than the degraded blur recorded by a large telescope, is reconstituted by combining a number of photographs each taken with a very short exposure. The method has also been used to simulate the image quality of a much larger telescope than any now existing, by combining photographs taken using two linked instruments.

spectrograph An instrument for analysing and recording the light received from a particular object. It does this by spreading the light out into a coloured band, or spectrum, and recording the bright or dark lines that indicate the presence of different atoms.

spectroheliograph An instrument for photographing the Sun's visible surface, or photosphere, in the light emitted by a particular element, usually either hydrogen or calcium. The device, using slits to scan the whole disc, has been largely superseded by modern filters.

spectrohelioscope A version of the *spectroheliograph* adapted for visual use. The slit scans the solar disc so rapidly that persistence of vision in the eye gives the illusion of being able to view the entire Sun simultaneously.

spectrometer A device for measuring the *wavelength* of any particular line in the spread-out coloured band, or spectrum, of light from an object. Each line in the spectrum corresponds to a particular substance, which can be identified if its wavelength is known.

spectrophotometer An instrument for measuring the intensity of the lines appearing in the spread-out coloured band, or spectrum, of the light from an object. This intensity is a measure of the abundance of the particular element or compound causing the line.

spectroscope An instrument for spreading light into a coloured band, or spectrum. The early types incorporated a glass prism, which bends or refracts light through a different angle according to its *wavelength*. Modern spectroscopes and spectrum-analysing devices generally use a diffraction grating, a reflecting plate ruled with several hundred fine lines per millimetre. This gives a more even distribution of wavelength across the spectrum.

spectrum The coloured band produced when polychromatic light (i.e., light comprising a mixture of various colours or *wavelengths*) is spread out into its different wavelengths. In order to provide a pure spectrum, the light must pass through a narrow slit arranged at right angles to the direction in which the light is dispersed. The coloured band then consists of an infinite number of images of the slit, each one isolating a certain wavelength. A spectrum may be of the continuous type (like a rainbow) – this will be the case if the light source is an incandescent body like the surface of the Sun or the filament in an electric lamp. More usefully, if it is associated with a rarefied gas it will be crossed by bright lines (emission spectrum) or dark lines (absorption spectrum), each of which is unique to a particular element in the gas.

speculum metal A highly reflective, very hard alloy of copper and tin, often with arsenic added. It was used for reflecting telescope mirrors before the discovery, round about 1850, that silver could be deposited chemically on to the surface of glass.

spherical aberration The failure of different zones of a lens or mirror to focus the image of an object in the same plane. This fault can be remedied either by polishing the optical surface to a non-spherical shape or by combining two or more optical components whose aberrations cancel one another out.

star A self-luminous, roughly spherical gaseous body which creates energy through nuclear reactions at its core. By contrast, a planet's interior is not sufficiently hot to set nuclear reactions going and its surface must depend upon a nearby star for light and heat.

steady-state theory An idea proposed by Hermann Bondi and Fred Hoyle in the 1940s to explain how the universe could be both expanding and essentially unchanging – i.e., without a finite beginning. To do this, they postulated that matter was being continuously created so as to occupy the growing space between the galaxies. Later work has, however, shown conclusively that the universe must have had a finite beginning some 20,000 million years ago in the *big bang*.

sunspot A cool, disturbed area on the Sun's surface that appears dark by contrast, although it is still much hotter and brighter than the surface of many cooler stars. This local coolness is the result of surface material losing energy at the centre of an intense magnetic field. Sunspot frequency varies with a period of about eleven years.

superior planets The name given to the planets Mars, Jupiter, Saturn, Uranus, Neptune and Pluto, which are all farther from the Sun than is the Earth.

supernova The explosion of a giant star, one that is several times as massive as the Sun. Such a star uses up its available store of hydrogen relatively quickly and becomes fantastically hot, temperatures of hundreds of millions of degrees being reached in the core. Eventually the centre collapses and the blast of energy released blows most of the star's material out into space. For a few days a supernova will shine as brightly as an entire galaxy of stars.

telescope An instrument for forming a bright and magnified image of a distant object. Astronomical telescopes are divided into refractors, which use a lens to receive and focus the light, and reflectors, which use a mirror instead. Some radio telescopes work on the reflecting principle, employing a metal bowl or dish to focus the radiation on to a receiver, while others use bare aerials tuned to the appropriate *wavelength*.

transit The passage of a celestial body across the north-south direction due to the Earth's rotation. Transit of the Sun across the meridian occurs at noon. Timing the transit of a star whose position in the sky is accurately known makes it possible to check a clock against the rotation of the Earth.

transit circle A telescope mounted on rigidly orientated pivots so that it can swing only in a vertical north-south plane. By its means, the transit of a celestial body across the meridian or north-south line can be timed. The transit circle replaced the older *mural arc* and has now been supplanted by the zenith tube, which observes the position of a star as it passes near the zenith or overhead point.

X-rays Very high-energy *electromagnetic radiation* whose wavelength or distance between pulses is in the range of one-millionth to one ten thousand-millionth of a millimetre. X-rays cannot penetrate the Earth's atmosphere and astronomical sources emitting this radiation (such as *supernovae*) must be studied by satellite-borne instruments.

ultraviolet radiation The name given to the band of invisible radiation beyond the violet end of the spectrum whose wavelength or distance between pulses ranges from 0·0004 millimetre down to about a millionth of a millimetre. Like X-rays, ultraviolet radiation is absorbed by the Earth's atmosphere.

wavelength The effective distance between successive intensity peaks in the waves of *electromagnetic* energy that are emitted by a source. The wavelength of visible light ranges from about 0·00045 millimetre (violet) to 0·0007 millimetre (red). The range of wavelength of all known electromagnetic radiation is from less than a million millionth of a millimetre (gamma rays) to thousands of metres (radio waves).

white dwarf Almost the final stage in the development of a normal star. Its material collapses gravitationally to form a very dense hot sphere with the mass of a star like the Sun but the diameter of a moderate planet. The star slowly cools during this white-dwarf stage, eventually becoming a non-luminous black dwarf.

variable stars Stars whose light output varies, either because of mutual covering and uncovering by the components of an orbiting pair (eclipsing binary) or because of real changes in their luminosity, such as *Cepheids*.

zenith The point in the sky which is exactly overhead.

zenith tube A telescope fixed permanently in a vertical plane, for observing the positions of stars near the zenith as a check on the rotation of the Earth. The zenith is chosen for precise work because the atmosphere has least effect on the star images.

zodiac A band around the sky 18° wide, centred on the projection of the Earth's orbital plane. Since most of the planetary orbits lie in almost the same plane as the Earth's, the planets are always found within the zodiac, with the occasional exception of Pluto. The word means "circle of animals", since most of the constellations through which the zodiac passes are connected with real or imaginary creatures – Taurus the Bull, Leo the Lion etc.

INDEX

Numbers in *italics* refer to illustrations

T

ACKNOWLEDGEMENTS

Managing Editor	Lionel Bender
Text Editor	Nigel Henbest
Assistant Editor	Zuza Vrbova
Designer	Nigel Partridge
Picture Research Manager	Celia Dearing
Picture Researcher	Anne-Marie Erhlich
Art Director	Nick Eddison
Production Manager	Kenneth Cowan
Production Editor	Fred Gill

Harrow House Editions would like to thank the following people for their assistance in producing this book: Gwynneth Learner and Professor A H Meadows, Peter Laurie, Jane Greening, Sue Gallacher and Nick Jackson.
Index compiled by Helen Baz
Glossary compiled by James Muirden

ARTISTS
All diagrams and illustrations by
Graham Marlowe, Folio Artists Agents,
excepting pages;
1, 2–3, 4–5, 6–7, Nigel Partridge
39, 78, Michael Mann Studios
46–47, Andy Farmer
178–179, Hugh Dixon, Spectron Artists Ltd.
184, Gerry Banks
Retouching, Michael Mann Studios
Front cover photography, Les Weiss

PICTURE CREDITS
A: Above B: Below C: Centre L: Left
R: Right

9(AL,BR): Ann Ronan Picture Library 11(A): Scala/Prado Madrid (BR): Scala/Bibliothéque National Firenze 12(AL): Ann Ronan Picture Library 13(A,B): Lowell Observatory photograph 14(A): Cooper-Bridgeman Library/De Unger Collection 15: Archiv für Kunst und Geschichte 17(AL,AR,CL): Archiv für Kunst und Geschichte (CR): Ann Ronan Picture Library 18(AL,AR): Ann Ronan Picture Library (C): Archiv für Kunst und Geschichte (B): Science Museum/photo A.C. Cooper 19(A): Space Frontiers (CR): Ann Ronan Picture Library 20(A): Scala/Uffizi Firenze (B): Ann Ronan Picture Library 23(B): Michael Holford 25(AC,AR): Michael Holford (BR): Cooper-Bridgeman Library 26(CR): Cooper-Bridgeman Library (BL): The John Hillelson Agency Ltd/Eric Lessing 28(BL): BBC Hulton Picture Library 29(AR): Mary Evans Picture Library 30(CL): The Mansell Collection 31(BL): Trustees of the British Museum (BR): Mount Wilson and Las Campanas Observatories Carnegie Institution of Washington 32(CR): The Mansell Collection 34(BL): The Mansell Collection 35: Michael Holford 36(AC): Ann Ronan Picture Library (BL): Archiv für Kunst und Geschichte 38(BL): Crown Copyright Science Museum London (BC): Archiv für Kunst und Geschichte (BR): Mary Evans Picture Library 39(B): Photo Science Museum London 40(AL): Archiv für Kunst und Geschichte 42–43(BL): by permission of Manchester City Council (AR): British Library/photo Ray Gardner (BL): Archiv für Kunst und Geschichte 44(AL): The Mansell Collection (BL): Jean-Loup Charmet 48(AL): National Portrait Gallery London 48–49(B): Fotmas Index 50(AL): R.A.S./photo A.C. Cooper 51(AR): R.A.S./photo A.C. Cooper 53(A): The National Maritime Museum London (B): National Portrait Gallery London 55(AR): Ann Ronan Picture Library (BR): The National Maritime Museum London 56(AR): Hamlyn/National Library of Australia, Canberra 56–57(C): Hamlyn/The National Maritime Museum London 56(B): British Library/photo Ray Gardner 57(BL): Royal Society/Science Museum London 58(L): Ann Ronan Picture Library 60(AL): BBC Hulton Picture Library (AC): Ann Ronan Picture Library (BL): National Portrait Gallery London 60–61(C): Ann Ronan Picture Library 61(AR): Crown Copyright Science Museum London 63(R): Ann Ronan Picture Library 64(RC): Dolland and Aitchison Museum 65(R): Mount Wilson and Las Campanas Observatories Carnegie Institution of Washington 66(C): Paul Brierley 67: The National Maritime Museum London 68(L): R.A.S./photo A.C. Cooper 69(A): The National Maritime Museum London 70: Deutsches Museum 71(AC,BL,BR): Deutsches Museum 74 (L): The Mansell Collection 75(L): Courtesy the Earl of Rosse (CR): US Naval Observatory (BR): photo Science Museum London 76(L): British Library/photo Ray Gardner 77(BR): photo Science Museum London 78: Courtesy the Earl of Rosse 79(R): R.A.S./photo A.C. Cooper 80(L): Ann Ronan Picture Library 82(R): The Mansell Collection 84(L): The Mansell Collection 85(AR): Harvard College Observatory (B): The Mansell Collection 86(AL): Jean-Loup Charmet (BL): Crown Copyright Science Museum London (BR): photo Science Museum London 87(B): Gernsheim Collection Humanities Research Center University of Texas at Austin 88(L): Jean-Loup Charmet 89(L): photo Science Museum London (R,BL): Ann Ronan Picture Library 91(B): © AURA The Kitt Peak National Observatory (A): Ann Ronan Picture Library 92: British Library/photo Ray Gardner 97: The Mansell Collection 98: Yerkes Observatory photograph 99: R.A.S./photo A.C. Cooper 100(AL): A.I.P. Niels Bohr Library (AR,BL): Ann Ronan Picture Library 103(AR): Lick Observatory photograph (BR): Ann Ronan Picture Library 104 (AL): R.A.S./photo A.C. Cooper (B): Yerkes Observatory photograph 107(A): Ann Ronan Picture Library 108(A): Ann Ronan Picture Library (B): Mount Wilson and Las Campanas Observatories Carnegie Institution of Washington 110: Mount Wilson and Las Campanas Observatories Carnegie Institution of Washington 113: Popperfoto 114(AL): NASA 114–115(C): Mount Wilson and Las Campanas Observatories Carnegie Institution of Washington 116(AL): Harvard College Observatory 118(A): Royal Observatory Edinburgh (B): Popperfoto 120(L): Popperfoto 123(AL): Mount Wilson and Las Campanas Observatories Carnegie Institution of Washington (AR): Dr. R. Spear (B): Michael Maunder 124(BR): S.P.L./Mount Wilson and Las Campanas Observatories Carnegie Institution of Washington 125(A): French Government Tourist Office 127: US Naval Observatory 128(B): Mount Wilson and Las Campanas Observatories Carnegie Institution of Washington 129: Space Frontiers 131(A): Ann Ronan Picture Library (BR): The Marconi Company Ltd 132(AL): Courtesy of Bell Laboratories (CL): National Radio Astronomy Observatory 135(AL): IWM/E.T. Archive (CL): Huygens Laboratorium (BL): Sefton Photo Library 136(BR): CSIRO 137(CR): Mullard Radio Astronomy Observatory Cambridge (BR): Professor Boksenberg UCL 139(AR): Keystone Press Agency (B): The Australian Information Service London 140(BL): Radiosterrenwacht Westerbork 142(AL): Mount Wilson and Las Campanas Observatories Carnegie Institution of Washington (B): The National Maritime Museum London 143(B): Black Star/Ralph Crane 148(AL): Novosti Press Agency (BL): The Kitt Peak National Observatory 151: Aspect Picture Library 152(CL): Royal Greenwich Observatory 152–153(AR,BR): Professor Boksenberg UCL 154 (A): Keystone Press Agency (B): The John Hillelson Agency Ltd/George Rodger 155(A): The John Hillelson Agency Ltd/Georg Gerster 156(BL): Jet Propulsion Laboratory 157(AR, BR): Lick Observatory photograph 158(B): S.P.L./Royal Observatory Edinburgh 159(AL): Susan Griggs Agency/Jonathan Blair (BL): The John Hillelson Agency/Georg Gerster 161 (A): Popperfoto (B): U.P.I. 162–163(B): University of Manchester 164(A,BL,BR): NASA 165(AL,BL,BR): NASA (AR): S.P.L./NASA 166: British Aerospace 167: Princeton University 168(A,B): NASA 170(AL,AR): NASA (BR): Smithsonian Institution 172(AL): NASA (AR): Perkin-Elmer Corps/New Scientist London 176(B): The Chatterton Astronomy Dept. University of Sydney 177(A): The Kitt Peak National Observatory 178(A,C): © AURA The Kitt Peak National Observatory 179(B): Rutherford and Appleton Laboratories 180 (A): CERGA/R. Futaully (B): The McMath-Hulbert Observatory of E.C.D. Inc 181(A): Princeton University 182(A): NASA (B): Brookhaven National Laboratory 186: Dr. R. D. Joseph 187(A): Dr. R. Spear (B): Royal Observatory Edinburgh 189(AR): Mount Wilson and Las Campanas Observatories Carnegie Institution of Washington 190(BL): Mount Wilson and Las Campanas Observatories Carnegie Institution of Washington 191(BR): Mount Wilson and Las Campanas Observatories Carnegie Institution of Washington 193(AR): Popperfoto 195(CR): Times Newspapers Ltd (BR): Michael Maunder 196(B): Michael Maunder 197(AL): Ben Mayer (B): Popperfoto 199(A): NASA (B): Royal Observatory Edinburgh 201: Royal Observatory Edinburgh 202(BL,BR): © AURA The Kitt Peak National Observatory (BC): The Kitt Peak National Observatory 203: © AURA The Kitt Peak National Observatory 205: Dennis di Cicco 208: Lawrence Berkeley Laboratory